Universitext

P. Hájek T. Havránek

Mechanizing Hypothesis Formation

Mathematical Foundations for a General Theory

Springer-Verlag
Berlin Heidelberg New York 1978

Petr Hájek
Mathematical Institute, Czechoslovak Academy of Sciences
Praha, Czechoslovakia

Tomáš Havránek
Department of Biomathematics, Czechoslovak Academy of Sciences
Praha, Czechoslovakia

The authors are members of the Society of
Czechoslovak Mathematicians and Physicists

AMS Subject Classification (1970): 02-02, 02 C 05, 68-02, 68 A 20, 68 A 45

ISBN-13: 978-3-540-08738-0 e-ISBN-13: 978-3-642-66943-9
DOI: 10.1007/978-3-642-66943-9

2141/3140-543210

To Marie and Marie

Preface

Hypothesis formation is known as one of the branches of Artificial Intelligence, The general question of Artificial Intelligence,"Can computers think?" is specified to the question,"Can computers formulate and justify hypotheses?" Various attempts have been made to answer the latter question positively. The present book is one such attempt . Our aim is not to formalize and mechanize the whole domain of inductive reasoning. Our ultimate question is: Can computers formulate and justify scientific hypotheses? Can they comprehend empirical data and process them rationally, using the apparatus of modern mathematical logic and statistics to try to produce a rational image of the observed empirical world?

Theories of hypothesis formation are sometimes called logics of discovery. Plotkin divides a logic of discovery into a logic of induction : studying the notion of justification of a hypothesis, and a logic of suggestion : studying methods of suggesting reasonable hypotheses. We use this division for the organization of the present book: Chapter I is introductory and explains the subject of our logic of discovery. The rest falls into two parts: Part A - a logic of induction, and Part B - a logic of suggestion.

In Part A we define and investigate formal calculi appropriate for formalizing (fragments of) observational and theoretical languages of scientific theories based on empirical data. The definitions are motivated by statistical considerations, which seem to be unjustly neglected in contemporary Artificial Intelligence. Our calculi are modified generalized predicate calculi and are related to calculi proposed by Suppes. The following are emphasized:

(i) explicit semantics in Tarski's style,

(ii) use of generalized quantifiers and abstract truth values,

(iii) relation to effective computability and complexity of computations.

As a result, we obtain (a) mathematical logic of observational and theoretical calculi and (b) logical foundations of computational statistics, i.e. theoretical investigation of the interplay of logical computability conditions and statistical measurability conditions. To avoid misunderstanding, let us mention the fact that we are not interested in foundations of probability; we use Kolmogorov's classical notion of probability.

In Part B, we give a formal definition of GUHA methods of automated formation of hypotheses and construct and study sufficiently complex particular GUHA methods capable of machine realization. The acronym "GUHA" (General Unary Hypotheses Automaton) was introduced ten years ago in the paper describing the first such method (Hájek, Havel and Chytil 1966a); at present, the word "GUHA" should be understood simply as an artificial name for a certain class of methods (procedures) based on the principle of mechanized construction of all interesting hypotheses justifiable on the basis of given empirical data. This principle is clearly independent of the statistical approach to the problem of induction; but the statistical approach is very natural and gives inspiration for several particular methods.

As a matter of fact, the whole conception presented here was inspired by the aim of finding general mathematical foundations of methods similar to that presented in the first GUHA-papers. In the meantime, various other versions of the GUHA methods have been developed and there exist up-to-date implemented FORTRAN programs, e.g. for IBM 370 (cf. Hájek et al. 1976). Both a historical note and an example of practical application are contained in the appendices. But we feel that Part A is not merely a necessary background for Part B but also presents a logic of induction of independent interest. On the other hand, Part B is no mere collection of some particular methods but contains theoretical foundations and impulses for further GUHA-methods and other similar methods.

In our opinion the following readers could profit from this book:
(a) Mathematical logicians interested in applications in computer science, in particular in hypothesis formation;
(b) Students of theories of inference in connection with computer applications;
(c) Computer scientists interested in hypothesis formation and its complexity properties, especially students of Artificial Intelligence;
(d) Statisticians who understand computation.

We assume some knowledge of the very basic parts of recursion theory, predicate logic and probability theory. Yasuhara [1972] and Burril [1972] can serve as completely satisfactory references. Every reader should read Chapter I; at the end of that chapter he will find some suggestions on reading only selected parts of the rest of the book. All the chapters are divided into sections. Each section concludes with a summary of key words; this summary serves as a check on whether the reader has memorized all the important notions of the section just concluded, by rapid repetition before passing on to the next section. Each chapter concludes with several problems and supplements.

Aknowledgements. We wish to thank Professors J. Bečvář and D.S. Scott who read parts of this book and whose comments have significantly influenced the final form of the text. We appreciate very much the continued interest and encouragement of Professor Gert H. Müller. We are indebted to Metoděj K. Chytil for numerous discussions of the GUHA method in which he formulated several general suggestions which were of relevance also to the theory presented here. A great number of valuable suggestions emerged from discussions with our colleagues Mrs. K. Bendová, Michal Chytil, I.M. Havel and F.N. Springsteel; in particular, they detected many mistakes. Most of all, we thank our wives, Marie Hájková and Marie Havránková, for their understanding and patience.

Průvodčí moji a každého, kdož v světě tápá, vpravdě jsou dva: Drzost mysli, všeho ohledující, a zastaralý při věcech Zvyk, pravdy barvu šalbám světa dávající ...

Jan Ámos Komenský, Labyrint světa a ráj srdce, Amsterodam 1663

(My guides, and indeed those of everyone who gropes through this world, are two: Insolence of the mind, which inquires into everything, and inveterate custom with regard to all things, which gives the colour of truth to the deceits of the world ...

Jan Amos Komenský Comenius, The labyrinth of the world and the paradise of the heart, edited and translated by Count Lützow, London 1901)

Contents

Chapter I. Introduction: What is a Logic of Discovery

"Logic" and "Discovery" are certainly very familiar notions. The term "logic of discovery" belonged originally to the philosophy of science; "philosophers of science have repeatedly mentioned the process of discovery of scientific hypotheses and the possibility or impossibility of formulating a logic for that process" (Buchanan 1966). The problem of the possibility of a logic of discovery takes on a new meaning as a problem of Artificial Intelligence (cf. the preface and Buchanan 1966). In the present chapter we shall outline some basic notions of the philosophy of science in a form which will lead us on the one hand to a certain notion of logic of discovery and on the other to several mathematical notions.

I.1. Informal considerations

1.1.1 Science can be regarded as a cognitive activity sui generis. Scientific procedures as elements of scientific cognitive activity can be characterized as operations with data (cf. Tondl 1972) . The aims of science are scientific explanation, prediction, verification, constitution, reduction etc. It must be stressed that science does not produce infallible truths; in science one formulates hypotheses and tries to justify them (or reject them) . The word "logic" in the term "logic of discovery" refers not only to the analysis of various scientific languages but also to a rational body of methods for finding and evaluating certain propositions (cf. Plotkin).

1.1.2 Consider three examples of hypothesis formation:

(1) This crow is black.
That crow is black.

All observed crows are black.

All crows are black.

(2) This crow is black.
That crow is black.

Many crows have been observed;
relative frequency of black
ones is high.

Crows have a considerable change of
being black.

(3)

rat no.	weight g	weight of the kidney mg
1	362	1432
2	372	1601
3	376	1436
4	407	1633
5	411	2262

The observed weights of the kidneys have the same order as the weights of the rats with one exception.

The weight of rat's kidney is positively dependent on the weight of the rat.

1.1.3 We shall stress some important features of these examples of "inductive inference". Each example consists of three parts. The first part describes our evidence ; it can have the form of simple sentences "This crow is black", or, equivalently, of a table or other similar form (Example 3). The second part is an observational statement: it is a more or less complicated sentence which can be asserted on the basis of the data. Finally, the third part is a theoretical statement, the inductive generalization (cf. Carnap 1936). The theoretical statement is not a consequence of the observational one; in Example 1, the observational statement is a logical consequence of the theoretical one, but in Examples 2, 3 the situation is more complicated. Nevertheless, we feel that the transition from the observational statement to the theoretical statement is justified by some rules of rational inductive inference, even if they are not formulated explicitly. As a matter of fact, some philosophers reject any possibility of formulating such rules and probably there is nobody who believes that there can be universal rules for rational inductive inference of theoretical statements from observational statements. On the other hand, one can show that there are non-trivial rules of inductive inference applicable under some well described circumstances and that some of them are useful in mechanized inductive inference.

1.1.4 The scheme of inductive inference is as follows:

$$\frac{\text{theoretical assumptions, observational statement(s)}}{\text{theoretical statement}}$$

This means that having accepted the theoretical assumptions and having verified the observational statement(s) in question, we accept the theoretical statement forming the conclusion. Having said "verified" we touched on semantics: observational statements are statements about something, about our data.

The question of the semantics of theoretical statements is dealt with in later sections. Let us stress the very important fact that it is an intelligent observation of the data (observational statement) that leads to theoretical conclusions, not the data themselves. One chooses the conceptual apparatus, i.e., one chooses both the observational and the theoretical language one wants to use. For instance, in Example 3 above, one could choose the notion of linear functional dependence; one could assert: "The weight of the kidney of the observed rats is not a linear function of the weight of the rat." and try to make a theoretical inference.

This will be important in our attempt to formalize the whole situation. We formulate our task into five questions (L0)- (L4) (our (L1) - (L4) are directly analogous to Plotkin's (H1) - (H4)).

1.1.5 The questions of the logic of discovery:

(L0) In what languages does one formulate observational and theoretical statements? (What is the syntax and semantics of these languages? What is their relation to the classical first order predicate calculus?)

(L1) What are rational inductive inference rules bridging the gap between observational and theoretical sentences? (What does it mean that a theoretical statement is justified?)

(L2) Are there rational methods for deciding whether a theoretical statement is justified (on the basis of given theoretical assumptions and observational statements ?)

(L3) What are the conditions for a theoretical statement or a set of theoretical statements to be of interest (importance) with respect to the task of scientific cognition?

(L4) Are there methods for suggesting such a set of sentences which is as interesting (important) as possible ?

Answers to (L0) - (L2) constitute a logic of induction; answers to (L3)- (L4) constitute a logic of suggestion. Answers to (L0)- (L4) constitute a logic of discovery. (Cf. Plotkin.)

Our aim is to develop a mathematical logic of discovery. This is certainly necessary from the point of view of Artificial Intelligence: only mathematically precise notions and results can be used as a basis for computer procedures. But let us mention at this point that there are important extramathematical problems, e.g. concerning the ways in which scientific data attain the mathematical form assumed below; such questions are beyond the scope of this book.

1.1.6 We shall now give some preliminary answers to (L0) - (L4); these answers indicate main standpoints, on which this book is based.

(L0) Our calculi will reflect the difference between observational and theoretical sentences; we develop observational and theoretical calculi. We elaborate both the syntax and semantics of these two kinds of calculi and, to some extent, develop their autonomous logic. A typical feature of observational calculi is effective calculability of the (truth) value of each sentence in each observational structure. A typical feature of (statistically motivated) theoretical calculi is their "modal" character: theoretical sentences refer to systems of "possible worlds" and probability is understood as a measure on such a system of possible worlds.

(L1) Observational and theoretical calculi are interrelated by inductive inference rules; from a theoretical frame assumption (background knowledge) and observational statement (describing data) one can infer a theoretical hypothesis. Rationality criteria for such rules can be formally expressed in accordance with some accepted notions of statistical hypothesis testing.

(L2) To answer (L2) statistical measurability conditions must be reconciled with notions concerning computability. This leads to a sort of computational statistics. Such considerations are quite unusual for statisticians but, fortunately, most statistical procedures pass the computability test well and hence can serve for definitions of particular observational quantifiers.

(L3) It is a very important fact that in many important inductive inference rules, hypotheses (succedents) are in one-one correspondence with some specific observational statements occuring in the corresponding antecedents. Thus in many cases the search for hypotheses can be reduced to the search for appropriate observational statements in a satisfactorily rich observational language. This leads to the formal notion of an observational research problem and its solution. The observational research problem itself specifies (among other things) a set of relevant observational questions; a solution is a representation of true relevant observational statements, serving as "logical patterns" - codes of theoretical hypotheses.

(L4) GUHA methods are methods for construction of solutions to observational research problems. The output of a GUHA procedure is not the (single) most interesting hypothesis but an important set of hypotheses. Kemeny writes: "I am convinced that the formation of possible theories will forever remain a job for the creative genius of the scientist. The choice may be aided by rules but no rules will replace original thinking." (cf. Buchanan, p. 66). GUHA procedures are procedures for aiding the choice of hypotheses, i.e. GUHA guarantees that we know where to look for the hypothesis. Statistical properties of the output set of hypotheses as a whole can be satisfactorily clarified.

The reader is advised to return to these answers later when he has learned some particular GUHA methods.

1.1.7 In most works on hypothesis formation, hypotheses are identified with some formulae of the classical predicate calculus. There are at least two arguments in favour of the predicate calculus from the point of view of logics of discovery:

(i) it has clear semantics and (ii) there are well-developed theorem proving methods. The second argument is particularly important if induction is understood deterministically, as inverse deduction. What are our reasons for modifications and generalizations of the classical predicate calculus? First, note that our calculi will satisfy (i) and the significance of (ii) will be minimized by the fact that we shall not equate induction with inverse deduction. The main reason is the fact that observational sentences are useful as "logical patterns" in statistical inference rules, and also that theoretical sentences expressing hypotheses are either only cumbersomely expressible or even not expressible in the classical predicate calculus.

1.1.8 Let us make some remarks on the structure of Part A (Chapters II-V). Chapters II and III are logical in character; in Chapter II we introduce step by step various observational and theoretical calculi and formulate their basic properties. In Chapter III we develop the logic of observational calculi;

in particular, we introduce and study some important classes of observational generalized quantifiers (associational and implicational quantifiers) and study calculi with models with incomplete information. Chapters IV and V have a more statistical flavour; in Chapter IV the usual theory of statistical hypothesis testing is presented in the logical framework of the previous chapters and various particular statistical associational and implicational quantifiers are exhibited. Chapter V is devoted to modern rank tests; various classes of observational rank quantifiers are described and their logical properties are investigated.

1.1.9 The next section contains various mathematical notions used throughout the book. We shall not summarize the contents of Part B in more detail here; but the reader is invited, after having read the rest of Chapter I, to read Section 1 of Chapter VI, where basic notions of our logic of suggestion are presented using only the apparatus of Chapter I. The reader interested in the logic of induction but not in methods of mechanized hypothesis formation may read only Chapters I - V. On the other hand, the reader wanting to comprehend quickly the basic theory of GUHA methods may read Chapter I, Chapter II Sections 1, 2, Chapter III Section 2, Chapter VI Section 1 and Chapter VII Sections 1 - 3 with the omissions indicated there . In this case he will be informed on GUHA methods as methods generating interesting observational statements, but will not know where these statements come from and in what sense GUHA generates hypotheses.

1.1.10 Key words: Logic of induction, logic of suggestion, observational and theoretical statements.

1.2 Some mathematical notions.

In this section we shall be interested in several mathematical notions concerning sentences. For the time being, we shall not analyse the structure of sentences, but we shall deal with sentences as abstract entities. Our aim is the following:

(i) We shall ask what is meant by the inference of sentences from other sentences and what is the meaning of sentences.

(ii) We shall relate these notions to notions concerning computability, recursiveness and other notions .

(iii) We shall introduce a formal notion of an observational language.

(iv) We shall be more specific on inductive inference rules relating theoretical languages to observational languages.

1.2.1 Definition. Let Sent be a non-empty set; call its elements sentences. An inference rule I on Sent is a relation consisting of some pairs $\langle\varphi, e\rangle$, where φ is a sentence and e is a finite possibly empty set of sentences. We often write

$$\frac{e}{\varphi} \in I$$

instead of $\langle\varphi, e\rangle \in I$, in agreement with the usual convention in expressing deduction rules; elements of e are antecedents and φ is the succedent of the pair $\langle\varphi, e\rangle$. We say that φ is inferred from e by I if $\langle\varphi, e\rangle \in I$. More generally, we call φ an (immediate) conclusion from $A \subseteq$ Sent (by I) if either $\varphi \in A$ or there is an $e \subseteq A$ such that $\langle\varphi, e\rangle \in I$. The set of all conclusions from A by I is denoted by I(A).

1.2.2 Remark. (1) Note that at this stage we say nothing about the truth or falsehood of the sentences or about the preservation of truth by inference rules. But if we deal with a relation I as with a rule of inference we may (and shall) ask what are rationality conditions for I, i.e. conditions guaranteeing that if $\langle\varphi, e\rangle \in I$ then having accepted e we may rationally accept φ.

(2) One could define inference rules in a more general way, considering mappings $I: Sent \times \mathcal{P}_{fin}(Sent) \rightarrow V$, where V is a set of values. ($\mathcal{P}_{fin}(Sent)$ denotes the set of finite subsets of Sent). Then the values $I(\varphi, e)$ could mean e.g. the degree of our belief in φ provided that e has been accepted. Clearly, our notion of rules of inference introduced in 1.2.1 corresponds to the case $V = \{0,1\}$. We limit ourselves to this particular case.

1.2.3 Definition. Let I be an inference rule on Sent and let $A \subseteq$ Sent. A finite sequence $\varphi_1, \ldots, \varphi_k$ of sentences is called a derivation (I-derivation) from A if for each $i = 1, \ldots, k$ either $\varphi_i \in A$ or there is an $e \subseteq \{\varphi_1, \ldots, \varphi_{i-1}\}$ such that $\frac{e}{\varphi_i} \in I$ i.e. φ_i is inferred from some preceding sentences. A sentence φ is I-derivable from A if there is an I-derivation from A containing φ (notation $A \vdash_I \varphi$).

1.2.4 So far, we have not inquired whether sentences have some meaning or whether they are true or false.

It is due to Frege that the two questions can be identified (see Church). According to Frege we consider sentences to be special names. When speaking of names we mean names of something (of some extralinguistic entity). Notice that a single name can denote different entities of truth and falsehood; consequently, a sentence is true iff its meaning (value, denotate) is truth, and is false iff its meaning is falsehood. We have to distinguish the meaning of a name from the sense of that name, for instance "Walter Scott" and "the author of Waverley" have the same meaning but a different sense. Similarly, the sentences "2+2=4" and "Prague is the capital of Czechoslovakia" have the same meaning - namely the truth - but a different sense. The problem of the sense of sentences will however not be discussed here.

We shall follow Frege in treating sentences as names of abstract values, but we shall assume neither that we necessarily have exactly two abstract values, nor that the abstract values are necessarily truth-values, i.e., that they are in some sense condernedwithvalidity, or, possibly, with the degree of our conviction concerning validity.

However, we shall take into account the fact that the value of a sentence depends, on the one hand, on the sentence itself and, on the other hand, on the extralinguistic entity which the sentence speaks of.

In accordance with the terminology usual in Mathematical Logic, the extralinguistic entities in which sentences are interpreted will be called models. Consequently, the value (meaning) is a function of two arguments: sentences and models. The theory of the relations between sentences and their meaning is called semantics.

1.2.5 Definition. A semantic system is determined by a non-empty set Sent of sentences, a non-empty set $\mathfrak{M}$ of models, a non-empty set V of abstract values and an evaluating function $\mathrm{Val}:(\mathrm{Sent} \times \mathfrak{M}) \rightarrow V$. If $\varphi \in$ Sent and $\underline{M} \in \mathfrak{M}$ then $\mathrm{Val}(\varphi, \underline{M})$ is the value of φ in $\underline{M}$; it is often denoted by $\|\varphi\|_{\underline{M}}$.

1.2.6 Examples and Comments. We begin with an abstract example; in (2) we offer a possible interpretation.

(1) For each number $n > 1$, let $\mathcal{S}_n$ be a semantic system defined as follows: Models are matrices of zeros and ones with n columns (and finitely many rows). If such a model $\underline{M}$ has m rows we regard $\underline{M}$ as the result of the evidence of n properties $P_1, \ldots, P_n$ on m observed objects: If the element in the i-th row and the j-th column is 1 then the i-th object has the property P_j. For each non-empty $e \subseteq \{1, \ldots, n\}$ the partial model $\underline{M}/e$ results from $\underline{M}$ by omitting all the columns c_i for which $i \notin e$.
With each such e we associate a sentence φ_e and define $\|\varphi_e\|_{\underline{M}} = 1$ iff each row in $\underline{M}/e$ contains at least one zero (i.e., no row consists only of ones); otherwise $\|\varphi_e\|_{\underline{M}} = 0$. The sentence φ_e can be read "the properties P_i, for $i \in e$, are incompatible in the observed material" hence, e.g., for $e = \{3,4,7\}$: "P_3, P_4, P_7 are incompatible in the observed material". If e is a singleton, $e = \{i\}$, we read φ_e "P_i is absent in the observed material". Thus, $\mathfrak{M}$ is the set of all matrices as described, Sent is the set of all φ_e and $V = \{0,1\}$.

	1	2	3	4	5	5	6	7 ...
1			1	0				1
2			0	1				1
3			1	1				0
4			0	1				0
5			1	0				0
6			0	0				1
7			0	0				0
8			0	1				1
9			1	0				1
.			.	.				.
.			.	.				.
.			.	.				.

(2) Consider, as a more concrete example, a research concerning mutational changes caused by gamma rays. Here models can describe populations of plants (e.g. marigolds) the seeds of which were subjected to gamma rays (cf. Zindel). Such a population of m plants can be described by the matrix having 1 in the i-th row and j-th column if the i-th plant in the population exhibits the mutation P_j. The sentence $\varphi_{\{3,4,7\}}$ can be read "The mutations P_3, P_4, P_7 are incompatible in the observed population" or "The mutations P_3, P_4 and P_7 do not occur simultaneously in the observed population". Similarly, $\varphi_{\{i\}}$ could be read "The mutation P_i does not occur in the observed population". A possible inductive generalization is: "Gamma rays do not cause the mutation P_i".

(3) Remember semantic systems in the first order predicate calculus: sentences are (some) closed formulae, models are relational structures (of the appropriate type) and $\|\varphi\|_{\underline{M}} = 1$ if φ is true in $\underline{M}$ in the sense of Tarski. The reader will meet this notion and its modifications in Chapter II and will see the relation of this notion to Example (1).

(4) We consider an example where V is the set of rational numbers. The semantic system $\mathcal{E}_n$ is defined as follows: models are rational matrices with n columns and finitely many rows. With each $i \in \{1,\ldots,n\}$ we associate a sentence φ_i whose value is the mean of the i-th column, i.e. if

$$\underline{M} = \left(r_k^i\right)_{\substack{i=1,\ldots,n \\ k=1,\ldots,m}} \text{ then } \mathrm{Val}(\varphi_i, \underline{M}) = \frac{1}{m} \sum_{k=1}^{m} r_k^i \, .$$

(In 1.2.8 we shall say in what sence such a sentence can be asserted.)

1.2.7 Definition. (1) Let $\mathcal{S} = \langle \text{Sent}, \mathfrak{M}, V, \text{Val} \rangle$ be a semantic system and let $V_o \subseteq V$ be a set of designated values. A sentence φ is V_o-true in a model $\underline{M}$ if $\|\varphi\|_{\underline{M}} \in V_o$ (notation $\underline{M} \models_{V_o} \varphi$). A sentence φ is a V_o-tautology if $\|\varphi\|_{\underline{M}} \in V_o$ for each $\underline{M} \in \mathfrak{M}$ (notation $\models_{V_o} \varphi$).

A sentence φ is a logical V_o-consequence of a set A of sentences if the following holds for each $\underline{M} \in \mathfrak{M}$: if each element of A is V_o-true in M then φ is V_o-true in $\underline{M}$. The set of all sentences V_o-true in a model $\underline{M}$ is denoted by $Tr_{V_o}(\underline{M})$.

(2) Let, moreover, I be an inference rule on Sent. I is V_o-sound w.r.t. $\mathcal{S}$ if the following holds for each $\underline{M} \in \mathfrak{M}$ and each $\langle \varphi, e \rangle \in I$: if each element of e is V_o-true in $\underline{M}$ then φ is V_o-true in $\underline{M}$. In symbols: $e \subseteq Tr_{V_o}(M)$ implies $\varphi \in Tr_{V_o}(\underline{M})$. One sees immediately that if I is V_o-sound w.r.t. $\mathcal{S}$ then the following holds: If φ is I-derivable from A then φ is a logical V_o-consequence of A. If I is a V_o-sound inference rule w.r.t. $\mathcal{S}$ then we call also I a V_o-deduction rule for $\mathcal{S}$ and say I-proof, I-provable or simply proof, provable instead of I-derivation, I-derivable.

Let I be a V_o-sound inference rule for $\mathcal{S}$. I is V_o-complete w.r.t. $\mathcal{S}$ if the following holds for each $A \subseteq$ Sent and $\varphi \in$ Sent: if φ is a logical V_o-consequence of A then φ is V_o-provable from A.

1.2.8 Remark. One V_o-asserts a sentence φ if one wants to say that φ is V_o-true (in the model one is speaking about). For example, one asserts (i.e. $\{1\}$-asserts) "the properties P_3, P_4 and P_7 are incompatible" if one wants to say that, in the model one is speaking about, the properties mentioned are incompatible. We can easily imagine a situation (in a research centre) where sentences are V_o-asserted e.g. for V_o being the set of all positive rational numbers. One has data $\underline{M} = (r^i_k)^i_k$ as in Example 1.2.6 (4) and one is interested in columns (quantities) with positive average. Then one V_o-asserts "the average of the third quantity " if one wants to say that average is indeed positive.

1.2.9 Examples and Comments. (1) Remember Example 1.2.6 (1) and (2) with sentences of the form φ_e "the properties P_i for $i \in e$ are incompatible". We have a $\{1\}$-sound deduction rule I_n on Sent defined as follows:

$$I_n = \left\{ \frac{\varphi_e}{\varphi_{e'}} \; ; \; \emptyset \neq e \subseteq e' \right\}$$

. I.e., whenever $e \subseteq e'$ we can infer (deduce) φ'_e from φ_e. Indeed, if for example $e = \{3,4,7\}$ and $e' = \{1,3,4,5,7\}$ and if, in a model $\underline{M}$, P_3, P_4 and P_7 are incompatible then a fortiori P_1, P_3, P_4, P_5 and P_7 are incompatible. (Think, for example, of mutations.)

(2) For Example 1.2.6. (3) , we have well known deduction rules, for example modus ponens:

$$\left\{ \frac{\varphi, \varphi \rightarrow \psi}{\psi} \; ; \quad \varphi, \psi \in \text{Sent} \right\}$$

($\rightarrow$ is the connective of implication, see Chapter II) .

(3) Concerning Example 1.2.6 (4) , it is easy to show that if V_o is a non-empty proper set of rationals then the only V_o-sound rule is the identity:

$$\left\{ \frac{\varphi}{\varphi} \; ; \quad \varphi \in \text{Sent} \right\}.$$

1.2.10 Discussion . (i) We now turn our attention to notions concerning computability. As we stated in the preface, we assume some knowledge of elementary recursion theory.

Partial recursive functions are particular natural-valued functions; the domain of a k-ary partial recursive function consists of some k-tuples of natural numbers and its range is included in $\mathbb{N}$ ($\mathbb{N}$ denotes the set of all natural numbers) . One defines partial recursive functions by specifying initial partial recursive functions and describing operations over functions that preserve partial recursiveness. (Partial) recursive functions can be identified with the (partial) functions mechanically computable in principle, i.e. without any restrictions as to the time and space necessary for the calculation of a particular value, except that they are requested to be finite. This identification is the content of the well-known extended Church's thesis.

(2) An alternative well-known approach is based on the notion of a Turing machine. Here one works with words in a finite alphabet rather than with natural

numbers. Given a Turing machine T with the tape alphabet $\{a_1, \ldots, a_n\}$ and a word x in this alphabet, one defines the computation of T with the input (initial tape inscription) x as a certain sequence of steps; a tape inscription is associated with each step. One can ask whether the computation halts or not; if it halts, one has the output computed by T, i.e. the final tape description.

(3) The two approaches are known to be equivalent. First, one can code (enumerate) words by some natural numbers (Gödel numbering). Conversely one can represent numbers by words, e.g. one can represent n by $\underbrace{1, \ldots, 1}_{(n+1) \text{ times}}$. One can code tuples of numbers in a similar way. For our purposes, it suffices to state the following fact: A function F from $\mathbb{N}^k$ into $\mathbb{N}$ is a partial recursive function iff there is a Turing machine T such that the following holds: for each k-tuple $\langle n_1, \ldots, n_k \rangle$ of natural numbers, the computation of T with (the code of) $\langle n_1, \ldots, n_k \rangle$ as input halts and gives the answer $F(n_1, \ldots, n_k)$ whenever $F(n_1, \ldots, n_k)$ is defined; the computation does not halt if $\langle n_1, \ldots, n_k \rangle \notin \mathrm{dom}(F)$. This equivalence is one of the arguments supporting Church's thesis.

(4) However, one objection to Church's thesis is that we do not always know ahead of time how many steps will be required to compute $F(n)$ for a recursive function F. This has led to various theories of computational complexity. We shall make use of the following definition:

Let T be a terminating Turing machine (halting for all inputs). T is said to operate in polynomial time if there is a polynomial p such that, for every word x in the tape alphabet, the number of steps of the computation with the input x is bounded by p(length x). It is a reasonable working hypothesis, by now widely accepted, that a problem concerning words in a finite alphabet can be regarded as tractable iff there is an algorithm (Turing machine) for its solution operating in polynomial time. See Karp [1972] for information. We shall pay attention to questions of polynomial complexity in two directions: in Chapter III we show that some problems concerning certain observational calculi are closely connected with open problems of complexity theory and in Part B we shall state various properties of the described methods and of related notions in terms of polynomial complexity. But, knowledge of complexity theory is not assumed for the main body of the text.

(5) One is often interested in computational properties of elements of domains D more general than the set of all natural numbers and/or the set of all words in

a finite alphabet. For instance, one considers functions whose arguments and values are finite graphs, matrices, finite rational structures etc. In this case one uses a simple <u>encoding</u> of elements of D by natural numbers or by words; then the whole theory is shifted to D, naturally in dependence on the chosen encoding. Note that Gödel numbering is an example of coding of words by numbers. Hence each encoding e of elements of D by words determines an encoding by numbers, namely the composition of e with Gödel numbering.

(6) Let us go over some notions and facts. A set $X \subseteq \mathbb{N}$ is a <u>recursive set</u> if its characteristic function is a total recursive function. Similarly for relations, i.e. subsets of $\mathbb{N}^k$ for some k.
A set $X \subseteq \mathbb{N}$ is <u>recursively enumerable</u> if there is a recursive relation R such that $X = \{a; (\exists b) R(a,b)\}$.

Facts: X is recursively enumerable iff X is the range of a partial recursive function. X is recursive iff both X and the complement of X are recursively enumerable. Clearly one calls a set Y of words recursively enumerable iff the set of Gödel numbers of elements of Y is recursively enumerable. Fact: A set Y of words in an alphabet Σ is recursively enumerable iff there is a Turing machine T whose computation halts iff the input is in Y. These and similar facts will be freely used in the sequel.

1.2.11 <u>Remark.</u> Let us return to our notions concerning sentences. It is natural to assume that sentences are finite objects, e.g. some finite words in a fixed alphabet. Furthermore, it is natural to assume that sentences form a recursive set; if inference rules are considered as <u>rules</u> it is natural to restrict oneself to <u>recursive</u> inference rules. It makes sense to speak of recursive inference rules since finite sets of natural numbers can be naturally coded by natural numbers (finite sets of words can be naturally coded by words). We can now relate the notions of sound and complete deduction rules with notions concerning recursiveness.

1.2.12 <u>Definition.</u> Let Sent be a recursive set, let $\mathcal{S} = \langle \mathrm{Sent}, \mathcal{M}, V, \mathrm{Val}\rangle$ be a semantic system and let $V_o \subseteq V$.
(1) $\mathcal{S}$ is V_o<u>-decidable</u> if the set of all V_o-tautologies is recursive. $\mathcal{S}$ is <u>strongly</u> V_o<u>-decidable</u> if the relation
$\{\langle\varphi,e\rangle ; \varphi \in \mathrm{Sent}, e \in \mathcal{P}_{fin}(\mathrm{Sent}), e \models_{V_o} \varphi\}$ of semantic consequence is recursive.

(2) $\mathcal{S}$ is V_o-axiomatizable if the set of all V_o-tautologies is recursively enumerable; $\mathcal{S}$ is strongly V_o-axiomatizable if the relation $\{\langle\varphi,e\rangle ; \varphi\in \text{Sent}, e\in \mathcal{P}_{fin}(\text{Sent}), e\models_{V_o}\varphi\}$ of semantic consequence is recursive enumerable.

1.2.13 Remark. (1) Having shown that a certain $\mathcal{S}$ is V_o-undecidable (for a V_o we are interested in), we can see that there is no mechanical procedure for deciding whether a sentence is a V_o-tautology or not. If we show $\mathcal{S}$ to be V_o-decidable, then we are faced with the important question whether there is a procedure deciding $\mathcal{S}$ simple enough to be realizable (tractable).

(2) The term "axiomatizable" is justified by the following theorem, due in essence to Craig (for strong axiomatizability see Problem (4)).

1.2.14 Theorem. Let Sent be a recursive set and let A be a recursively enumerable subset of Sent. Then there is a recursive inference rule I on Sent such that $A = \{\varphi\in \text{Sent} ; \vdash_I \varphi\}$ (A consists of all sentences I-provable from the empty set of assumptions).

Proof. Assume for simplicity that $\text{Sent}\subseteq \mathbb{N}$. Our assertion is obvious if A is finite: then $I = \{\frac{\emptyset}{\varphi} ; \varphi\in A\}$. Assume A be infinite and let $A = \{x\in \text{Sent}; (\exists n)R(x,n)\}$ where R is a recursive relation. Let B be an infinite recursive subset of A. Put $I = \{\frac{\emptyset}{\varphi} ; \varphi\in B\}\cup\{\frac{\varphi}{\psi} ; (\exists n<\varphi) R(\psi,n)\}$. Obviously, I is recursive and putting $D = \{\varphi ; \emptyset\vdash_I \varphi\}$ we have $D\subseteq A$. On the other hand, if $\psi\in A$ and if $R(\psi,n)$ holds then there is a $\varphi\in B$ such that $\varphi> n$ since B is infinite ; the two element sequence φ,ψ is an I-derivation of ψ from $\emptyset$. We have found $A\subseteq D$.

1.2.15 Corollary. Let Sent be a recursive set, let $\mathcal{S} = \langle \text{Sent}, \mathfrak{M}, V, \text{Val}\rangle$ be a semantic system and let $V_o\subseteq V$. $\mathcal{S}$ is V_o-axiomatizable iff there is a recursive V_o-sound deduction rule on Sent such that I is V_o-complete, i.e. V_o-tautologies coincide with sentences I-provable from the empty set of assumptions.

1.2.16 Example. For each n, the semantic system $\mathcal{S}_n$ of Example 1.2.6 (1) is strongly $\{1\}$-decidable since the set of its sentences is finite, hence the relation $\models_{\{1\}}$ is also finite and therefore recursive. For another example see Problem (5).

1.2.17 We shall now formulate a definition of basic importance, namely of an observational semantic system. The definition reflects a very important aspect of the informal notion of observationality, namely that an observational model is a finite set of observed data (hence the whole model is a finite object) and that one has to be able to determine mechanically (compute) the value of each formula in each observational model. (Note that we suppose data to be digital.)

1.2.18 Definition. A semantic system $\mathcal{S} = \langle \mathrm{Sent}, \mathcal{M}, V, \mathrm{Val} \rangle$ is an observational semantic system if Sent, $\mathcal{M}$, V are recursive sets and Val is a partial recursive function.

1.2.19 The definition assumes that Sent, $\mathcal{M}$, V are subsets of a countable domain D that has been encoded by natural numbers (see the following example). Note that Val is a partial recursive funtion whose domain is a recursive set; hence Val is the restriction of a total recursive function to Sent $\times \mathcal{M}$.

1.2.20 Example. With Example 1.2.16 in mind we show how to encode Sent, $\mathcal{M}$ and V in cases (1) and (4); then it is obvious that the systems $\mathcal{S}_m$ and $\mathcal{E}_m$ are observational. We use coding by words.

(1) The alphabet is $\{0,1,\square,*\}$. Each sentence φ_e is coded by the word $\square, \varepsilon_1, \ldots, \varepsilon_n$ of the length n+1 where $\varepsilon_i = 1$ iff $i \in e$. A matrix

$M = (r_i^j)_{i=1,\ldots,m}^{j=1,\ldots,n}$ is coded by the word

$$*, r_{11}, \ldots, r_{1n}, *, \ldots, *, r_{m1}, \ldots, r_{mn}, * \qquad (*)$$

and V is the set $\{0,1\}$.

(4) The alphabet is $\{0, \ldots, 9, \square, *, +, -, /\}$.

Natural numbers are coded by their usual decimal expansion; rational numbers are treated as signed fractions, i.e. words of the form +x/y or -x/y where x and y are codes of natural numbers, y not zero. For instance +3/17 is a word of length 5. Sentences are coded as $\square x$ where x is the code of a natural number between 1 and n. Models are coded as in (1); r_i^j is now a code of a rational number and the expression $(*)$ is to be understood as the juxtaposition of the respective parts. For instance, the code of

$$\begin{pmatrix} \frac{3}{17} , & \frac{2}{3} \\ 0 , & 5 \end{pmatrix}$$

is $*+3/17*-2/3*+0/1*+5/1$.

1.2.21 Theorem. Let $\mathcal{S} = \langle \mathrm{Sent}, \mathfrak{M}, V, \mathrm{Val} \rangle$ be an observational semantic system and let V_o be a recursive subset of V.

(1) $\mathcal{S}$ is V_o-axiomatizable iff $\mathcal{S}$ is V_o-decidable.

(2) $\mathcal{S}$ is strongly V_o-axiomatizable iff $\mathcal{S}$ is strongly V_o-decidable.

(3) The set Sf_{V_o} of all V_o-satisfiable sentences (i.e. sentences φ such that there is an $\underline{M}$ such that $\|\varphi\|_{\underline{M}} \in V_o$) is recursively enumerable. Similarly, the complement of $\models_{V_o}$, i.e. the relation

$$\{\langle\varphi, e\rangle \; ; \; \varphi \text{ is not a logical consequence of } e, \; e \in \mathcal{P}_{fin}(\mathrm{Sent})\}$$

is recursively enumerable.

Proof. Let us first show (3). We have

$Sf_{V_o} = \{\varphi \; ; \; (\exists \underline{M})(\varphi \in \mathrm{Sent} \; \& \; \underline{M} \in \mathfrak{M} \; \& \; \|\varphi\|_{\underline{M}} \in V_o\}$ and the relation $\varphi \in \mathrm{Sent} \; \& \; \underline{M} \in \mathfrak{M} \; \& \; \|\varphi\|_{\underline{M}} \in V_o$ is recursive; hence Sf_{V_o} is recursively enumerable. Similarly for the more general case.

We show that (3) implies (1). Evidently, decidability implies axiomatizability. Conversely, if $\mathcal{S}$ is V_o-axiomatizable then the set $Taut_{V_o}$ of all V_o-tautologies is recursively enumerable. Its complement (w.r.t. Sent) is the set Sf_{V-V_o} and $V-V_o$ is recursive; by (3) Sf_{V-V_o} is recursively enumerable. Hence $Taut_{V_o}$ is recursive. Similarly for the second case.

1.2.22 Remark. (1) The reader familiar with the first order predicate calculus sees that observational semantic systems differ from the semantic systems of the first order predicate calculus (cf. Example 1.2.6 (3)); the semantic system of each predicate calculus with at least one binary predicate is axiomatizable (i.e. $\{1\}$-axiomatizable) but not decidable. See the next chapter for more details.

(2) We can replace the conditions of recursiveness in the definition of an observational semantic system by stronger conditions (primitive recursiveness, computability in polynomial time etc.) . In this way we obtain similar, more

restrictive notions. We shall pay attention to this fact; but we shall consider the definition of an observational semantic system as one of our basic definitions.

(3) In the rest of this section we introduce some formal notions concerning observational and theoretical semantic systems and their relations; in particular, we introduce a formal notion of inductive inference rules.

1.2.23 Suppose we have a recursive set Sent whose elements are called sentences and which is a union of recursive (not necessarily disjoint) sets $Sent_0$, $Sent_T$; elements of $Sent_0$ are called observational sentences and elements of $Sent_T$ are called theoretical sentences. An inference rule I on Sent will be called <u>inductive</u> if it consists of some pairs of the form

$$\frac{\Gamma, \Delta}{\Psi} \qquad (*)$$

where Γ is a finite set of theoretical sentences, Δ is a non-empty finite set of observational sentences and Ψ is a theoretical sentence. (Cf. 1.1.4). Now we want to indicate the way in which we shall try to answer the question (L1), namely what are the criteria of rationality (measures of rationality) of inferences using I. We shall pay attention to the semantic properties of the sentences involved. This means that the notions of rationality will be defined with respect to two sematic systems:

an observational semantic system $\mathcal{S}^0 = \langle \mathrm{Sent}^0, \mathfrak{M}^0, V^0, \mathrm{Val}^0 \rangle$ and a "theoretical" system $\mathcal{S}^T = \langle \mathrm{Sent}^T, \mathfrak{M}^T, V^T, \mathrm{Val}^T \rangle$. Note that we shall have to answer our question (L0) ; the answer will consist in a detailed theory of the appropriate structure of observational and theoretical systems.

Observational models will be considered as possible "parts" of partial information on theoretical models; for the time being let us have an abstract relation $\prec$ - a subset of $\mathfrak{M}^0 \times \mathfrak{M}^T$; $\underline{M}_0 \prec \underline{M}_T$ is read "$\underline{M}_0$ is a part of $\underline{M}_T$". The pattern of an act of inductive inference is as follows: one has an observational model $\underline{M}_0$ - evidence - which is, in some sense, a part of a theoretical universe $\underline{M}_T$. One is interested in sentences true in $\underline{M}_T$ but does not have $\underline{M}_T$ as a totality at one's disposal. Let

$$\frac{\Gamma, \Delta}{\Psi}$$

be an element of I as above (cf. (*)). One has accepted that Γ is true in $\underline{M}_T$ (V^T_o-true for some $V^T_o \subseteq V^T$) and one has verified that Δ is true in $\underline{M}_0$. Then one accepts the hypothesis that Ψ is true in $\underline{M}_T$. The question of the rationality of I is a question about the properties of this sort of reasoning.

1.2.24 We shall pause here for a simple example. Here the "part of" - relation is inclusion; let us stress that this is not the only possibility. As a matter of fact, our main attention will be paid to more general "part of" - relations.

Remember Example 1.2.6 (1) (cf. 1.2.20 (1)). We fix an n and let $\mathcal{S}_n$ be an observational semantic system $\mathcal{S}^0 = \langle \mathrm{Sent}^0, \mathfrak{M}^0, V^0, \mathrm{Val}^0 \rangle$. Hence models are matrices of zeros and ones with n columns; if $e \subseteq \{1, \ldots, n\}$ then the sentence φ_e is read "the properties P_i for $i \in e$ are incompatible in the observed model".

Our theoretical system $\mathcal{S}^T$ has for each $e \subseteq \{1, \ldots, n\}$ a sentence Ψ_e which reads "the properties P_i ($i \in e$) are incompatible in the theoretical model". Theoretical models are matrices of zeros and ones with n columns and countably many rows (the rows form a sequence indexed by all natural numbers). The evaluation function Val^T is the obvious modification of Val^0. $\prec$ is defined by: $\underline{M}_0$ is a part of $\underline{M}_T$ if $\underline{M}_0$ results from $\underline{M}_T$ by omitting all but finitely many rows. The inference rule is

$$I = \left\{ \frac{\varphi_e}{\Psi_e} ; \quad e \subseteq \{1, \ldots, n\} \right\} .$$

The reader realizes that this is an inference rule related to Example 1.1.2 (1) (All crows are black) : If we verify that P_1, P_3, P_4 are incompatible in the observed model, we accept the hypothesis that those properties are incompatible in the universe. Concerning the rationality of this inference rule, we can say that I is a sort of inverse deduction: for each $\underline{M}_0 \prec \underline{M}_T$ and each e, $\|\psi_e\|_{\underline{M}_T} = 1$ implies $\|\varphi_e\|_{\underline{M}_0} = 1$ and (φ_e is the "logically weakest" element of Sent^T with this property).

1.2.25 A general definition would be as follows. Let $\mathcal{S}^0$, $\mathcal{S}^T$, be as above, let I be an inductive inference rule and let V_o^0, V_o^T be some sets of observational and theoretical values respectively. I is deterministic (w.r.t. to the things just named) if the following holds for each $\underline{M}_0 \prec \underline{M}_T$ and each $\frac{\Gamma, \Delta}{\Psi} \in$ I : the V_o^T - truthfulness of (each element of) Γ and of Ψ in $\underline{M}_T$ implies the V_o^0 - truthfulness of all elements of Δ in M_0. Let us stress now that we shall be mainly interested in rules that are not deterministic. We want to formalize inferences like those in Example 1.1.2 (2),(3), where the inferred theoretical sentences express something about chance or belief. In particular, our aim is to analyse statistical inferences in the present terms. Our task will be to find some appropriate structure of observational and theoretical systems, including appropriate "part of" - relations, for this purpose. Our main notion will be the notion of a state-dependent structure, related to the semantics of some modal calculi. Cf. (Scott and Krauss) and references given there.

1.2.26 Key words: Inference rule, derivation, semantic system (sentences + models + values + evaluation), soundness and completeness (of an inference rule w.r.t. a semantic system), axiomatizability, decidability, observational semantic system, inductive inference rules (deterministic or otherwise).

PROBLEMS AND SUPPLEMENTS TO CHAPTER I

(1) Let I be an inference rule on Sent. I is <u>transitive</u> if, for each $X \subseteq$ Sent, $I(X) = I(I(X))$. (I is transitive iff I-derivability coincides with being an (immediate) I-conclusion). I is <u>regular</u> if the following holds for each $\varphi \in$ Sent and each $e_1, e_2 \in \mathcal{P}_{fin}(\text{Sent})$:

(i) $$\frac{\varphi}{\varphi} \in I\ ,$$

(ii) $$\left(\frac{e_1}{\varphi} \in I \text{ and } e_1 \subseteq e_2 \quad \text{implies} \quad \frac{e_2}{\varphi} \in I\right).$$

(a) Show that for each I there is a regular inference rule I on Sent such that, for each $A \subseteq$ Sent, $I(A) = I'(A)$. (b) Show that for each I, the rule $I' = \{\langle \varphi, e\rangle \ ; \ \varphi$ is I-derivable from $e\}$ is a regular transitive inference rule and I-derivability coincides with I′- derivability.

(2) Let $\mathcal{S} = \langle \text{Sent}, \mathfrak{M}, V, \text{Val}\rangle$ be a semantic system and let $V_o \subseteq V$. The relation $\{\langle\varphi, e\rangle \ ; \ \varphi \in \text{Sent} \ \& \ e \in \mathcal{P}_{fin}(\text{Sent}) \ \& \ e \models_{V_o} \varphi\}$ is a regular transitive inference rule.

(3) The following is a generalization of Craig's theorem:

<u>Theorem</u>. Let Sent be a recursive set and let K be a recursively enumerable regular transitive inference rule of Sent. Suppose that for each finite set e of sentences, the set $K(e) = \{\varphi \ ; \ \frac{e}{\varphi} \in K\}$ is infinite. Then there is a recursive inference rule I on Sent such that, for each e and φ ,

$$\frac{e}{\varphi} \in K \text{ iff } \varphi \text{ is I-derivable from } e\ .$$

<u>Hint</u>: Use the following fact from recursion theory:

Let $X, Y \subseteq \mathbb{N}$ be recursive sets, let $R \subseteq X \times Y$ be recursively enumerable and suppose that for each $n \in Y$, the set $\{m \in X; \ R(m,n)\}$ is infinite. Then there is a recursive relation $R_o \subseteq R$ such that for each $n \in Y$ the set $\{m \in X \ ; \ R_o(m,n)\}$ is infinite.

Let $\frac{e}{\varphi} \in K$ iff $(\exists n)S(\varphi, e, n)$ where S is a recursive relation. Let K_o be a subrelation of K such that, for each e, $K_o(e)$ is infinite. Put

$$I = K_o \cup \left\{ \frac{e, \varphi}{\psi} \; ; \; (\exists n < \varphi)\, S(\psi, e, n) \right\}.$$

(4) Corollary. Let $\mathcal{S} = \langle \mathrm{Sent}, \mathfrak{M}, V, \mathrm{Val} \rangle$ be a semantic system and let $V_o \subseteq V$. Assume that for each $\varphi \in \mathrm{Sent}$ the set $\{\psi \; ; \; \varphi \models_{V_o} \psi\}$ is infinite. Then the following holds:

$\mathcal{S}$ is strongly V_o-axiomatizable iff there is a recursive V_o-sound inference rule I on Sent such that I is strongly V_o-complete. (i.e. $e \models_{V_o} \varphi$ iff $e \vdash_I \varphi$ for each φ and each e).

Remark. The infinity condition is quite natural: it suffices if we can express each formula in infinitely many mutually V_o-equivalent ways. For example, in the predicate calculus

φ, $\varphi \& \varphi$, $\varphi \& \varphi \& \varphi$, ... are equivalent.

(5) We modify the example 1.2.6 (1) of an observational semantic system (cf. 1.2.20 (1)).

The semantic system $\mathcal{S}^*$ is defined as follows: $V = \{0,1\}$; models are arbitrary matrices of zeros and ones (with finitely many rows and columns) . With each finite set e of positive natural numbers we associate a sentence φ_e (read: the properties P_i for $i \in e$ are incompatible). $\|\varphi_e\|_{\underline{M}}$ is defined as follows: if $\underline{M}$ has n columns and $e \subseteq \{1, \ldots, n\}$ then the definition is as in 1.2.6 (1) . If e is not included in $\{1, \ldots, n\}$ and if $m = \max(e)$ then we define $\|\varphi_e\|_{\underline{M}}$ as $\|\varphi_e\|_{\underline{M}'}$ where $\underline{M}'$ results from $\underline{M}$ by adding to $\underline{M}$ an (n+1)- th , ... ,m-th column, all added columns consisting only of zeros - non-interpreted properties are always assumed not to hold.

(a) Show that the system $\mathcal{S}^*$ is an observational semantic system.

(b) Consider the rule

$$I = \left\{ \frac{\varphi_{e_1}}{\varphi_{e_2}} \; ; \; e_1 \subseteq e_2 \right\}.$$

Show that I is transitive, $\{1\}$ -sound and $\{1\}$-complete. (Easy.)

(6) Remark. There is an alternative approach to recursion theory, in which recursive functions are defined as functions on hereditarily finite sets

(finite sets of finite sets of finite sets ...). Put $V_0 = \emptyset$ and for each n let V_{n+1} be the set of all subsets of V_n. Then the set HF of hereditarily finite sets is the union $\bigcup_{n \in \mathbb{N}} V_n$ (cf. Rödding, Jensen and Karp).

This approach would be very useful for our purposes; unfortunately, it is relatively little known. We also mention the informal treatment of recursive functions in Shoenfield [1971] based on an informal notion of a finite object; Shoenfield's treatment can easily be formalized using recursion on hereditarily finite sets by identifying finite objects with hereditarily finite sets. But we decided in favour of an approach that is well known.

Part A
A Logic of Induction

Chapter II. A Formalization of Observational and Theoretical Languages

Recall our notion of a semantic system as consisting of sentences, models, abstract values and an evaluation function assigning to each sentence and each model M the value $\|\varphi\|_{\underline{M}}$ of φ in M. In the present chapter we are going to analyse possible structures of sentences and of models and the dependence of $\|\varphi\|_{\underline{M}}$ on the structure of φ and of $\underline{M}$. Our aim is to generalize and modify the classical predicate calculus in various ways, in particular by admitting generalized quantifiers. (The following are preliminary examples of sentences containing generalized quantifiers: (i) For sufficiently many x, P(x). (ii) The property Q(x) is associated with R(x).)

Our plan is to divide our generalizations and modifications into several easy steps. We obtain various formal calculi similar to the classical predicate calculus and find conditions under which they can be naturally called observational. We also describe calculi that will be used as our formalization of theoretical languages. We shall illustrate defined notions by examples and state basic facts about them. At the end of the chapter we shall be able to offer our answer to the questions of the logic of induction ((L0)-(L2) of 1.1.5.) Systematic mathematical theory of observational and theoretical calculi is postponed to Chapters III and IV; but at the end of each section (except Section 1) the reader will be informed which parts of Chapters III and IV can be read as an immediate continuation.

II.1 Structures

2.1.1 We are going to present the definition of V-valued structures as a familiar generalization of relational structures. V-valued structures will play the role of models in the sense of semantic systems A relational structure consists of a non-empty domain M and some relations $R_1, \ldots, R_n$ on M of various arities. In symbols,

$$\underline{M} = \langle M, R_1, \ldots, R_n \rangle \quad .$$

Obviously, relations can be replaced by two-valued functions on M of the appropriate arity, i.e. characteristic functions of the relations. This can be generalized by allowing functions with values in an abstract set V. For example, the question "are x,y related?" defines a two-valued binary function (binary relation) ; the question "what is the degree of relationship of x,y?" defines a binary function with more general values. We make the following definition.

2.1.2 Definition. (1) A type is a finite sequence $\langle t_1, \ldots, t_n \rangle$ of positive natural numbers. A V-structure of the type $t = \langle t_1, \ldots, t_n \rangle$ is a tuple

$$\underline{M} = \langle M, f_1, \ldots, f_n \rangle$$

where M is a non-empty set called the domain (or field) of $\underline{M}$ and each f_i is a mapping from M^{t_i} into V. A V-structure $\underline{N} = \langle N, g_1, \ldots, g_n \rangle$ of type t is a substructure of $\underline{M}$ if $N \subseteq M$ and each g_i is the restriction of f_i to N^{t_i}. A one-one mapping j of M onto N is an isomorphism of $\underline{M}$, $\underline{N}$ if it preserves the structure, i.e. for each i and $o_1, \ldots, o_{t_i} \in M$ we have $f_i(o_1, \ldots, o_{t_i}) = g_i(j(o_1), \ldots, j(o_{t_i}))$.

2.1.3 Examples (1) Let V be the set of non-negative reals. A metric space is a V-structure $\langle M, d \rangle$ of type $\langle 2 \rangle$ satisfying the well known assumptions.

(2) Let $\mathbb{N}$ denote the set of natural numbers. The arithmetical structure on $\mathbb{N}$ (addition and multiplication) can be characterized

(a) as a $\{0,1\}$ - structure $\langle \mathbb{N}, ad, mt \rangle$ of type $\langle 3,3 \rangle$ where $ad(i,j,k) = 1$ iff $i+j=k$ and $mt(i,j,k) = 1$ iff $i.j=k$, or

(b) as an $\mathbb{N}$-structure $\langle \mathbb{N}, a, m \rangle$ of type $\langle 2,2 \rangle$ where $a(i,j) = i+j$ and $m(i,j) = i.j$.

(3) Let M be a finite set and let $\leq$ be a linear ordering of M. The set M ordered by $\leq$ can be expressed

(a) as a $\{0,1\}$-structure $\langle M, f \rangle$ of type $\langle 2 \rangle$ such that $f(o_1, o_2) = 1$ iff $o_1 \leq o_2$, or

(b) as a $\mathbb{N}$-structure $\langle M, r \rangle$ of type $\langle 1 \rangle$ where $r(o)$ is the rank of o w.r.t. $\leq$, i.e. $r(o) = 0$ iff o is the least element, $r(o) = 1$ iff o is the immediate successor of the least element etc.

2.1.4 Denote by $\mathfrak{M}_t^V$ the set of all V-structures $\underline{M}$ of type t such that the domain of M is a finite set of natural numbers. Thus each V-structure $\underline{M}'$ of type t with a finite domain is isomorphic to a member of $\mathfrak{M}_t^V$.

In addition, let V be a recursive set of natural numbers. It is easy to define a natural coding of $\mathfrak{M}_t^V$ by some natural numbers. For example, remember that there is a natural coding of all tuples of natural numbers by natural numbers. A type is a tuple of natural numbers; a V-valued function on a finite set $M \subseteq \mathbb{N}$ can be represented as a tuple of tuples of natural numbers, using the natural ordering of $\mathbb{N}$; a structure can be represented as a tuple consisting of the domain, the type and the respective V-valued functions.

Elements of $\mathfrak{M}_t^V$ where V is a recursive set can be called <u>observational V-structures</u> of type t. Note that V may become a subset of $\mathbb{N}$ using a coding; thus it makes sense to speak for example of observational $\mathbb{Q}$-structures where $\mathbb{Q}$ is the set of all rationals. Note also that it makes sense to speak of a recursive function some arguments of which vary over observational V-structures of a given type. This remark will be used in the definition of observational calculi.

2.1.5 A further generalization, also well known, consists in the "parametrization" of the V-valued function by a new argument ranging over an abstract set Σ of "states". This corresponds to the classical idea of a system of "possible worlds" rather than a single "world". We shall define Σ-state dependent V-structures. Such structures have been used in modal logic (Kripke) and also in robotics (Mc Carthy-Hayes). Our treatment of state dependent structures will differ from that in the literature.

2.1.6 Definition. A Σ-state dependent V-structure of type $t = \langle t_1, \ldots t_n \rangle$ is a tuple

$\underline{U} = \langle U, f_1, \ldots, f_n \rangle$

where $U \neq \emptyset$ and each f_i, maps $U^{t_i} \times \Sigma$ into V.

Let V, U be fixed. Any mapping of Σ into V is called a state dependent variate. Obviously, for each i, each t_i-tuple $o_1, \ldots, o_{t_i}$ of elements of U determines a state dependent variate $\vartheta^i_{o_1, \ldots, o_{t_i}}(\sigma) \overset{i}{=} f_i(o_1, \ldots, o_{t_i}, \sigma)$ called the variate determined by $o_1, \ldots, o_{t_i}$. On the other hand, each σ determines a V-structure $\underline{U}_\sigma = \langle U, f_1(-, \sigma)^i, \ldots, f_n(-, \sigma) \rangle$ called the structure determined by σ. A sample is a finite (non-empty) subset M of U. The sample structure $\underline{M}^{\underline{U}}_\sigma$ determined by a sample M and a state σ is the substructure of $\underline{U}$ whose domain is M. If there is no danger of a misunderstanding we write $\underline{M}_\sigma$ instead of $\underline{M}^{\underline{U}}_\sigma$.

2.1.7 One often assumes, using state dependent structures, that the set of states is endowed with a structure. For example, if we have a linear ordering on Σ we may understand elements of Σ as moments of time; $\underline{U}_\sigma$ is the state of $\underline{U}$ in the moment σ.

Our idea is to understand a state dependent structure as a structure $\underline{U} = \langle U, f_1, \ldots, f_n \rangle$, where the value of each f_i for $o_1, \ldots, o_{t_i}$ is not determined by $o_1, \ldots, o_{t_i}$ themselves but depends on some random factors. Suppose we have a system $\mathcal{E}$ of subsets of Σ such that $X \subseteq Y \in \mathcal{E}$ imples $X \in \mathcal{E}$; call elements of $\mathcal{E}$ small subsets. Then we say that the value of f_i for $o_1, \ldots, o_{t_i}$ has little chance of belonging to $V_o \subseteq V$ if

$\{\sigma ; f_i(o_1, \ldots, o_{t_i}, \sigma) \in V_o\} \in \mathcal{E}$.

In Chapter IV we shall study random structures: Suppose we have a probability measure P on Σ, i.e. P maps some subsets of Σ into the real interval $[0,1]$ and satisfies the usual conditions. Write $\mathcal{R}$ for the domain of P and call

$\underline{\Sigma} = \langle \Sigma , \mathcal{R} , P \rangle$ a probability space (see Chapter IV, Section 1 for details). Any Σ-state dependent structure may be called a $\underline{\Sigma}$- random V-structure. The probability measure on $\underline{\Sigma}$ defines various notions of small sets. Take an $\alpha \in [0,0.5]$ and define $\mathcal{E}_\alpha = \{X \subseteq \Sigma ;$ for some $Y \supseteq X, Y \in \mathcal{R} , P(Y) \leq \alpha\}$. This is a typical example of a system of small subsets of Σ .

2.1.8 We are now able to be more specific on rationality criteria for inductive inference rules. Let V be a set of abstract values; for simplicity, assume V to be a recursive set. Let Σ be a set of states; let $\mathcal{E}$ be a system of small subsets of Σ . Assume we have a theoretical semantic system $\mathcal{S}^T$ with abstract values V and with Σ-state dependent V-structures of type t as models. Furthermore, let us have an observational semantic system $\mathcal{S}^O$ with abstract values V and with finite V-structures of type t as models. (More precisely, the set of models of $\mathcal{S}^O$ is $\mathfrak{M}_t^V$, cf. 2.1.4). Note that saying "observational" we assume that the evaluation function of $\mathcal{S}^O$ is recursive . The "part of" - relation $\prec$ of 1.3.1 is now defined as follows: $\underline{M} \prec \underline{U}$ iff $\underline{M}$ is a sample structure from $\underline{U}$.

Let I be an inductive inference rule w.r.t. $Sent^T$ and $Sent^O$ and assume for simplicity that I is formed by triples $\frac{\Phi, \varphi}{\Psi}$ $(\Phi, \Psi \in Sent^T, \varphi \in Sent^O)$.

We may consider the following rationality criteria concerning V_o-truthfulness $(V_o \subseteq V)$, just to mention two possibilities:

(a) I is rational if for each state dependent structure $\underline{U}$ and each finite non-empty $M \subseteq U$ we have the following: $\frac{\Phi, \varphi}{\Psi} \in I$, $\|\Phi\|_{\underline{U}} \in V_o$ and $\|\Psi\|_{\underline{U}} \notin V_o$ implies $\{\sigma ; \|\varphi\|_{\underline{M}_\sigma} \in V_o\} \in \mathcal{E}$.

(b) I is rational if for each $\underline{U}$ and M we have: $\frac{\Phi, \varphi}{\Psi} \in I$, $\|\Phi\|_{\underline{U}} \in V_o$ and $\|\Psi\|_{\underline{U}} \in V_o$ implies $\{\sigma ; \|\varphi\|_{\underline{M}_\sigma} \notin V_o\} \in \mathcal{E}$.

The criterion (a) says: If Ψ is inferred from Φ and φ then V_o-falsehood of Ψ implies that φ is V_o-true in $\underline{M}_\sigma$ for only a few σ. Hence if we accept Φ and have verified φ in the observed sample structure then we accept Ψ since if φ were V_o-false then our observation would be unlikely. This criterion will be used for various statistical inference rules in Chapter IV. Note that if Σ is a one-element set $(\Sigma = \{\sigma\})$ and "a few" means "no"

($\mathcal{E} = \{\emptyset\}$) then (a) guarantees that I is sound inverse deduction: If $\frac{\Phi, \varphi}{\Psi} \in I$ and Φ, Ψ are V_o-true in $\underline{U}$ then φ is V_o-true in $\underline{M}$. (Each M determines a unique $\underline{M}$.)

2.1.9 Remark. If V is not recursive, e.g. if $V = \mathbb{R}$ (reals) then the observational system must have another set of values V^0 which is (coded by) a recursive set (e.g. $V^0 = \mathbb{Q}$ -rationals). Then we have to approximate sample structures by V^0-structures. See Chapter IV, Section 2.

2.1.10 Key words: type, structure, isomorphism, observational structures; state dependent structures, state dependent variates, sample structures; systems of small subsets of states, rationality criteria for inductive inference rules.

II.2 Observational predicate calculi

2.2.1 By "the classical first order predicate calculus of type t $t = \langle t_1, \ldots, t_n \rangle$ " we mean the following: Formulae are built up from predicates $P_1, \ldots, P_n$ of arity $t_1, \ldots, t_n$ respectively, variables, logical connectives and quantifiers in the usual way. $\{0,1\}$ -structures of type t serve as models: satisfaction and truth are defined inductively; for each closed formula φ (i.e. formula without free variables) and each model $\underline{M}$, we define $\|\varphi\|_{\underline{M}} = 1$ iff φ is true in $\underline{M}$ and $\|\varphi\|_{\underline{M}} = 0$ otherwise. Note that we obtain a semantic system in this way: sentences are closed formulae, models are as described and the evaluation function $\|\varphi\|_{\underline{M}}$ is defined. By the classical results of Gödel and others, this semantic system is axiomatizable (completeness theorem), but if it is rich enough (more precisely, if $t_i > 1$ for some i) then it is not decidable. These are commonly known facts. We now start our generalizations and modifications.

We shall use the term "a predicate calculus" for each calculus similar to the classical predicate calculus provided it uses two abstract values 0,1, i.e. formulae are interpreted in (some) $\{0,1\}$ - structures. But we admit more quantifiers than $\forall, \exists$. The study of generalized quantifiers was initiated by

Mostowski and continued by many scholars (cf. Lindström 1966, 1969, Tharp 1973, Keisler 1970). We shall investigate generalized quantifiers in predicate calculi from a point of view different from that of Mostowski and his followers, (who are interested mainly in the behaviour of formulae with generalized quantifiers in infinite models) since we shall study predicate calculi called observational. The main point is that we shall admit only finite structures as models and make other assumptions such that the following will be true: Closed formulae, models and the evaluation function of an observational predicate calculus form an observational semantic system (see 2.2.5 below). The reader will see later (especially in Chapter IV) that some generalized quantifiers are very natural in formalizing observational languages. Let us make a precise definition.

2.2.2 Definition. A predicate language of type $t = \langle t_1, \dots t_n \rangle$ consists of the following:

predicates $P_1, \dots, P_n$ of arity $t_1, \dots, t_n$ respectively,

an infinite sequence $x_o, x_1, x_2, \dots$ of variables.

junctors $\underline{0}, \underline{1}$ (nullary), $\neg$ (unary), $\&, \vee, \rightarrow, \leftrightarrow$ (binary), called falsehood, truth, negation, conjuction, disjunction, implication and equivalence,

quantifiers $q_o, q_1, q_2, \dots$ of types $s_o, s_1, s_2, \dots$ respectively. The sequence of quantifiers is either infinite or finite (non-empty). Each quantifier type is a sequence $\langle 1, 1, \dots, 1 \rangle$. If there are infinitely many quantifiers then the function associating the type s_i with each i is recursive.

A predicate language with identity contains furthermore an additional binary predicate $=$ distinct from $P_1, \dots, P_n$ (the equality predicate).

Formulae are defined inductively, the notion of atomic formulae and the induction step for connectives being as usual.

Each expression $P_i(u_1, \dots, u_{t_i})$ where $u_1, \dots, u_{t_i}$ are variables is an atomic formula (and $u_1 = u_2$ is an atomic formula); $\underline{0}$ and $\underline{1}$ are formulae; if φ is a formula then $\neg\varphi$ is; if φ, ψ are formulae then $\varphi \& \psi$, $\varphi \vee \psi$, $\varphi \rightarrow \psi$, $\varphi \leftrightarrow \psi$ are formulae.

If q_i is a quantifier of type $\langle 1^{s_i} \rangle$, if u is a variable and if $\varphi_1, \dots, \varphi_{s_i}$ are formulae then $(qu)(\varphi_1, \dots, \varphi_{s_i})$ is a formula. This completes the inductive definition.

Free and bound variables are defined as usual. The induction step for $(qu)(\varphi_1,\dots,\varphi_s)$ is as follows: a variable is free in $(qu)(\varphi_1, \dots, \varphi_s)$ iff it is free in one of the formulae $\varphi_1, \dots, \varphi_s$ and is distinct from u. A variable is bound in $(qu)(\varphi_1, \dots, \varphi_s)$ iff it is bound in one of the formulae $\varphi_1,\dots,\varphi_s$ or it is u.

Formulae of this language can be coded (Gödel numbered) by natural numbers in the same manner as formulae of the classical predicate calculus. We fix such a coding; note that then the set of all codes for closed formulae becomes recursive.

Example. The classical quantifiers $\forall, \exists$ have type $\langle 1 \rangle$. The quantifier "is associated with" has type $\langle 1,1 \rangle$ (until now we have not said anything about the semantics of quantifiers; this is our next task).

2.2.3 Associated functions of the junctors of a predicate calculus show how the value of a composed formula depends on the value of its components. The associated function of a nullary junctor is simply its value: the value of $\underline{0}$ is 0, the value of $\underline{1}$ is 1. The following tables define the usual associated functions of other junctors:

¬	
0	1
1	0

&	0	1
0	0	0
1	0	1

∨	0	1
0	0	1
1	1	1

→	0	1
0	1	1
1	0	1

↔	0	1
0	1	0
1	0	1

2.2.4 What determines the meaning of a formula beginning with a quantifier? Consider $(\forall x)\, P(x)$ where P is unary. Let P be interpreted in a model M by a function f. The truth value of $(\forall x)P(x)$ is fully determined by the model $\langle M,f \rangle$: we have $\|(\forall x)P(x)\|_{\underline{M}} = 1$ iff f is identically 1 on M. Similarly for $\exists$.
Let $Asf_{\forall}$ be the function defined on all models of type $\langle 1 \rangle$ such that $Asf_{\forall}(\langle M,f \rangle) = 1$ iff f is identically 1 on M. Call this function the associated function of $\forall$ and note that it completely determines the semantics of $\forall$.
More generally, the associated function of a quantifier of type $\langle 1^s \rangle$ will be a mapping from all models of type $\langle 1^s \rangle$ into $\{0, 1\}$ which is invariant under isomorphism.

We shall give some examples. In accordance with the convention at the end of 2.2.1, "model" means "a finite $\{0,1\}$-structure". (This is important for example (f) below.)

(a) Universal quantifier $\forall$ - type $\langle 1\rangle$. $Asf_{\forall}(\langle M,f\rangle) = 1$ iff $f(o) = 1$ for all $o \in M$; otherwise $Asf_{\forall}(\langle M,f\rangle) = 0$.

(b) Existential quantifier $\exists$ - type $\langle 1\rangle$. $Asf_{\exists}(\langle M,f\rangle) = 1$ iff $f(o) = 1$ for some $o \in M$; otherwise $Asf_{\exists}(\langle M,f\rangle) = 0$.

These quantifiers are called classical quantifiers.

(c) Plurality quantifier W (Rescher 1962) - type $\langle 1\rangle$. $Asf_{W}(\langle M,f\rangle) = 1$ iff the cardinality of $\{o \in M; f(o) = 1\}$ is larger than the cardinality of $\{o \in M, f(o) = 0\}$ (most objects have the value 1).

(d) The quantifier of implication $\Rightarrow$ (Church 1951), type $\langle 1,1\rangle$ (not to be confused with the junctor ("logical connective") of implication).
$Asf_{\Rightarrow}(\langle M,f,g\rangle) = 1$ iff for each $o \in M$ such that $f(o) = 1$ we have $g(o) = 1$ i.e., there is no object $o \in M$ with $f(o) = 1$ and $g(o) = 0$.

(e) The quantifier of simple association $\sim$ - type $\langle 1,1\rangle$.
For $\underline{M} = \langle M,f,g\rangle$ put $a_{\underline{M}} = \text{card}\{o \in M;\ f(o) = g(o) = 1\}$,
$b_{\underline{M}} = \text{card}\{o \in M;\ f(o) = 1 \text{ and } g(o) = 0\}$,
$c_{\underline{M}} = \text{card}\{o \in M;\ f(o) = 0 \text{ and } g(o) = 1\}$,
$d_{\underline{M}} = \text{card}\{o \in M;\ f(o) = g(o) = 0\}$. $Asf_{\sim}(\underline{M}) = 1$ iff
$a_{\underline{M}} d_{\underline{M}} > b_{\underline{M}} c_{\underline{M}}$ (i.e., coincidence predominates over difference - in the present simple sense).

(f) The quantifier of founded p-implication $\Rightarrow_{p,a}$ ($a \in \mathbb{N}$, p rational, $0 \leq p \leq 1$. $Asf_{\Rightarrow_{p,a}}(\underline{M}) = 1$ iff $a_{\underline{M}} \geq p(a_{\underline{M}} + b_{\underline{M}})$ and $a_{\underline{M}} \geq a$.

2.2.5 Definition. An observational predicate calculus OPC of type t is given by the following:

(i) a predicate language L of type t,

(ii) for each quantifier q_i of L, its associated function Asf_{q_i} , mapping the set $\mathfrak{M}_{s_i}^{\{0,1\}}$ of all models of type s_i whose domain is a finite subset of $\mathbb{N}$ into $\{0,1\}$ such that the following is satisfied:

(iia) Each Asf_{q_i} is invariant under isomorphism, i. e., if $\underline{M}, \underline{N} \in \mathfrak{M}_{s_i}^{\{0,1\}}$ are isomorphic then $Asf_{q_i}(\underline{M}) = Asf_{q_i}(\underline{N})$,

(iib) $Asf_{q_i}(\underline{M})$ is a recursive function of two variables q_i, $\underline{M}$.

The OPC of type t <u>with classical quantifiers</u> is the OPC of type t having two quantifiers $\forall, \exists$ with their usual associated functions.

2.2.6 <u>Definition</u>. (Values of formulae). Let $\mathcal{P}$ be an OPC, let $\underline{M} = \langle M, f_1, \dots, f_n \rangle$ be a model and let φ be a formula; write $FV(\varphi)$ for the set of free variables of φ. An M-<u>sequence</u> for φ is a mapping e of $FV(\varphi)$ into M. If the domain of e is $u_1, \dots, u_n$ and if $e(u_i) = m_i$ then we write

$$e = \frac{u_1, \dots, u_n}{m_1, \dots, m_n} .$$

We define inductively $\|\varphi\|_{\underline{M}}[e]$ - the M-value of φ for e.

(a) $$\| P_i(u_1, \dots, u_k) \|_{\underline{M}} \left[\frac{u_1, \dots, u_k}{m_1, \dots, m_k} \right] = f_i(m_1, \dots, m_k) ;$$

$$\| u_1 = u_2 \|_{\underline{M}} \left[\frac{u_1, u_2}{m_1, m_2} \right] = 1 \quad \text{iff} \quad m_1 = m_2 .$$

(b) $\| \underline{0} \|_{\underline{M}}[\emptyset] = 0$, $\| \underline{1} \|_{\underline{M}}[\emptyset] = 1$; $\| \neg\varphi \|_{\underline{M}}[e] = 1 - \| \varphi \|_{\underline{M}}[e]$.

If $FV(\varphi) \subseteq \mathrm{dom}(e)$ then write e/φ instead of $e \upharpoonright FV(\varphi)$ (restriction). If ι is $\&, \vee, \rightarrow, \leftrightarrow$, then

(c) $$\| \varphi \iota \psi \|_{\underline{M}}[e] = \mathrm{Asf}_{\iota}(\| \varphi \|_{\underline{M}}[e/\varphi], \| \psi \|_{\underline{M}}[e/\psi]).$$

If $\mathrm{dom}(e) \supseteq FV(\varphi) - \{x\}$ and $x \notin \mathrm{dom}(e)$ then letting x vary over M we obtain a unary function on M

$$\| \varphi \|^{e}_{\underline{M}}[m] = \| \varphi \|_{\underline{M}}[(e \cup \frac{x}{m})/\varphi]$$

($\|\varphi\|_{\underline{M}}$ can be viewed as a k-ary function, k being the number of free variables of φ. Now all variables except x are fixed according to e : x varies over M .)

(d) $$\| (qx)(\varphi_1, \dots, \varphi_k) \|_{\underline{M}}[e] = \mathrm{Asf}_q(\langle M, \|\varphi_1\|^{e}_{\underline{M}}, \dots, \|\varphi_k\|^{e}_{\underline{M}} \rangle)$$

2.2.7 Example

Let R be a binary predicate. In the following formulas (denoted $\varphi_1, \varphi_2, \varphi_3$) x is free and y is bound.

φ_1: $(\forall y) R(x,y)$

φ_2: $(\exists y) R(x,y)$

φ_3: $(W y) R(y,x)$.

The following are then closed formulas:

ψ_1: $\varphi_1 \overset{x}{\Rightarrow} \varphi_2$

ψ_2: $\varphi_1 \overset{x}{\sim} \varphi_2$

We should write $(\Rightarrow x)(\varphi_1, \varphi_2)$; however, if there is no danger of misunderstanding we simply write $\varphi_1 \Rightarrow \varphi_2$ and similarly for the other quantifiers of type $\langle 1,1 \rangle$.

Let $\underline{M} = \langle M, f \rangle$ be the $\{0,1\}$ - structure of type $\langle 2 \rangle$ with six objects 0, ..., 5 and with the function defined by the following table:

	0	1	2	3	4	5
0	0	0	0	0	0	0
1	1	1	1	1	1	1
2	0	0	0	1	0	0
3	1	1	1	1	1	1
4	0	0	0	1	0	1
5	1	1	1	1	1	1

	$\Vert\varphi_1\Vert_{\underline{M}}$	$\Vert\varphi_2\Vert_{\underline{M}}$	$\Vert\varphi_3\Vert_{\underline{M}}$
0	0	0	0
1	1	1	0
2	0	1	0
3	1	1	1
4	0	1	0
5	1	1	1

In the right-hand table we have the functions $\Vert\varphi_i\Vert_{\underline{M}}$; verify that $\Vert\psi_1\Vert_{\underline{M}} = \Vert\psi_2\Vert_{\underline{M}} = 1$, i.e., that both ψ_1 and ψ_2 are true in $\underline{M}$; $\Vert\varphi_1 \Rightarrow \varphi_3\Vert_{\underline{M}} = 0$.

2.2.8 Theorem. Let $\mathcal{P}$ be an OPC of type t and let $\mathcal{S}$ be the semantic system whose sentences are closed formulas of $\mathcal{P}$, whose models are elements of $\mathfrak{M}_t^{\{0,1\}}$ and whose evaluation function is defined by

$$\mathrm{Val}(\varphi , \underline{M}) = \Vert\varphi\Vert_{\underline{M}}[\emptyset].$$

Then $\mathcal{S}$ is an observational semantic system.

Proof. The only thing to be proved is that the function Val is recursive in $\mathcal{P}$ and $\underline{M}$. (Remember that Sent and $\mathfrak{M}_t^{\{0,1\}}$ are recursive infinite sets.) This follows from the fact that $\|\varphi\|_{\underline{M}}[e]$ is a recursive function of φ, $\underline{M}$, e. The last fact follows from the inductive definition of $\|\varphi\|_{\underline{M}}[e]$; details are left to the reader. (Hint : Let $G(\underline{M},e,i) = f_i(e_1,\dots,e_{t_i})$ if $e = \langle e_1,\dots,e_{t_i}\rangle$, $\underline{M} = \langle M, f_1,\dots,f_n\rangle$, $1 \leq i \leq n$ and f_i is t_i-ary; let $G(\underline{M},e,i) = 0$ otherwise. G is recursive.)

2.2.9 Definition and Remark. Let $\mathcal{P}$ be an OPC.

Suppose that φ, ψ are two formulae such that $FV(\varphi) = FV(\psi)$ (φ and ψ have the same free variables). φ and ψ are said to be logically equivalent if $\|\varphi\|_{\underline{M}} = \|\psi\|_{\underline{M}}$ for each $\underline{M} \in \mathfrak{M}$.

Note that in general the last equality is an equality of functions (with the same domain; if both φ and ψ are closed then $\|\varphi\|_{\underline{M}} = \|\psi\|_{\underline{M}}$ expresses the equality of two values). The definition can be easily generalized for arbitrary pairs of formulae.

2.2.10 We are going to summarize facts not involving quantifiers that are true for each OPC and are proved exactly as for the classical predicate calculus. Assume an OPC to be given, $\Leftrightarrow$ is used as the symbol of logical equivalence.

(1) $\varphi \,\&\, \psi \Leftrightarrow \psi \,\&\, \varphi$, (2) $\varphi \vee \psi \Leftrightarrow \psi \vee \varphi$ (commutativity),

(3) $\varphi \,\&\, \varphi \Leftrightarrow \varphi$, (4) $\varphi \vee \varphi \Leftrightarrow \varphi$ (idempotence),

(5) $\varphi \,\&\, (\psi \,\&\, \chi) \Leftrightarrow (\varphi \,\&\, \psi) \,\&\, \chi$, (6) $\varphi \vee (\psi \vee \chi) \Leftrightarrow (\varphi \vee \psi) \vee \chi$ (associativity),

(7) $\varphi \,\&\, \underline{1} \Leftrightarrow \varphi \vee \underline{0} \Leftrightarrow \varphi$,

(8) $\varphi \,\&\, \underline{0} \Leftrightarrow \underline{0}$, $\varphi \vee \underline{1} \Leftrightarrow \underline{1}$,

(9) $(\varphi \to \psi) \Leftrightarrow (\neg\varphi \vee \psi) \Leftrightarrow \neg(\varphi \,\&\, \neg\psi)$,

(10) $\varphi \,\&\, (\psi \vee \chi) \Leftrightarrow (\varphi \,\&\, \psi) \vee (\varphi \,\&\, \chi)$ (distributivity),

(11) $\varphi \vee (\psi \,\&\, \chi) \Leftrightarrow (\varphi \vee \psi) \,\&\, (\varphi \vee \chi)$ (distributivity),

(12) $\neg\neg\varphi \Leftrightarrow \varphi$

(13) $\neg(\varphi \,\&\, \psi) \Leftrightarrow \neg\varphi \vee \neg\psi$ (de Morgan law),

(14) $\neg(\varphi \vee \psi) \Leftrightarrow \neg\varphi \,\&\, \neg\psi$ (de Morgan law),

(15) $\varphi \,\&\, \neg\varphi \Leftrightarrow \underline{0}$, (16) $\varphi \vee \neg\varphi \Leftrightarrow \underline{1}$ (complementation).

2.2.11 If $B = \{\varphi_1, \ldots, \varphi_n\}$ then we write $\bigwedge B$ or $\bigwedge_{i=1}^{n} \varphi_i$ with the usual meaning: the conjunction of $\varphi_1, \ldots, \varphi_n$; if B is empty then $\bigwedge B$ means $\underline{1}$. Similarly, $\bigvee_{i=1}^{n} \varphi_i$ or $\bigvee B$ is the disjunction of $\varphi_1, \ldots, \varphi_n$; $\bigvee \emptyset$ is $\underline{0}$.

Let A, B be disjoint finite sets of formulae.

Then (17) $\bigwedge(B \cup A) \Leftrightarrow \bigwedge B \,\&\, \bigwedge A$, (18) $\bigvee(B \cup A) \Leftrightarrow \bigvee B \vee \bigvee A$,

(19) $\bigwedge B \vee \bigwedge A \Leftrightarrow \bigvee\{\varphi \,\&\, \psi ;\ \varphi \in B \text{ and } \psi \in A\}$,

(20) $\bigvee B \,\&\, \bigvee A \Leftrightarrow \bigwedge\{\varphi \vee \psi ;\ \varphi \in B \text{ and } \psi \in A\}$,

(21) $\neg\bigwedge B \Leftrightarrow \bigvee\{\neg\varphi ; \varphi \in B\}$, (22) $\neg\bigvee B \Leftrightarrow \bigwedge\{\neg\varphi ; \varphi \in B\}$.

2.2.12. A formula is <u>open</u> if it contains no quantifiers. Particular open formulae: <u>atomic formulae; literals</u>, i.e., atomic formulae or negated atomic formulae; <u>elementary conjunctions</u>, i.e., formulae of the form $\bigwedge B$ where B is a non-empty set of literals such that for no atomic formula both $\varphi \in B$ and $\neg\varphi \in B$; <u>elementary disjunctions</u> (analogous) ; <u>formulae in conjunctive normal form</u> (multiple (possibly empty) conjunctions of elementary disjunctions); formulae in <u>disjunctive normal form</u> (disjunctions of elementary conjunctions) . <u>Fact:</u> For each open formula φ different from $\underline{0}$ there is a logically equivalent formula in conjunctive normal form containing only predicates and variables which occur in φ . The same is true for open formulas different from $\underline{1}$ and disjunctive normal form.

2.2.13 For each closed formula φ and each finite set B of closed formulae, φ is a logical $\{1\}$ - consequence of B (in the sense of 1.2.7) iff $(\bigwedge B) \rightarrow \varphi$ is a tautology. "Closed" means "having no free variables".

2.2.14 <u>Corollary.</u> Let $\mathcal{P}$ be an OPC and let I be a $\{1\}$ -sound deduction rule. I is strongly $\{1\}$-complete (w.r.t. the semantic system given by all closed formulae) iff I is $\{1\}$-complete and for each closed formula φ and each finite set B of closed formulae we have

$$B, (\bigwedge B) \rightarrow \varphi \vdash_I \varphi . \qquad (*)$$

Proof. ($\Rightarrow$) is evident. Conversely, assume the condition and let B be a finite set of closed formulae and φ a closed formula such that $B \models_{\{1\}} \varphi$. Then $(\bigwedge B) \rightarrow \varphi$ is a tautology and hence $\vdash_I (\bigwedge B) \rightarrow \varphi$. This yields $B \vdash_I \varphi$ by (*) .

2.2.15 In accordance with 1.2.12 we call an OPC $\mathcal{P}$ decidable if the set $Taut_{\mathcal{P}}$ of all closed formulae that are $\{1\}$-tautologies is recursive. We could define $\mathcal{P}$ to be axiomatizable if $Taut_{\mathcal{P}}$ is recursively enumerable; but it follows immediately from 1.2.15 that $Taut_{\mathcal{P}}$ is recursive iff it is recursively enumerable. This shows that properties of OPC's may differ considerably from properties of the classical predicate calculus which is axiomatizable but not decidable (provided its type is rich enough, cf. 2.2.1). We shall study OPC's in Chapter III; here we only present a classical result due to Trachtenbrot concerning OPC's with classical quantifiers. The theorem is the counterpart of Gödel's completeness theorem (for the classical predicate calculus) in the logic of OPC's.

2.2.16 Theorem (Trachtenbrot 1950). There is a type t such that the OPC of the type t whose only quantifiers are classical quantifiers $\forall, \exists$ is not decidable (and hence not axiomatizable).

The original proof is rather complicated. We outline a proof using later results (in particular, Matijasevič's result on Diophantine sets) in Chapter III, Section 5.

2.2.17 Remarks. (1) One can show that if t is not monadic ($t_i > 1$ for some i) then the OPC of type t with classical quantifiers is not decidable. In Chapter III we show that each monadic OPC (without equality) with classical quantifiers (and, moreover, each OPC with finitely many quantifiers) is decidable. We also prove other facts concerning (un) decidability. (Cf. Ivánek.)

(2) Section 1,2,5 of Chapter III may be read immediately after the present section.

2.2.18 Key words: Predicate language (with equality), formulae, free and bound variables; associated functions of junctors and of (generalized) quantifiers, observational predicate calculi, the value of a formula in a model for a sequence; open formulae, literals, elementary conjunctions and disjunctions.

II. 3 Function calculi

2.3.1 The aim of the present section is to generalize the classical predicate calculus in order to obtain formal languages appropriate for expressing statements on V-structures for any set V of abstract values. There is no doubt that we need such languages. In most cases our V will be a subset of the set $\mathbb{R}$ of real numbers, or a subset of $\mathbb{R} \cup \{\times\}$ where $\times$ is an abstract value for missing information (cf. III, Section 3). There are various means of constructing languages for V-structures. In many-valued logics, one considers elements of V as generalized truth values (e.g. degrees of certainty); one generalizes associated functions of junctors and quantifiers to appropriate V-valued functions and often makes use of a structure given on V. (Cf. Rosser-Turquette 1952,Chang-Keisler 1966.) Suppes works with finite real-valued structures but has only three truth-values meaning "true", "false" and "meaningless".The reader is recommended to read Suppes's paper; but we shall choose another way.

We shall allow formulae to have arbitrary values from V,(i.e. we shall work with V-valued associated functions of various junctors and quantifiers) but we shall not deal with values from V as truth values. Instead, we shall work with various subsets $V_o \subseteq V$ and investigate the notions of V_o-truth and V_o-consequence. (Cf. the discussion 1.2.8 on V_o-assertions.)

2.3.2 Definition. Let t be a type. A language of type t consists of the following:

function symbols $F_1, \ldots, F_n$ of arities $t_1, \ldots, t_n$ respectively;

variables $x_o, x_1, \ldots$ (infinite sequence);

junctors $\iota_o, \iota_1, \ldots$ of arities $j_o, j_1, j_2, \ldots$

respectively; the sequence of junctors is finite or infinite and if it is infinite then the sequence $j_o, j_1, \ldots$ as a function over the natural numbers is recursive;

quantifiers $q_o, q_1, \ldots$ of types $s_o, s_1, \ldots$ respectively.

The sequence of quantifiers is finite (non-empty) or infinite; each quantifier type is a tuple of ones. If there are infinitely many quantifiers then the function $i \longrightarrow s_i$ is recursive.

Atomic formulae have the form $F_i(u_1, \dots, u_{t_i})$ where the u_j are variables. If ι is a k-ary junctor and if $\varphi_1, \dots, \varphi_k$ are formulae then $\iota(\varphi_1, \dots, \varphi_n)$ is a formula. If q is a quantifier of type $\langle 1^k \rangle$ and $\varphi_1, \dots, \varphi_k$ are formulae and if u is a variable then $(qu)(\varphi_1, \dots, \varphi_k)$ is a formula.

The definition of free and bound variables generalizes trivially for the present notion of formulae. Given a language, we fix a Gödel numbering of formulae by natural numbers.

2.3.3 Definition. Let t be a type and let V be an abstract set of values. A V-valued functor calculus $\mathcal{F}$ of type t consists of the following:

a language of type t,

a non-empty class $\mathfrak{M}$ of V-structures of type t, called models of $\mathcal{F}$;

for each k-ary junctor ι, a mapping $Asf_\iota : V^k \rightarrow V$ called the associated function of ι;

for each quantifier q of type $\langle 1^k \rangle$, a mapping Asf_q, called the associated function of q.

Say that a V-structure $\langle M, f_1, \dots, f_k \rangle$ of type $\langle 1^k \rangle$ belongs to $\mathfrak{M}$ if there is a structure $\langle M, g_1, \dots, g_n \rangle$ having the same domain. Asf_q maps all V-structures of type $\langle 1^k \rangle$ belonging to $\mathfrak{M}$ into V.

For example, if $\mathfrak{M}$ consists of all finite V-structures of type t then $dom(Asf_q)$ consists of all finite V-structures of the type $\langle 1^k \rangle$.

2.3.4 Examples. (1) Each OPC is a function calculus whose class $\mathfrak{M}$ of models is $\mathfrak{M}_t^{\{0,1\}}$ of all $\{0,1\}$-structures of type t whose domain is a finite set of natural numbers.

(2) The classical predicate calculus is a function calculus whose class of models is formed by all $\{0,1\}$-structures of type t.

(3) We give a very simple example of natural valued calculi. For each $n \geq 1$ we denote by $\mathcal{F}^n$ the function calculus defined as follows: the set of abstract values is $\mathbb{N}$, $\mathfrak{M}$ is $\mathfrak{M}_{\langle 1^n \rangle}^{\mathbb{N}}$ (the set of all $\mathbb{N}$-structures of type $\langle 1^n \rangle$ whose domain is a finite set of natural numbers). The language is specified as follows: it has n unary function symbols, junctors +, . (binary) and z (unary), quantifiers Σ and Π of type $\langle 1 \rangle$. The associated function of + and $\cdot$ is addition and multiplication respectively;

$Asf_{\leq}(p,q) = 1$ iff $p \leq q$ and $= 0$ otherwise. $Asf_z(p) = 1$ if $p=0$, $= 0$ otherwise. $Asf_{\Pi}(\langle M,f\rangle) = \prod_{o \in M} f(o)$, $Asf_{\Sigma}(\langle M,f\rangle) = \sum_{o \in M} f(o)$ (sum and product over the model). Since $\mathcal{F}^n$ is a very simple but natural calculus call it the <u>pocket calculus</u> of type $\langle 1^n \rangle$.

(4) $\mathcal{E}_n$ is a real-valued calculus with n unary function symbols, $\mathcal{M} = \mathcal{M}^{\mathbb{R}}_{\langle 1^n \rangle}$, there are no junctors and there is one quantifier ρ of type $\langle 1,1\rangle$ called the <u>correlation coefficient</u>; Asf_{ρ} is defined as follows: If $\langle M,f,g\rangle$ is a model let $\bar{f}$ denote the arithmetic mean of $\{f(o);\ o \in M\}$ and similarly for $\bar{g}$.

$$Asf_{\rho}(\langle M,f,g\rangle) = \frac{\sum_{o \in M}(f(o)-\bar{f})(g(o)-\bar{g})}{\sqrt{\sum_{o \in M}(f(o)-\bar{f})^2 \sum_{o \in M}(g(o)-\bar{g})^2}} \qquad (*)$$

(and $Asf_{\rho}(\langle M,f,g\rangle) = 0$ if the denominator of $(*)$ is zero).

(5) We can restrict ourselves to rational-valued models in the above example, i.e. take $\mathcal{M} = \mathcal{M}^{\mathbb{Q}}_{\langle 1^n \rangle}$. If we want to declare $\mathbb{Q}$ as our set of abstract values we must modify the definition of the associated function of the quantifier in order to guarantee that its values will be rational, hence, we may work with the quantifier ρ^* defined by

$$Asf_{\rho^*}(\langle M,f,g\rangle) = \operatorname{sgn}\left(\sum_{o \in M}(f(o)-\bar{f})(g(o)-\bar{g})\right)\frac{\left(\sum_{o \in M}(f(o)-\bar{f})(g(o)-\bar{g})\right)^2}{\sum_{o \in M}(f(o)-\bar{f})^2 \sum_{o \in M}(g(o)-\bar{g})^2}$$

(signed square of the correlation coefficient), which gives equivalent information.

2.3.5 <u>Definition.</u> (Values of formulas). The definition is fully analogous to 2.2.6. For an atomic formula $F_i(u_1,\ldots,u_k)$,

$$\| F_i(u_1,\ldots,u_k) \|_{\underline{M}} \left[\frac{u_1,\ldots,u_k}{m_1,\ldots,m_k}\right] = f_i(m_1,\ldots,m_k) \ ;$$

if ι is a k-ary junctor then

$$\| \iota(\varphi_1,\ldots,\varphi_k) \|_{\underline{M}}[e] = Asf_{\iota}(\|\varphi_1\|_{\underline{M}}[e/\varphi_1], \ldots, \|\varphi_k\|_{\underline{M}}[e/\varphi_k]) \ ;$$

if q is a quantifier of type $\langle 1^k \rangle$ then

$$\| (qu)(\varphi_1, \dots, \varphi_k) \|_{\underline{M}}[e] = \mathrm{Asf}_q (\langle M, \|\varphi_1\|^e_{\underline{M}}, \dots, \|\varphi_k\|^e_{\underline{M}} \rangle).$$

2.3.6 Remark. (1) If V contains 0 and 1 (and perhaps other values) then sentences taking only values 0,1 i.e. $\{0,1\}$- tautologies can be called proper sentences. Given a $V_o \subseteq V$, we may introduce a new unary junctor ι whose associated function is the characteristic function of V_o over V ($\mathrm{Asf}_\iota(v) = 1$ if $v \in V_o$, $= 0$ if $v \in V - V_o$). Then for each sentence φ and each $\underline{M}$, $\|\varphi\|_{\underline{M}} \in V_o$ iff $\|\iota\varphi\|_{\underline{M}} = 1$; $\iota\varphi$ is a proper sentence.

(2) We could construct function calculi with equality as an extra binary functor = such that $\| u_1 = u_2 \|_{\underline{M}} \left[\frac{u_1, u_2}{m_1, m_2} \right] = 1$ iff $m_1 = m_2$, otherwise $= 0$ (provided $0,1 \in V$).

2.3.7 Definition. A function calculus $\mathcal{F}$ is observational (OFC) if the following holds:

(a) V is a recursive set,

(b) $\mathrm{Asf}_\iota(\underline{v})$ is a recursive function of ι and $\underline{v}$,

(c) $\mathrm{Asf}_q(\underline{M})$ is a recursive function of q and $\underline{M}$.

2.3.8 Pedantically we should say: "$\mathrm{Asf}_\iota(\underline{v})$ is a partial recursive function of ι and $\underline{v}$" and similarly for $\mathrm{Asf}_q(\underline{M})$. But under our recursiveness assumptions concerning V and the coding of formulae (e.g. we quietly assume that the set of all junctors is recursive etc.), the domain of Asf is a recursive set; hence we could equivalently say "there is a (total) recursive function whose restriction is Asf" . Thus there is no danger of confusion. The following theorem is then obvious (cf. 2.2.8) :

2.3.9 Theorem. Let $\mathcal{F}$ be an OFC, then the semantic system $\mathcal{S}$ whose sentences are closed formulas of $\mathcal{F}$, whose models are models of $\mathcal{F}$ and whose evaluation function is

$$\mathrm{Val}(\varphi, \underline{M}) = \|\varphi\|_{\underline{M}}[\emptyset]$$

is an observational semantic system.

2.3.10 Remark. (1) Trachtenbrot's theorem gives an example of undecidable OFC's; in Chapter III we shall show other results. Note that, for reasonable V_o, the pocket calculi are undecidable (Hájek 1973).

(2) The definition 2.2.9 can be used for arbitrary function calculi: φ and ψ are logically equivalent if $\|\varphi\|_{\underline{M}} = \|\psi\|_{\underline{M}}$ for each $\underline{M} \in \mathfrak{M}$.

(3) Sections 2 and 3 of Chapter III can be read immediately after this section provided the reader has already read Chapter III, Section 1 .

2.3.11 Key words: language, formulae, free and bound variables; V-valued function calculus, values of formulae; observational function calculi.

II.4 Function calculi with state dependent models (state dependent calculi)

2.4.1 In the present short section we generalize function calculi to calculi whose models are state-dependent structures (cf.2.1.6-7). According to 2.1.8 semantic system with state-dependent models are useful as possible formalizations of theoretical languages since they make it possible to speak about chance. We shall now analyse the structure of sentences interpretable in state-dependent structures in more details.

Generalizing function calculi we first introduce a new variable, say s, for states; and we modify the definition of an atomic formula as follows : if F_i is k-ary and if $u_1,\ldots,u_k$ are object variables then $F_i(u_1,\ldots,u_k,s)$ is a formula. Naturally, the value of such a formula in a Σ-state dependent V-structure $\underline{U} = \langle U, f_1,\ldots,f_n\rangle$ for a sequence

$$e = \frac{u_1,\ldots u_k, s}{o_1,\ldots,o_k,\sigma} \quad \text{is}$$

$$\| F_i(u_1,\ldots,u_k,s)\|_{\underline{U}}[e] = f_i(o_1,\ldots,o_k,\sigma).$$

There is no problem concerning junctors; but we must be careful about quantifiers. We distinguish quantifiers of three kinds: object-quantifiers binding an object variable, state-quantifiers binding the state variable and mixed quantifiers binding both an object variable and the state variable. We turn now to exact definitions; they will be followed by examples.

2.4.2 Definition. Let $t = \langle t_1,\ldots t_n\rangle$ be a type.

(1) A state dependent language of type t consists of the following :

function symbols $F_1,\ldots, F_n$ of arities $t_1,\ldots, t_n$ respectively,

object variables $x_o, x_1,\ldots$, a state variable s ,

junctors $\iota_o, \iota_1, \ldots$ of arities $j_o, j_1, \ldots$

respectively satisfying the usual recursiveness condition,

quantifiers $q_o, q_1,\ldots$. With each quantifier q_i we associate its kind $k_i \in \{$ ob, st, mx $\}$ and its quantifier type (a tuple of ones). The function $i \mapsto k_i, s_i$ is a recursive function if the sequence of quantifiers is infinite.

(2) Atomic formulae are defined according to 2.4.1; if ι is a k-ary junctor and if $\varphi_1, \dots, \varphi_k$ are formulae then $\iota(\varphi_1, \dots, \varphi_k)$ is a formula. If q is an object quantifier of type $\langle 1^k \rangle$, if u is an object variable and if $\varphi_1, \dots, \varphi_k$ are formulae then

(i) $(qu)(\varphi_1, \dots, \varphi_k)$

is a formula. Similarly for q a state quantifier; then

(ii) $(qs)(\varphi_1, \dots, \varphi_k)$

is a formula. Finally, if q is a mixed quantifier then

(iii) $(qu,s)(\varphi_1, \dots, \varphi_k)$

is a formula.

The definition of free and bound variables is clear when we postulate that q binds u in (i), q binds s in (ii) and q binds u,s in (iii).

2.4.3 Definition. Let Σ be a fixed abstract set of states, let V be an abstract set of values and let t be a type. A v-valued function calculus $\mathcal{F}$ with Σ-state dependent models of type t (briefly, a s.d. function calculus) is determined by the following:

a s.d. language of type t;

a non-empty class $\mathfrak{M}$ of Σ-state dependent V-structures called models of $\mathcal{F}$;

for each k-ary junctor ι, its associated function $Asf_\iota : V^k \longrightarrow \hat{V}$;

for each quantifier q of type $\langle 1^k \rangle$, its associated function Asf_q with the following properties:

Say that a structure (or a state dependent structure) $\underline{M}$ belongs to $\mathfrak{M}$ if there is a state dependent structure in $\mathfrak{M}$ having the same domain as $\underline{M}$.

(i) If q is an object quantifier then Asf_q maps the class of all state dependent structures of type $\langle 1^k \rangle$ belonging to $\mathfrak{M}$ into V.

(ii) If q is a mixed quantifier then Asf_q maps the class of all Σ-state dependent structures of type $\langle 1^k \rangle$ belonging to $\mathfrak{M}$ into V.

(iii) If q is a state quantifier then Asf_q maps the class of all k-tuples of Σ-state dependent variates into V. (Hence $Asf_q(\langle g_1, \dots, g_k \rangle)$ is defined iff each g_i maps Σ into V.)

2.4.4 Examples. Assume $V = \{0,1\}$.

(1) Object quantifiers of type $\langle 1 \rangle$: $\forall$ with the usual associated function; $\exists^{\infty}$ (Mostowski's quantifier) ; $\mathrm{Asf}_{\exists^{\infty}}(\langle M, f \rangle) = 1$ iff $\{o;\ f(o) = 1\}$ is infinite.

(2) Object quantifier of type $\langle 1,1 \rangle$: H (Härtig's quantifier)

$\mathrm{Asf}_H(\langle M, f, g \rangle) = 1$ iff $\{o;\ f(o) = 1\}$ has the same cardinality as $\{o;\ g(o) = 1\}$.

(3) Mixed quantifier of type $\langle 1^k \rangle$: Full.
For each Σ-state dependent structure $\underline{U} = \langle U, f_1, \ldots, f_k \rangle$, $\mathrm{Asf}_{\underline{\mathrm{Full}}}(\underline{U}) = 1$ iff for each finite $\{0,1\}$-structure $\underline{N}$ of type $\langle 1^k \rangle$ there is a finite $M \subseteq U$ and a state σ such that $\underline{N}$ is isomorphic to $\underline{M}^{U}_{\sigma}$ (each finite structure can be obtained as a sample from U).

(4) Let $\mathcal{E}$ be a system of small sets on Σ. We have the state quantifier few of type $\langle 1 \rangle$ defined as follows:

$$\mathrm{Asf}_{\underline{\mathrm{few}}}(\langle g \rangle) = 1 \text{ iff } \{\sigma\ ;\ g(\sigma) = 1\} \in \mathcal{E} .$$

2.4.5 Examples with real values. Since we are often forced to work with associated functions that are not always defined, put $V = \mathbb{R} \cup \{\mathrm{undef}\}$ where undef is the value "undefined". Let $\mathfrak{M}$ be the class of all V-structures whose domain is $\mathbb{N}$ (the set of all natural numbers).

(1) Object quantifier lim of type $\langle 1 \rangle$:

$$\mathrm{Asf}_{\underline{\lim}}(\langle M, f \rangle) = \lim_{n \to \infty} f(n), \text{ if defined}$$

$$= \mathrm{undef} \text{ otherwise} .$$

(2) The mixed quantifier Full (2.4.4 (3)) makes sense for all V-valued models and $\mathrm{Asf}_{\underline{\mathrm{Full}}}$ is always either 1 or 0 .

(3) The state quantifier E (expectation): let P be a probability measure on $\langle \Sigma, \mathfrak{A} \rangle$ then

$$\mathrm{Asf}_E(\langle g \rangle) = \int g\,dP \quad \text{if defined,} \ = \mathrm{undef} \text{ otherwise}.$$

2.4.6 Remark and Convention. Let φ be an open formula in a state dependent function calculus and suppose that the free variables of φ are $u_1, \ldots, u_n, s$. If $\underline{U}$ is a model (i.e. state dependent structure) then $\underline{U}$ and φ determine a mapping of $U^k \times \Sigma$ into V associating with each $o_1, \ldots, o_n \in U$ and $\sigma \in \Sigma$ the value $\|\varphi\|_{\underline{U}} \left[\frac{u_1, \ldots, u_n, s}{o_1, \ldots, o_n, \sigma} \right]$. Without any danger of misunderstanding this mapping can be denoted by $\|\varphi\|_{\underline{U}}$. If $\varphi_1, \ldots, \varphi_k$ are open formulae and if $\underline{U}$ is a model with domain U then we have the Σ-state dependent structure

$$\underline{U}_{\varphi_1, \ldots, \varphi_k} = \langle U, \|\varphi_1\|_{\underline{U}}, \ldots, \|\varphi_k\|_{\underline{U}} \rangle$$

we say that $\underline{U}_{\varphi_1, \ldots, \varphi_k}$ is derived from $\underline{U}$ with the help of $\varphi_1, \ldots, \varphi_k$

2.4.7 Remark. Let Φ be a sentence (closed formula) of a s.d. calculus $\mathcal{F}$ of type $t = \langle 1^k \rangle$. We may extend $\mathcal{F}$ to a.s.d. calculus $\mathcal{F}'$ having exactly one more mixed quantifier q of type t such that

$$\mathrm{Asf}_q(\underline{U}) = \|\Phi\|_{\underline{U}}.$$

Observe that for each k-tuple of open formulas $\varphi_1, \ldots, \varphi_k$ containing exactly one free object variable u and the state variable s we have

$$\|\Phi\|_{\underline{U}_{\varphi_1, \ldots, \varphi_k}} = \|(q\, u, s)(\varphi_1, \ldots, \varphi_k)\|_{\underline{U}}$$

Thus Φ says about $\underline{U}_{\varphi_1, \ldots, \varphi_k}$ the same as $(q\, u, s)(\varphi_1, \ldots, \varphi_k)$ about $\underline{U}$. This fact will be used in Chapter IV.

2.4.8 Let us now discuss the state of the questions (L0)-(L2) of the logic of induction (see 1.1.5).

(L0) We shall use state dependent function calculi as our formalization of theoretical languages, since they make it possible to express (and interpret) statements concerning chance. We shall use observational function calculi as our formalization of observational languages, since they have recursive syntax and semantics and, therefore, sentences can be generated and evaluated by a machine (in principle).

(L1) The notion "a theoretical hypothesis is justified by some (true) observational statement (in a certain theoretical context)" is formalized by our notion of inductive inference rules; we gave a criterion of rationality of such a rule, which is an extract of statistical inference rules as we shall see. (We shall formulate further rationality criteria in Chapter IV).

(L2) The question concerning methods of deciding whether a hypothesis is justified by some observational statement reduces to the requirement of recursiveness of the inductive inference rule chosen or, better, to an easy (e.g. polynomial) recognizability of the rule. This requirement will be trivially fulfilled for the rules of Chapter IV.

2.4.9 Remark. The reader may read Chapter IV Sections 1-5 as the immediate continuation of the present section without reading Chapter III.

2.4.10 Key words: function calculi with state dependent models (s.d. function calculi); the structure derived from another structure with the help of open formulae.

PROBLEMS AND SUPPLEMENTS TO CHAPTER II.

(1) Prove the following lemma (on renaming free variables):

Let $\mathfrak{F}$ be a function calculus.

Let φ be a formula with free variables $x_1, \ldots, x_n$ and let $y_1, \ldots, y_n$ be a sequence of variables such that, for each $i = 1, \ldots, n$, either y_i is x_i or y_i does not occur in φ. Let φ' be the result of replacing all occurrences of x_i in φ by y_i $(i = 1, \ldots, n)$ and, for each M-sequence e for φ,

if $e = \dfrac{x_1, \ldots, x_n}{m_1, \ldots, m_n}$, let e' be the sequence $\dfrac{y_1, \ldots, y_n}{m_1, \ldots, m_n}$. Then

$$\|\varphi\|_{\underline{M}}[e] = \|\varphi'\|_{\underline{M}}[e'] .$$

(2) Prove the following theorem (on renaming bound variables):

Let $\mathfrak{F}$ be a function calculus, let q be a quantifier and let $\varphi = (qu)(\varphi_1, \ldots, \varphi_n)$ be a formula. Finally let y be a variable not occuring in φ. For $i = 1, \ldots, k$ denote by ψ_i the formulae resulting from φ_i by replacing each occurrence of u in φ_i by y. Then $(qy)(\psi_1, \ldots, \psi_n)$ is logically equivalent to $(qu)(\varphi_1, \ldots, \varphi_n)$.

(3) (Rescher) Prove that the following are tautologies (W is Rescher's plurality quantifier):

(i) $(\forall x)\varphi(x) \rightarrow (Wx)\varphi(x)$, (ii) $(Wx)\varphi(x) \rightarrow (\exists x)\varphi(x)$,

(iii) $(Wx)\varphi_1(x)$ & $(Wx)\varphi_2(x) \rightarrow (\exists x)(\varphi_1(x)$ & $\varphi_2(x))$,

(iv) $(Wx)\varphi(x) \rightarrow \neg(Wx)\neg\varphi(x)$,

(v) $((\forall x)\varphi(x)$ & $(Wx)(\varphi(x) \rightarrow \psi(x))) \rightarrow (Wx)\psi(x)$,

(vi) $((Wx)\varphi(x)$ & $(\forall x)(\varphi(x) \rightarrow \psi(x))) \rightarrow (Wx)\psi(x)$.

Consider a model with three elements a, b, c and the binary relation R such that a is in relation R to nothing, b only to a and b, and c only to b and c. Show that the sentence $(Wx)(Wy)R(x,y) \rightarrow (Wy)(Wx)R(x,y)$ is <u>not</u> true in this model.

(4) (Chytil 1975). Let $\equiv$ be the junctor of equivalence

Asf $(u,v) = 1$ iff $u = v$ and, for each $n \geq 2$, let E_n and O_n be n-ary junctors defined as follows:

$$\mathrm{Asf}_{E_n}(u_1, \ldots, u_n) = 1 \text{ iff } \sum_1^n u_i \text{ is even}$$

$$\mathrm{Asf}_{O_n}(u_1, \ldots, u_n) = 1 \text{ iff } \sum_1^n u_i \text{ is odd.}$$

Consider a predicate calculus whose junctors are those just defined, negation, 0 and 1. Show that each open formula built up from atomic formulas using $\equiv$ and $\neg$ is logically equivalent to a formula of one of the following forms:

$$\underline{0}, \underline{1}, E_n(\varphi_1, \ldots, \varphi_n), \quad O_n(\varphi_1, \ldots, \varphi_n),$$

where $\varphi_1, \ldots, \varphi_n$ are distinct atomic formulas.

(5) Show that (2) remains valid if we replace "function calculus" by "state dependent calculus", "variable" by "object variable" and "quantifier" by "object quantifier" (Modify appropriately (1)).

(6) Show that the following formulas are not logically equivalent

$$(\forall x)(\underline{\text{few}}\ s)\varphi(x,s),$$
$$(\underline{\text{few}}\ s)(\forall x)\varphi(x,s).$$

(7) (Fraissé) Consider $\{0,1\}$-structures of a type t; let p be a natural number. Put $\underline{M} \simeq_p \underline{N}$ if each substructure $\underline{M}_o$ of $\underline{M}$, M_o of cardinality $\leq p$ is isomorphic to a substructure of N and vice versa. (Then call M and N p-equivalent).

(a) $\simeq_p$ is an equivalence relation with finitely many equivalence classes.

(b) There is a natural number $r(t,p)$ such that each structure has a p-equivalent substructure of cardinality $\leq r(t,p)$.

(c) There is a natural number $s(t,p)$ such that for each finite structure $\underline{M}$ of cardinality $\geq s(t,p)$ there is a p-equivalent countably infinite structure $\underline{M}'$.

(8) Let Σ be a set of states and let $\preceq$ be a linear quasiordering on a field $\mathcal{R}$ of subsets of Σ (i. e. a transitive relation such that for all X, Y ($X \preceq Y$ or $Y \succeq X$) and suppose that $X \subseteq Y$ implies $X \preceq Y$ for each $X, Y \in \mathcal{R}$).

Define a state quantifier More of type $\langle 1,1 \rangle$ for state dependent predicate calculi putting

$$\mathrm{Asf}_{\underline{\mathrm{More}}}(\langle g_1, g_2 \rangle) = 1 \text{ iff } \{\sigma \in \Sigma; g_1(\sigma) = 1\} \preceq \{\sigma \in \Sigma; g_2(\sigma) = 1\}$$
$$= 0 \text{ otherwise.}$$

Read a formula $(\underline{\mathrm{More}}\ s)(\varphi(s), \psi(s))$ " ψ is more likely than φ ".
Find some tautologies for More. (For example, if P is a probability measure on Σ, one can define

$X \preceq Y$ iff $P(X) \leq P(Y)$.)

Chapter III. The Logic of Observational Functor Calculi

Observational calculi are logical systems similar to the first order predicate calculus; thus it is possible to consider them from the logical point of view. In particular, questions concerning decidability, axiomatizability, and definability can be naturally asked. It appears that observational calculi could and should be studied also in the "pure" symbolic logic; but our question is what is the importance of the logic of observational calculi for AI and, more generally, for computer science. We claim three things:

(1) Questions concerning decidability of observational calculi are relevant for Hypothesis Formation. Reason: nobody would call a tautology an intelligent observation concerning particular data, if he knew that it is a tautology. Can we recognize tautologies of an observational calculus we are using? Naturally, the decidability question is only the beginning. If the answer is yes, the next question concerns the complexity of the decision problem. If the answer is no, the next task is to find natural subclasses $Sent_o$ of the set Sent of all sentences such that the tautology problem restricted to $Sent_o$ is decidable.

(2) The notion of an immediate consequence can be used for optimized representation of sets of observational statements. This is typical for GUHA methods (see Part B): one finds simple sound deduction rules and uses them in a non-iterative way to represent relevant observational truths.

(3) There are close relations between logical notions concerning observational calculi and notions concerning recognizability of languages in polynomial time. Hence the logic of observational calculi is related to and can be used in the theory of computational complexity.

The chapter is arranged as follows: Section 1 deals with monadic observational predicate calculi, i.e. OPC's all of whose predicates are unary (except the equality predicate, if present). The reader will see that these simple observational calculi do have non-trivial theory thanks to generalized quantifiers. Note also that the particular methods of suggestion described in Part B are based on monadic OPC's and their generalizations-monadic observational function calculi. In Section 2, we investigate a very important class of observational quantifiers in OPC's, called associational quantifiers and its subclass of implicational quantifiers. These are natural classes of quantifiers and we shall see in Chapter IV that various important statistically motivated quantifiers are associational or even implicational. Section 3 is devoted to the problem of incomplete information: we describe a uniform way of extending each OFC to an OFC having one more additional value x-unknown (missing information). In Section 4, we shall investigate calculi having finitely many abstract values without any preferred structure on the values; such calculi are called calculi with nominal or qualitative values. Results of the first four sections will be utilized in Chapter IV and in Part B.

Section 5 surveys abstract model theory of the OPC's and describes its connection with the well known problems of complexity theory. It shows how OPC's differ from the predicate calculi with both finite and infinite models in questions concerning the interpolation theorem and related problems.(Full treatment of this matter will be published elsewhere.) Section 5 may be omitted on a first reading.

III.1 Monadic observational predicate calculi

3.1.1 Definition and conventions. Observational predicate calculi (OPC's) were defined in Chapter II, Section 2. An OPC is monadic if all its predicates are unary, i.e. if its type is $\langle 1, \ldots, 1\rangle$. We write MOPC for "monadic observational predicate calculus". A MOPC whose only quantifiers are the classical quantifiers $\forall, \exists$ is called a classical MOPC or CMOPC. Similarly for a MOPC with equality, in particular a CMOPC with equality.

3.1.2 Definition. Let $\mathcal{P}$ be a MOPC. The first variable x_o is called the designated variable. Open (= quantifier free) formulas containing no variable distinct from the designated variable x are called designated open formulas. Let $\langle P_i, i < n\rangle$ be the sequence of predicates of $\mathcal{P}$. An n-ary card is a sequence $\langle u_i; i < n\rangle$ of zeros and ones. If $\underline{M} = \langle M, \langle p_i ; i < n\rangle\rangle$ is a model (a $\{0,1\}$-structure of type $\langle 1^n\rangle$) and if $o \in M$ then the M-card of o is the tuple $C_{\underline{M}}(o) = \langle p_i(o); i < n\rangle$; it is evidently an n-ary card.

3.1.3 Lemma. Let $\varphi(x)$ be a designated open formula, let $\underline{M}$ be a model and let $o \in M$. Then the value $\|\varphi\|_{\underline{M}}[o]$ depends only on $C_{\underline{M}}(o)$, i.e., whenever M' is a model and $C_{\underline{M}}(o) = C_{\underline{M}'}(o)$ then $\|\varphi\|_{\underline{M}}[o] = \|\varphi\|_{\underline{M}'}[o]$. Moreover, if $P_{i_1}, \ldots, P_{i_k}$ are the predicates occurring in φ then $\|\varphi\|_{\underline{M}}[o]$ depends only on the i_1-th, ..., i_k-th members of $C_{\underline{M}}(o)$, i.e., whenever $\underline{M}'$ is a model, $o' \in M'$ and $C_{\underline{M}}(o)$ coincides with $C_{\underline{M}'}(o')$ on the i_1-th, ..., i_k-th place then $\|\varphi\|_{\underline{M}}[o] = \|\varphi\|_{\underline{M}'}[o']$.

Proof: obvious.

3.1.4 Notation. If u is an m-card and if φ is a designated open formula then $\|\varphi\|[u]$ is defined as $\|\varphi\|_{\underline{M}}[o]$ for each M such that $C_{\underline{M}}(o) = u$.

3.1.5 Definition. Let $\mathcal{P}$ be a MOPC of type $\langle 1^n \rangle$ and let q be a quantifier of type $\langle 1^k \rangle$, $k \le n$. q is definable in $\mathcal{P}$ if there is a sentence Φ of $\mathcal{P}$ not containing q such that the sentence $(qx)(P_1(x), \dots, P_k(x))$ is logically equivalent to Φ.

3.1.6 Lemma. Let $\mathcal{P}$ and q be as in 3.1.5. q is definable in $\mathcal{P}$ iff each sentence of $\mathcal{P}$ is logically equivalent to a sentence not containing the quantifier q.

Proof. $\Leftarrow$ is trivial. To prove $\Rightarrow$, one shows by induction on the complexity of formulae that the following holds:

For each formula $\varphi(x_o, \dots, x_k)$ of $\mathcal{P}$ there is a formula $\hat{\varphi}(x_o, \dots, x_k)$ not containing q and logically equivalent to φ. The only non-trivial step concerns the case that $\varphi(\underline{x})$ has the form $(qy)(\varphi_1(x,y), \dots, \varphi_k(x,y))$. Let Φ be the sentence from the definition; by renaming bound variables, assume that no variable occurring in $(qy)(\varphi_1, \dots, \varphi_k)$ occurs in Φ. Let $\Phi^*(\underline{x})$ be the formula resulting from Φ by replacing each occurrence of $P_i(z)$ by $\varphi_i(\underline{x}, z)$ $(i = 1, \dots, k)$; then evidently $\Phi^*(\underline{x})$ is logically equivalent to $\varphi(\underline{x})$.

In the sequel we shall study CMOPC's (first without equality, then with equality). We shall see that CMOPC's are uninteresting since their expressive power is too weak; hence it is reasonable to turn to MOPC's with non-classical quantifiers. Let $\mathcal{P}$ be a fixed CMOPC without equality.

3.1.7 Definition. A canonical sentence (of $\mathcal{P}$) is a sentence of the form $(\forall x)\varphi$ where x is the designated variable and φ is a designated elementary disjunction (i.e., an elementary disjunction containing only the designated variable).

3.1.8 Theorem (Normal form). Each sentence of $\mathcal{P}$ is logically equivalent to a Boolean combination of canonical sentences (i.e., to a sentence built up from canonical sentences using only the junctors &, $\vee$, $\neg$).

Proof: We prove the following slightly more general assertion concerning arbitrary formulas: For each formula φ (with the free variables $u, \dots, v$) there is a logically equivalent formula $\hat{\varphi}$ (with the same free variables) which is a boolean combination of canonical sentences and atomic formulae.

We prove the last assertion by induction on the complexity of φ. If φ is atomic then the assertion is trivial. The induction step for connectives is also trivial. Let φ be $(\forall z)$ and let $\psi \Leftrightarrow \hat{\psi}$ where $\hat{\psi}$ is a boolean combination of canonical and atomic formulae. Call a formula δ a quasielementary disjunction if there are pairwise distinct canonical and/or atomic formulae $\chi_1, \ldots, \chi_e$ $(e \geq 1)$ such that δ results from them as follows: One first negates some (possibly one) of them and then joins the resulting sequence by the sign $\vee$. Thus quantifier-free quasielementary disjunctions are just elementary disjunctions. We may assume that our $\hat{\psi}$ is a conjunction $\delta_1 \,\&\, \ldots \,\&\, \delta_k$ of quasielementary disjunctions. Then, evidently, $\varphi \Leftrightarrow (\forall z)\delta_1 \,\&\, \ldots \,\&\, (\forall z)\delta_k$ and it suffices to show that each $(\forall z)\delta_i$ can be reduced to the desired form. Thus, let i be fixed and suppose that δ_i is the disjunction $\alpha_1(z) \vee \ldots \vee \alpha_p(z) \vee \omega_1 \vee \ldots \vee \omega_q$ where the α's contain z free and the ω's do not.
Then $(\forall z)(\alpha_1(z) \vee \ldots \vee \alpha_p(z) \vee \omega_1 \vee \ldots \vee \omega_q) = (\forall z)(\alpha_1(z) \vee \ldots \vee \alpha_p(z)) \vee \omega_1 \vee \ldots \vee \omega_q$.
If $p = 0$ then there is nothing to prove since $\omega_1 \vee \ldots \vee \omega_p$ is of the desired form; if $p > 0$ then we **rename the bound variables to replace** $(\forall z)(\alpha_1(z) \vee \ldots \vee \alpha_p(z))$ by $(\forall x)(\alpha_1(x) \vee \ldots \vee \alpha_p(x))$, which is a canonical sentence, since $\alpha_1(x) \vee \ldots \vee \alpha_p(x)$ is an elementary disjunction.

3.1.9 Definition. Let $\underline{M}$ be a model of type $\langle 1^n \rangle$. The characteristic of $\underline{M}$ is the mapping $\chi_{\underline{M}}$ of the set of all n-cards into $\{0,1\}$ defined as follows: $\chi_{\underline{M}}(\underline{u}) = 1$ iff there is an $o \in M$ whose card $C_{\underline{M}}(o)$ is $\underline{u}$.

3.1.10 Theorem (on the characteristic). Let $\mathcal{P}$ be the CMOPC with predicates $P_1, \ldots, P_n$ and let $\underline{M}_1, \underline{M}_2$ be models. If $\chi_{\underline{M}_1} = \chi_{\underline{M}_2}$ then, for each sentence Φ, $\underline{M}_1 \models \Phi$ iff $\underline{M}_2 \models \Phi$.

Proof: By the Normal form theorem 3.1.8, it suffices to suppose Φ to be a canonical sentence $(\forall x)\delta$. Then $\underline{M}_i \models \Phi$ iff, for each object $o \in M_i$, $\|\delta\|_{\underline{M}_i}[o] = 1$. However by 3.1.2, the $\underline{M}_i$-value of δ for o depends only on the $\underline{M}_i$-card of o; hence if $\underline{M}_1, \underline{M}_2$ have the same cards and $\|\delta\|_{\underline{M}_1}[o] = 1$ for all $o \in M_1$ then $\|\delta\|_{\underline{M}_2}[o] = 1$ for each $o \in M_2$.

3.1.11 Corollary (stability). Let $\mathcal{P}$ be as above, let $\underline{M}$ be a model and Φ a sentence. $\underline{M} \models \Phi$ iff there is a submodel $\underline{M}_o \subseteq \underline{M}$ such that $\underline{M}_o$ has at most 2^n elements and, for each $\underline{M}_1$, $\underline{M}_o \subseteq \underline{M}_1 \subseteq \underline{M}$ implies $\underline{M}_1 \models \Phi$.

Proof: For each card occurring in $\underline{M}$ select an object with this card; all selected objects form $\underline{M}_o$.

3.1.12 Corollary (decidability). Each CMOPC is decidable.

Proof: By 3.1.11, if Φ is a sentence containing n predicates then Φ is a tautology iff Φ is true in all models having at most 2^n elements; thus, to decide whether Φ is a tautology it suffices to consider all (finitely many) models with the domain $\{0, \ldots, i\}$ (for all $i \leq 2^n - 1$) and verify whether Φ is true in all of them.

3.1.13 Lemma. For each function χ mapping the set of all n-cards into $\{0,1\}$ and not identically equal to 0:

(1) There is a model $\underline{M}$ such that $\chi = \chi_{\underline{M}}$;

(2) There is a sentence Φ_χ such that, for each $\underline{M}$, $\underline{M} \models \Phi_\chi$ iff $\chi = \chi_{\underline{M}}$.

Proof: (1) Let K be the set of all n-cards, let $M = \{\underline{u} \in K ; \chi(\underline{u}) = 1\}$, let, for $\underline{u} = \langle u_1, \ldots, u_n \rangle$, $p_i(\underline{u}) = u_i$. Then, for $\underline{M} = \langle M, p_1, \ldots, p_n \rangle$, we have $\chi = \chi_{\underline{M}}$.

(2) Let, for $\underline{u} = \langle u_1, \ldots, u_n \rangle \in K$, $u_i P_i$ be P_i if $u_i = 1$ and $u_i P_i$ be $\neg P_i$ if $u_i = 0$. Let $\kappa_{\underline{u}}$ be $u_1 P_1 \& \ldots \& u_n P_n$. Let Φ be the formula

$$(\bigwedge_{\chi(\underline{u})=1} (\exists x)\kappa_{\underline{u}}) \& (\forall x) \bigvee_{\chi(\underline{u})=1} \kappa_{\underline{u}}).$$

3.1.14 Theorem (definability of quantifiers). Let $\mathcal{P}$ be the CMOPC with the predicates $P_1, \ldots, P_n$ and let $\mathcal{P}'$ be the extension of $\mathcal{P}$ resulting from the addition of a new quantifier q of type $\langle 1^k \rangle$ $(k \leq n)$. q is definable in $\mathcal{P}'$ iff Asf_q is constant on models with equal characteristic i.e., iff $\chi_{\underline{M}} = \chi_{\underline{M}'}$ implies $\mathrm{Asf}_q(\underline{M}) = \mathrm{Asf}_q(\underline{M}')$.

Proof: The implication $\Rightarrow$ follows from Theorem 5.1.10 on the characteristic . Conversely, suppose $k = n$ (ignore $P_{k+1}, \ldots P_n$) and let X be the set of all characteristics $\chi_{\underline{M}}$ such that $Asf_q(\underline{M}) = 1$ iff $\chi_{\underline{M}} \in X$. Let Φ be $\bigvee_{\chi \in X} \Phi_\chi$; then $\|\Phi\|_{\underline{M}} = 1$ iff $\chi_{\underline{M}} \in X$, i.e., Φ defines q.

3.1.15 Remark. The preceding results concerning CMOPC's by no means constitute a novelty but the authors were not able to find appropriate references.

We shall now consider CMOPC's with the equality predicate. We shall see how this generalization increases the expressive power of CMOPC's; we show that the equality predicate can be replaced by infinitely many quantifiers $\exists^k$ (there are k objects such that. ...; k a natural number) and we shall argue that, from our point of view, such quantifiers yield a generalization of CMOPC's which is more natural than the equality predicate. Hence we are led again to MOPC's with arbitrary quantifiers. Our exposition is based on Slomson [1968] but the results (except 3.1.28) are older; cf. Jensen [1965].

3.1.16 Definition. (1) For each natural number $k > 0$, $\exists^k$ is a quantifier of type $\langle 1 \rangle$ whose associated function is defined as follows: For each finite model $\underline{M} = \langle M, f \rangle$, $Asf_{\exists^k}(\underline{M}) = 1$ iff there are at least k elements $o \in M$ such that $f(o) = 1$.

(2) If $\mathcal{P}$ is a CMOPC then $\mathcal{P}^=$ denotes the corresponding CMOPC with equality and $\mathcal{P}^*$ denotes the extension of $\mathcal{P}^=$ by adding all the quantifiers $\exists^k$ (k a natural number).

3.1.17 Lemma. Let k be a natural number and let $\mathcal{P}^k$ be the extension of $\mathcal{P}^=$ by $\exists^k$. Then $\exists^k$ is definable in $\mathcal{P}^k$ by the following formula

$$\Phi_k : (\exists x_1, \ldots, x_k)(\bigwedge_{i \neq j,\ 1 \leq i,\ j \leq k} x_i \neq x_j \ \& \bigwedge_{1 \leq i \leq k} P_1 x_i)$$

The proof is obvious.

3.1.18 Conventions and Definition. (1) In the next few paragraphs, we consider a fixed CMOPC $\mathcal{P}^{=}$ with equality and n unary predicates.

(2) K denotes the set of all n-cards. A genus is a mapping of K into natural numbers not identically equal to 0. With each model $\underline{M}$ we associate the genus $g_{\underline{M}}$ such that $g_{\underline{M}}(\underline{u}) = i$ iff the number of objects in M having the card $\underline{u}$ is i. Note that M is finite. If g is a genus and $p \in \mathbb{N}$ then g/p is the genus defined as follows: $g/p(\underline{u}) = \min(g(\underline{u}), p)$.

(3) Let $\underline{M}$ be a model and let $\underline{m} = \langle m_o, \ldots, m_{k-1} \rangle$, $\underline{n} = \langle n_o, \ldots, n_{k-1} \rangle$ be k-tuples of elements of M. $\underline{m}$ and $\underline{n}$ are M-similar (notation: $\underline{m} \simeq_{\underline{M}} \underline{n}$) if (a) $C_{\underline{M}}(\underline{m}_i) = C_{\underline{M}}(\underline{n}_i)$ for each $i < k$ and (b) $n_i = n_j$ iff $m_i = m_j$ for each $i,j < k$.

3.1.19 Lemma. Let $\underline{M}$ be a model, let $\underline{m},\underline{n}$ be k-tuples of elements from M such that $\underline{m} \simeq_{\underline{M}} \underline{n}$ and let φ be a formula with the free variables $x_o, \ldots, x_{k-1}$. Then

$$\|\varphi\|_{\underline{M}}[\underline{m}] = \|\varphi\|_{\underline{M}}[\underline{n}] \quad .$$

Proof: Let $\underline{m} = \langle m_o, \ldots, m_{k-1} \rangle$ and $\underline{n} = \langle n_o, \ldots, n_{k-1} \rangle$. Put $\iota(o) = 0$ for each $o \in M$ distinct from all the members of $\underline{m}$ and $\underline{n}$, and let $\iota(m_i) = n_i$ and $\iota(n_i) = m_i$ for i=1, ..., k-1. Then ι is an isomorphism between $\underline{M}$ and $\underline{M}$ (an automorphism of $\underline{M}$) and hence preserves the values of the formulae. Thus, $\|\varphi\|_{\underline{M}}[\underline{m}]$ is equal to $\|\varphi\|_{\underline{M}}[\underline{n}]$.

3.1.20 Lemma. Let $\underline{M}$ be a model and let $\varphi(v_o, \ldots, v_{k-1})$ be a formula containing less than p variables (both free and bound; the free ones are $v_o, \ldots, v_{k-1}$). Suppose that $\underline{N}$ is a submodel of $\underline{M}$ such that $g_{\underline{N}} = g_{\underline{M}}/p$. Then $\|\varphi\|_{\underline{M}}[\underline{m}] = \|\varphi\|_{\underline{N}}[\underline{m}]$ for each k-tuple $\underline{m}$ of elements of N.

Proof: We use induction on the complexity of φ. If φ is atomic then the assertion is obvious. The induction step for connectives is also obvious. Thus, suppose φ to be the formula $(\exists v_k)\, \psi(v_o, \ldots, v_k)$ and let the assertion hold for ψ.

Let $m_o, \dots, m_{k-1} \in N$, $\underline{m} = \langle m_o, \dots, m_{k-1} \rangle$. First, suppose $\|\varphi\|_{\underline{N}}[\underline{m}] = 1$. Then there is an $m_k \in N$ such that $\|\psi\|_{\underline{N}}[m_o, \dots, m_{k-1}, m_k] = 1$, hence $\|\psi\|_{\underline{M}}[m_o, \dots, m_{k-1}, m_k] = 1$ by the induction assumption. Consequently, $\|(\exists v_k)\, \psi(v_o, \dots v_k)\|_{\underline{M}}\, [m_o, \dots, m_{k-1}] = 1$.

Conversely, suppose $\|\varphi\|_{\underline{M}}\, \underline{m} = 1$. Then there is an $m_k \in M$ such that $\|\psi\|_{\underline{M}}[m_o, \dots, m_k] = 1$. If $m_k \in N$ we obtain $\|\psi\|_{\underline{N}}[m_o, \dots, m_k] = 1$ which implies $\|\varphi\|_{\underline{N}}[\underline{m}] = 1$.
If $m_k \in M - N$ and if $\underline{u}$ is the card of m_k in $\underline{M}$ then there are at least p objects in M with the card $\underline{u}$; exactly p of them belong to N. Thus, there is an $m_k \in N$ such that $\hat{m}_k \neq m_o, \dots, m_{k-1}$ and $C_{\underline{M}}(m_k) = C_{\underline{M}}(\hat{m}_k)$.
Then $\langle m_o, \dots, m_{k-1}, m_k \rangle \simeq_{\underline{M}} \langle m_o, \dots, m_{k-1}, \hat{m}_k \rangle$ and, by Lemma 3.1.19, we have $\|\psi\|_{\underline{M}}[m_o, \dots, m_k] = 1$. Then $\|\psi\|_{\underline{N}}[m_o, \dots, m_k] = 1$ and $\|\varphi\|_{\underline{N}}[\underline{m}] = 1$.

3.1.21 Remark. The preceding lemma can be generalized as follows: The assumption $g_{\underline{N}} = g_{\underline{N}}/p$ can be replaced by the assumption that, for each card $\underline{u}$, (i) whenever $g_{\underline{M}}(\underline{u}) < p$, $g_{\underline{M}}(\underline{u}) = g_{\underline{N}}(\underline{u})$ (all elements with the card $\underline{u}$ belong to N), (ii) whenever $g_{\underline{M}}(\underline{u}) \geq p$, $g_{\underline{N}}(\underline{u})$ is <u>at least</u> p (and, obviously, $g_{\underline{N}}(\underline{u}) \leq g_{\underline{M}}(\underline{u})$).

3.1.22 Theorem (stability). Let $\mathcal{P}^{=}$ be the CMOPC with equality and n unary predicates; let Φ be a sentence containing less than p variables. $\underline{M} \models \Phi$ iff there is a submodel $\underline{M}_o \subseteq \underline{M}$ such that M_o has at most $p.2^n$ elements and, for each $\underline{M}_1$, $\underline{M}_o \subseteq \underline{M}_1 \subseteq \underline{M}$ implies $\underline{M}_1 \models \Phi$.

Proof: Take for $\underline{M}_o$ a model with the genus $g_{\underline{M}_o} = g_{\underline{M}}/p$. Use 3.1.20.

3.1.23 Corollary (decidability). Each CMOPC with equality is decidable.

Proof: By 3.1.22 , if Φ is a sentence containing n predicates and p variables then Φ is a tautology iff Φ is true in all models having at most $p.2^n$ elements.

3.1.24 Remark and Definition. Let $\underline{u} = \langle u_o, \ldots, u_{n-1} \rangle$ be an n-card; for each $i < n$, let λ_i be $P_i(x)$ if $u_i = 1$ and let λ_i be $\neg P_i(x)$ if $u_i = 0$. Let $\kappa_{\underline{u}}$ be $\bigwedge_{i=0}^{n-1} \lambda_i$ the elementary conjunction given by $\underline{u}$. Let $\underline{M}$ be a model. Let $k > 0$. Then (a) $\underline{M} \models (\exists^k x)\kappa_{\underline{u}}$ iff at least k objects in $\underline{M}$ have the card $\underline{u}$;

(b) $\underline{M} \models (\exists^k x)\kappa_{\underline{u}} \ \& \ \neg(\exists^{k+1} x)\kappa_{\underline{u}}$ iff exactly k objects in $\underline{M}$ have the card $\underline{u}$;

(c) $\underline{M} = \neg(\exists^1 x)\ \kappa_{\underline{u}}$ iff no object in $\underline{M}$ has the card $\underline{u}$. Note that the formula $(\exists^k x)\kappa_{\underline{u}}$ contains only the designated variable x and does not contain the equality predicate. Each formula of the form $(\exists^k x)\kappa_{\underline{u}}$ (where $\underline{u}$ is a card) will be called a canonical sentence (for CMOPC's with equality).

3.1.25 Theorem (normal form). Let $\mathcal{P}^=$ be a CMOPC with equality and let $\mathcal{P}^*$ be the extension of $\mathcal{P}^=$ by adding the quantifiers $\exists^k$ (k a natural number). Let Φ be a sentence from $\mathcal{P}^=$. Then there is a sentence Φ^* from $\mathcal{P}^*$ logically equivalent to Φ (in $\mathcal{P}^*$) and such that Φ^* is a boolean combination of canonical sentences. (In particular, Φ^* contains neither the equality predicate nor any variable distinct from the canonical variable).

Proof: Suppose that Φ contains n predicates and p variables. Let $\underline{N}$ be a model (of type $\langle 1^n \rangle$) having $\leq p.2^n$ elements. For each card $\underline{u}$, let $\varphi_{\underline{N},\underline{u}}$ be a Boolean combination of canonical sentences such that
(i) if $g_{\underline{N}}(\underline{u}) < p$ then $\varphi_{\underline{N},\underline{u}}$ says "exactly $g_{\underline{N}}(\underline{u})$ objects have the card $\underline{u}$",
(ii) if $g_{\underline{N}}(\underline{u}) \geq p$ then $\varphi_{\underline{N},\underline{u}}$ says "at least p objects have the card $\underline{u}$".
(Use 3.1.24.) Let $\varphi_{\underline{N}}$ be $\bigwedge_{\underline{u}\text{ card}} \varphi_{\underline{N},\underline{u}}$ and let Φ^* be $\bigwedge_{\underline{N}\models\Phi} \varphi_{\underline{N}}$. (Observe that this is a disjunction of finitely many formulae.) We claim that Φ^* is logically equivalent to Φ. Indeed, if $\underline{M} = \Phi$ then, by 3.1.20, there is an $\underline{N} \subseteq \underline{M}$ such that $\underline{N} \models \Phi$ and $g_{\underline{N}} = g_{\underline{M}}/p$; then $\underline{M} \models \varphi_{\underline{N}}$. Conversely, if $\underline{M} \models \neg\Phi$ and $\underline{M} \models \varphi_N$ for some $\underline{N}$ then let $\underline{N}_o$ be a submodel of $\underline{M}$ with the genus $g_{\underline{M}}/p$. Then $\underline{N}_o$ can be considered to be a submodel of $\underline{M}$ and a submodel of $\underline{N}$; by 3.1.20, $\underline{M} \models \Phi$ iff $\underline{N}_o \models \Phi$ iff $\underline{N} \models \Phi$, hence $\underline{N} \models \neg\Phi$. We have proved $\underline{M} \models \neg\Phi^*$.

3.1.26 Theorem (definability of quantifiers). (Tharp 1973). Let $\mathcal{P}^{=}$ be a CMOPC with equality and unary predicates $P_1, \ldots, P_n$ and let $\mathcal{P}'$ be its extension by adding a quantifier q of type $\langle 1^k \rangle$ ($k \leq n$). q is definable in $\mathcal{P}'$ iff there is a natural number m such that the following holds for $\varepsilon = 0,1$ and each model $\underline{M}$ of type $\langle 1^k \rangle$:

$Asf_q(\underline{M}) = \varepsilon$ iff $(\exists \underline{M}_o \subseteq \underline{M})(\underline{M}_o$ has $\leq$ m elements and

$$(\forall \underline{M}_1)(\underline{M}_o \subseteq \underline{M}_1 \subseteq \underline{M} \text{ implies } Asf_q(\underline{M}_1) = \varepsilon)).$$

Proof: If q is definable then the condition follows immediately from the stability theorem.

Conversely, let the condition hold. Call a sentence Φ classically expressible (or expressible) if there is a sentence not containing q and logically equivalent to Φ. Our aim is to prove that $q(P_1, \ldots, P_\ell)$ is expressible. Assume that this is not the case. We shall construct more and more special non-expressible sentences and at the end arrive at a contradiction. Let $u_1, \ldots, u_{2^\ell}$ be all the cards. Remember the sentence $(\exists^k x)\,\kappa_{u_i}$ saying that there are at least k objects with the card u_i and the sentence $(\exists^k x)\,\kappa_{u_i} \,\&\, \neg(\exists^{k+1} x)\,\kappa_{u_i}$ saying that there are exactly k objects with the card u_i. We denote the former sentence by $|u_i| \geq k$ and the second by $|u_i| = k$. Note that both sentences are expressible. Put $\chi_o^+ = q(P_1, \ldots, P_k)$ and $\chi_o^- = \neg q(P_1, \ldots, P_k)$. By assumption, neither χ_o^+ nor χ_o^- is expressible. We proceed in steps $i = 1, \ldots, 2^k$. In step i we define the numbers k_i and formulae χ_i^+, χ_i^-. Let m be the number from our assumption.

In step 1 we consider u_1. (a) If there is a $k_1 < m$ such that $\chi_o^+ \& |u_1| = k_1$ is not expressible then choose one such k_1, and put $\chi_1^+ = \chi_o^+ \& |u_1| = k_1$, $\chi_1^- = \chi_o^- \& |u_1| = k_1$. Neither χ_1^+ nor χ_1^- is expressible (if χ_1^- were then χ_1^+ would also be expressible since χ_1^+ is equivalent to $\neg \chi_1^- \& |u_1| = k_1$).

(b) If there is no such k_1 we put $k_1 = m$ and

$\chi_1^+ = \chi_1^+ \,\&\, |u_1| \geq k_1$, $\chi_1^- = \chi_o^- \,\&\, |u_1| \geq k_1$. In the present case neither χ_1^+ nor χ_1^- is expressible. (Note that χ_o^+ is equivalent to the disjunction

$[(\chi_o^+ \,\&\, |u_1| = 0) \vee \ldots\ldots \vee (\chi_o^+ \,\&\, |u_1| = m-1)] \vee (\chi_o^+ \,\&\, |u_1| \geq m)$:

if all disjuncts were expressible, χ_o^+ would be too.)

Suppose that step (i-1) has been completed; in step i we consider u_i.

(a) If there is a $k_i < m$ such that $\chi_{i-1}^+ \,\&\, |u_1| = k_i$ is not expressible we choose such a k_i and put

$\chi_i^+ = \chi_i^+ \,\&\, |u_i| = k_i$, $\chi_i^- = \chi_{i-1}^- \,\&\, |u_i| = k_i$,

(b) otherwise, we put $k_i = m$ and put

$\chi_i^+ = \chi_{i-1}^+ \,\&\, |u_i| \geq k_i$, $\chi_i^- = \chi_{i-1}^- \,\&\, |u_i| \geq k_i$.

Neither χ_i^+ nor χ_i^- is expressible.

Put $\chi^+ = \chi_{2^\ell}^+$, $\chi^- = \chi_{2^\ell}^-$. Let $\underline{M} \models \chi^+$, $\underline{N} \models \chi^-$. **Here are some examples; otherwise** χ^+ and χ^- would be expressible. Let $\underline{M}_o$ be a submodel of $\underline{M}$ with at least m elements such that all models between $\underline{M}_o$ and $\underline{M}$ satisfy χ_o^+; the same holds for $\underline{N}_o$, $\underline{N}$ and χ_o^-. Then, we can find $\underline{M}_1$ and $\underline{N}_1$ such that $\underline{M}_o \subseteq \underline{M}_1 \subseteq \underline{M}$, $\underline{N}_o \subseteq \underline{N}_1 \subseteq \underline{N}$ and

(*) $$\underline{M}_1, \underline{N}_1 \models \bigwedge_{i=1}^{2^k} |u_i| = k_i$$

Note that $\underline{M}_1 \models \chi^+$ and $\underline{N}_1 \models \chi^-$; but (*) implies that $\underline{M}_1$ and $\underline{N}_1$ are isomorphic, so that, e.g., $\underline{N} \models \chi^+$, which is a contradiction.

3.1.27 <u>Remark.</u> q is not definable iff for each m there is an $\varepsilon = 0,1$ and a model $\underline{M}$ with $Asf_q(\underline{M}) = \varepsilon$ such that for each $\underline{M}_o \subseteq \underline{M}$ with at most m elements there is an $\underline{M}_1$ between $\underline{M}_o$, $\underline{M}$ such that $Asf_q(\underline{M}_1) \neq \varepsilon$.

Examples cf. 2.2.4 .

(1) The plurality quantifier W : Given m, take a model $\underline{M}$ with 2m + 3 objects, f(o) = 1 for m+2 objects, f(o) = 0 for the rest. Each submodel with m objects can be extended to a submodel with m ones and m+1 zeros. Thus, W is not definable.

(2) The simple **associational quantifier** $\sim$: For the sake of simplicity, assume $m = 2^e$. Let $\underline{M}$ be a model with 4e + 3 objects such that

$a_{\underline{M}} = b_{\underline{M}} = d_{\underline{M}} = e + 1$, $c_{\underline{M}} = e$. Each submodel with 2e elements can be extended to a submodel $\underline{M}_1$ with $a_{\underline{M}_1} = c_{\underline{M}_1} = d_{\underline{M}_1} = e$, $b_{\underline{M}_1} = e + 1$. Thus, $\sim$ is not definable.

(3) We know that $\exists^k$ is definable; we can take k for the constant m in 3.1.26. Indeed, if $\mathrm{Asf}_{\exists^k}(\langle M, f\rangle) = 1$ then take $\underline{M}_o \subseteq \underline{M}$ such that f is identically 1 on $\underline{M}_o$; if $\mathrm{Asf}_{\exists^k}(\langle M,f\rangle) = 0$ then take an arbitrary non-empty $\underline{M}_o \subseteq \underline{M}$ with at most m elements.

3.1.28 We shall now consider arbitrary MOPC's (without equality). We show that we have a normal form theorem and, under the assumption that the number of quanfitiers is finite, we prove decidability. However, we shall show that there are undecidable MOPC's with finitely many quantifiers and equality.

We shall consider a MOPC $\mathcal{P}$. Recall that the type of a quantifier q is a tuple of ones. Since the notation for the general case is somewhat cumbersome we shall use, in our proofs, a typical example of a quantifier q of type $\langle 1,1\rangle$; hence Asf_q is defined on models of the form $\langle M,f,g\rangle$, where f,g, are mappings of M into $\{0,1\}$. If φ , ψ are formulas and x is a variable then $(qx)(\varphi,\psi)$ is a formula. Assume that the free variables of φ and ψ are x, z and let e be the mapping $\frac{z}{a}$. Then

$$\|(qx)(\varphi(x,z),\psi(x,z))\|_{\underline{M}}[e] = \mathrm{Asf}_q(\langle M, \|\varphi\|^e_{\underline{M}}, \|\psi\|^e_{\underline{M}}\rangle) .$$

3.1.29 Definition. A closed formula is pure prenex if it begins with a quantifier and if the subformulae joined by this quantifier are open (do not contain any quantifiers). Hence, $(qx)(\varphi,\psi)$ is a pure prenex formula iff φ, ψ are open and contain only the variable x.

3.1.30 Lemma. Each formula φ is logically equivalent to a **Boolean** combination ψ of pure prenex formulae and atomic formulae such that φ and ψ have the same free variables.

Proof. We proceed by induction on the complexity of formulae. If φ is atomic or $\underline{0}$ or $\underline{1}$ then φ is itself a Boolean combination of the desired form; if φ is $\neg\varphi_o$ and the assertion holds for φ_o or if φ is $\varphi_1 \& \varphi_2$ or $\varphi_1 \vee \varphi_2$ or $\varphi_1 \rightarrow \varphi_2$ and the assertion holds for φ_1 and φ_2 then it evidently holds for φ. Thus, suppose that φ begins with a quantifier, say φ is $(qx)(\varphi_1, \varphi_2)$ and let the assertion hold for φ_1, φ_2. Thus, φ_1 is equivalent to the formula $\bigvee_{i=1}^{m}(\varphi_1 i(x) \& \psi_1 i)$, where each $\varphi_1 i(x)$ is an elementary conjunction built up from atomic formulae of the form $P_j(x)$ and ψ_{1i} is a (quasi) elementary conjunction built up from atomic formulas with variables other than x and from some prenex formulae; similarly for φ_2. A state is a system

$$\underline{\varepsilon} = \begin{pmatrix} \varepsilon_{11}, \ldots, \varepsilon_{1m} \\ \\ \varepsilon_{21}, \ldots, \varepsilon_{2n} \end{pmatrix} \quad \text{of m+n zeros and ones;}$$

the corresponding state description is the formula

$$S_{\underline{\varepsilon}} = (\varepsilon_{11}\psi_{11} \& \ldots \& \varepsilon_{1m}\psi_{1m}) \& (\varepsilon_{21}\psi_{21} \& \ldots \& \varepsilon_{2m}\psi_{2m})$$

where $\varepsilon_{11}\psi_{11}$ is ψ_{11} if $\varepsilon_{11} = 1$ and $\neg\psi_{11}$ if $\varepsilon_{11} = 0$ etc. Clearly, $\bigvee_{\underline{\varepsilon}} S_{\underline{\varepsilon}}$ is a tautology (by truth-tables); hence, we have the following equivalences ($\Leftrightarrow$ stands for "logically equivalent"):

$$\varphi \Leftrightarrow \varphi \& \bigvee_{\underline{\varepsilon}} S_{\underline{\varepsilon}} \Leftrightarrow$$

$$\Leftrightarrow \bigvee_{\varepsilon}[S_{\varepsilon} \,\&(qx)(\bigvee_i(\varphi_{1i}(x)\,\&\,\psi_{1i}), \bigvee_j(\varphi_{2j}(x)\,\&\,\psi_{2j}))] \Leftrightarrow$$

$$\Leftrightarrow \bigvee_{\varepsilon}[S_{\varepsilon} \,\&(qx)(\bigvee_i(\varphi_{1i}(x)\,\&\,\varepsilon_{1i}), \bigvee_j(\varphi_{2j}(x)\,\&\,\varepsilon_{2j}))] \Leftrightarrow$$

$$\Leftrightarrow \bigvee_{\varepsilon}[S_{\varepsilon} \,\&\, \underbrace{(qx)(\bigvee_{\varepsilon_{1j}=1} \varphi_{1j}(x), \bigvee_{\varepsilon_{2j}=1} \varphi_{2j}(x))}_{(*)}].$$

But each S_{ε} is a Boolean combination of the desired form and each $(*)$ is a pure prenex formula; thus, the induction step is concluded.

3.1.30 Corollary. (Pure prenex normal form). Each sentence is logically equivalent to a Boolean combination of pure prenex formulae.

3.1.31 Theorem (representation). If the MOPC $\mathcal{P}$ under consideration has finitely many quantifiers then there is a finite set S of sentences such that each sentence is logically equivalent to a sentence from S; there is a recursive function nf such that, for each sentence φ, $nf(\varphi) \in S$ and φ is logically equivalent to $nf(\varphi)$.

Proof: There is a recursive function dof associating with each designated open formula (i.e., open formula containing no variable except x) a logically equivalent formula from a finite set OF of designated open formulas; we may require that each predicate occurring in $dof(\varphi)$ occurs in φ and if the variable x occurs in $dof(\varphi)$ then it occurs in φ. (OF consists of designated open formulae in normal disjunctive form and, in addition, of $\underline{1}$). It is easy to show that OF has $1 + 2^{(3^n-1)}$ elements; one could find another set with exactly 2^{2^n} elements.

By 3.1.30 (using the proof of 3.1.29), there is a recursive function associating with each sentence Φ a logically equivalent sentence Φ' which is a Boolean combination of pure prenex sentences. Now, each pure prenex sentence can be effectively replaced by another pure prenex sentence Φ'' such that the open formulae joined in Φ'' by the quantifier of Φ'' are all in OF. Similarly for our Boolean combination Φ' of pure prenex formulae. Clearly, there are finitely many such pure prenex sentences (call them OF-pure prenex sentences);

each Boolean combination Φ'' of OF-pure prenex sentences can be effectively transformed to another combination Φ''', which is either $\underline{1}$ or a disjunction of (quasi) elementary conjunctions built up from OF-pure prenex sentences. The set of all sentences Φ''' is finite and can be taken as S; the mapping nf associating with each sentence Φ the sentence Φ''' is recursive.

3.1.32 Theorem (decidability). If $\mathcal{P}$ is a MOPC with finitely many quantifiers then $\mathcal{P}$ is decidable.

Proof. Let S and nf be as above, i.e., S is a finite set of sentences, nf is a recursive function and if Φ is a sentence, then $nf(\Phi) \in S$ and Φ is equivalent to $nf(\Phi)$. Let t be a function on S such that $t(\Phi) = 1$ iff Φ is a tautology and $t(\Phi) = 0$ otherwise. Note that the domain of t is finite; if we extend t to the set of all sentences putting $t(\Phi) = 0$ for $\Phi \notin S$ then the resulting function is recursive. Hence, the function $t(nf(\Phi))$ is recursive and it is the characteristic function of Taut.

3.1.33 Remark. (1) Theorem 3.1.32 is proved in [Mostowski] for the case that the type of each quantifier is $\langle 1 \rangle$.
(2) Our proof of 3.1.32 does not make decidability transparent. Even if we agree that the function is very reasonable, we see that the function t is recursive because it equals 0 except for finitely many exceptions. It would be difficult not to call such a function recursive: but it is possible that, for a $\varphi \in S$, we do not know whether φ is a tautology or not. Thus, we conclude that our decidability result does not give any result concerning transparency. The following theorem supports our last claim.

3.1.34 Theorem. (1) There is a MOPC $\mathcal{P}_1$ with one predicate and infinitely many quantifiers (without equality) which is not decidable.
(2) There is a MOPC $\mathcal{P}_2$ with one predicate, two quantifiers and equality, which is not decidable.

Proof: Let R be a binary (primitive) recursive relation such that the set $A = \{ n; (\exists m)R(m,n)\}$ is not recursive. We define the calculi $\mathcal{P}_1$, $\mathcal{P}_2$, and in each calculus we find a recursive sequence $\{\varphi_n; n \in \mathbb{N}\}$ of sentences such that φ_n has a model iff $n \in A$. Consequently, $n \in A$ iff $\neg\varphi_n \notin$ Taut and, hence, Taut is not recursive.

Both $\mathcal{P}_1$ and $\mathcal{P}_2$ will contain a quantifier $\mathfrak{R}$ such that $Asf_{\mathfrak{R}}(\langle M,f\rangle) = 1$ iff $R(a_{\underline{M}}, b_{\underline{M}})$, ($a_{\underline{M}}$ is the cardinality of $\{o \in M; f(o) = 1\}$, $b_{\underline{M}}$ the cardinality of $\{o \in M: f(o) = 0\}$. Furthermore, $\mathcal{P}_1$ will contain the quantifiers $\exists^{!k}$ (k a natural number, $\exists^{!k}$ of type <1>) saying "there are exactly k objects such that ...". Let P be the only predicate of $\mathcal{P}_2$; φ_n is $(\exists^{!n}x)Px \,\&(\mathfrak{R}x)Px$. Clearly, the sequence $\{\varphi_n, n \in \mathbb{N}\}$ satisfies our requirements.

$\mathcal{P}_2$ has one predicate, the quantifers $\mathfrak{R}$, $\exists$ and the equality predicate: we use the fact that each formula $(\exists^{!k}x)Px$ can be equivalently replaced by a formula containing only P, $\exists$, = variables and connectives.

3.1.35 Remark. (1) We can obtain an undecidability result for finitely many quantifiers, no equality and infinitely many predicates; for a precise formulation see Problem 4.

(2) Section 5 of the present Chapter, which is devoted to a general theory of OPC's, can be read as an immediate continuation of the present section.

3.1.36 Key words: Classical monadic observational predicate calculi (with or without equality), normal form theorems, definability of quantifiers, pure prenex normal form for MOPC's, decidability.

III.2 Associational and implicational quantifiers

In the present section, we are going to study MOPC's with some particular quantifiers of the type $\langle 1,1\rangle$, called associational quantifiers. The formal definition of associational quantifiers formalizes the following intuitive relation of two properties: Coincidence of the two properties predominates over difference. This can be made precise in many ways, both statistical and non-statistical. Some simple examples will be presented below: statistically motivated associational quantifiers will be obtained in Chapter IV.

Let an observational predicate calculus be given.

3.2.1 Definition. (1) Let $\underline{M}$ be a model, denote by $a_{\underline{M}}$, $b_{\underline{M}}$, $c_{\underline{M}}$, $d_{\underline{M}}$ the cardinality of the set of objects having the card $\langle 1,1\rangle$, $\langle 1,0\rangle$, $\langle 0,1\rangle$, $\langle 0,0\rangle$, respectively. Remember that the card of an object $o \in M$ in $\underline{M} = \langle M, f_1, f_2\rangle$ is $\langle f_1(o), f_2(o)\rangle$. We put

$$q_{\underline{M}} = \langle a_{\underline{M}}, b_{\underline{M}}, c_{\underline{M}}, d_{\underline{M}}\rangle .$$

(2) In the sequel, when saying "quadruple" we mean a quadruple $\langle a, b, c, d\rangle$ of natural numbers whose sum is positive so that there is an $\underline{M}$ such that $a = a_{\underline{M}}$, $b = b_{\underline{M}}$, $c = c_{\underline{M}}$, $d = d_{\underline{M}}$. If $q = \langle a,b,c,d\rangle$ then we put $\mathrm{Sum}(q) = a+b+c+d$. If it does not lead to a misunderstanding, we shall write m for Sum(q).

3.2.2 Definition. (1) A quadruple $q_2 = \langle a_2,b_2,c_2,d_2\rangle$ is a-better than a quadruple $q_1 = \langle a_1,b_1,c_1,d_1\rangle$ if $a_2 \geq a_1$, $b_2 \leq b_1$, $c_2 \leq c_1$, $d_2 \geq d_1$ a comes from "associational".

(2) A model $\underline{M}_2$ is a-better than $\underline{M}_1$ if $q_{\underline{M}_2}$ is a-better than $q_{\underline{M}_1}$

(3) A quantifier $\sim$ of type $\langle 1,1\rangle$ is associational if the following holds for all models $\underline{M}_1, \underline{M}_2$ of type $\langle 1,1\rangle$: If $\mathrm{Asf}_{\sim}(\underline{M}_1) = 1$ and if $\underline{M}_2$ is a-better than $\underline{M}_1$ then $\mathrm{Asf}_{\sim}(\underline{M}_2) = 1$.

3.2.3 Examples of associational quantifiers (cf. 2.2.4):
(a) the quantifier of simple association; the quantifier of implication; the quantifier of founded p-implication. (Further see Section 4 of Chapter IV.)

3.2.4 Lemma. Let $\underline{M}_1, \underline{M}_2$ be models of type $\langle 1,1\rangle$. If $\underline{M}_2$ is not a-better than $\underline{M}_1$ then one can define an associational quantifier $\sim$ such that $\mathrm{Asf}(\underline{M}_1) = 1$ and $\mathrm{Asf}(\underline{M}_2) = 0$.

Proof: For each model $\underline{M}$, put $\mathrm{Asf}_{\sim}(\underline{M}) = 1$ iff $q_{\underline{M}}$ is a-better than $q_{\underline{M}_1}$. By the transitivity of "a-better", this is an associational operator; $\mathrm{Asf}_{\sim}(\underline{M}_1) = 1$ and $\mathrm{Asf}_{\sim}(\underline{M}_2) = 0$.

We wish to introduce an "improvement" relation between 2-cards such that the following holds: If a 2-card $\underline{v}$ a-improves a 2-card $\underline{u}$ then changing the card $\underline{u}$ to $\underline{v}$ in a model $\underline{M}$ changes $\underline{M}$ into an a-better model. We need auxiliary notation also useful for the next sections. So we give the definition for arbitrary V-structures (of type $\langle 1,1\rangle$), but the generalization for type $\langle 1^k\rangle$ is evident.

3.2.5 Definition. Let $\underline{M} = \langle M, f_1, f_2\rangle$ be a V-structure, let $A \subseteq M$ and let $\underline{u} = \langle u_1, u_2\rangle \in V^2$. Then $\underline{M}(A:\underline{u})$ is the model $\langle M, g_1, g_2\rangle$, where $g_i(o) = f_i(o)$ for $o \notin A$ and $g_i(o) = u_i$ for $o \in A$ (cards of elements of A are changed to be $\underline{u}$). In particular, if $o \in M$ then $\underline{M}(o:\underline{u})$ means $\underline{M}(\{o\}:\underline{u})$ and if $\underline{v} \in V^2$ then $\underline{M}(\underline{v}:\underline{u})$ means $\underline{M}(A:\underline{u})$ for $A = \{o \in M;$ the card of o is $\underline{v}\}$.

3.2.6 Remark. If $o_1 \neq o_2$ then $\underline{M}(o_1:\underline{u})(o_2:\underline{v}) = \underline{M}(o_2:\underline{v})(o_1:\underline{u})$. If $A = \{o_1, \ldots, o_n\}$ then $\underline{M}(A:\underline{u}) = M(o_1:\underline{u})(o_2:\underline{u}) \ldots (o_n:\underline{u})$.

3.2.7 Definition. Let $\underline{u}, \underline{v} \in \{0,1\}^2$. $\underline{v}$ a-improves $\underline{u}$ (notation: $\underline{u} \leq_a \underline{v}$) if for each model M of type $\langle 1,1\rangle$ and each $o \in M$, we have the following: If the card of o is $\underline{u}$ then $\underline{M}(o:\underline{v})$ is a-better than $\underline{M}$.

3.2.8 Remark. (1) Evidently, $\leq_a$ is a quasiordering.
(2) An alternative definition reads as follows: $\underline{u} \leq_a \underline{v}$ iff for each model $\underline{M}$ of type $\langle 1,1 \rangle$ and each $o \in M$ we have: If the card of o is $\underline{v}$ then $\underline{M}$ is a-better than $\underline{M}(o:\underline{u})$.

3.2.9 Theorem. The relation $\leq_a$ is an ordering described by the following conditions:

(a) $\langle 1,0 \rangle <_a \langle 1,1 \rangle$, $\langle 1,0 \rangle <_a \langle 0,0 \rangle$,

$\langle 0,1 \rangle <_a \langle 1,1 \rangle$, $\langle 0,1 \rangle <_a \langle 0,0 \rangle$,

(b) $\{\langle 1,1 \rangle, \langle 0,0 \rangle\}$ and $\{\langle 1,0 \rangle, \langle 0,1 \rangle\}$

are incomparable pairs.

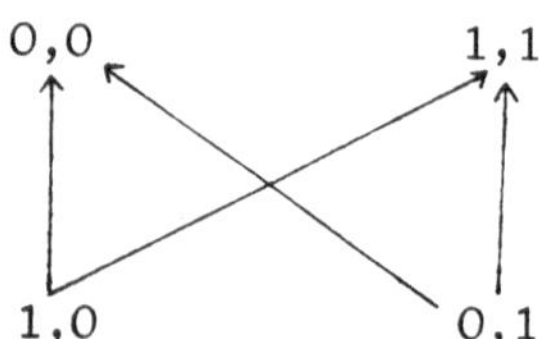

Proof: (a) Evidently, $\langle 1,1 \rangle \geq_a \langle 1,0 \rangle$. To prove that $\langle 1,1 \rangle \leq_a \langle 1,0 \rangle$ does not hold, it suffices to take a model $\underline{M}$ of type $\langle 1^2 \rangle$ with at least one card $\langle 1,0 \rangle = C_{\underline{M}}(o)$ and observe that for $\underline{N} = \underline{M}(o: \langle 1,1 \rangle)$ $q_{\underline{M}}$ is not a-better than $q_{\underline{N}}$.

(b) Let $C_{\underline{M}}(o) = \langle 1,1 \rangle$ in a model $\underline{M}$, put $\underline{N} = \underline{M}(o: \langle 0,0 \rangle)$. Then $q_{\underline{N}} = \langle a_{\underline{M}} - 1, b_{\underline{M}}, c_{\underline{M}}, d_{\underline{M}} + 1 \rangle$ so that neither is $q_{\underline{M}}$ a-better than $q_{\underline{N}}$ nor is $q_{\underline{N}}$ a-better than $q_{\underline{M}}$. This shows that $\langle 1,1 \rangle$, $\langle 0,0 \rangle$ are incomparable.

All the remaining cases are treated similarly.

We shall study some particular associational quantifiers called implicational quantifiers; they have some properties of the quantifier of implication.

3.2.10 Definition. (1) A quadruple $q_2 = \langle a_2, b_2, c_2, d_2 \rangle$ is i-better than a quadruple $q_1 = \langle a_1, b_1, c_1, d_1 \rangle$ if $a_2 \geq a_1$, $b_2 \leq b_1$.

(2) A model $\underline{M}_2$ is i-better than $\underline{M}_1$ if $q_{\underline{M}_2}$ is i-better than $q_{\underline{M}_1}$.
(3) A quantifier $\sim$ of the type $\langle 1,1 \rangle$ is implicational if the following holds for any models $\underline{M}_1, \underline{M}_2$ of type $\langle 1,1 \rangle$: If $Asf_{\sim}(\underline{M}_1) = 1$ and $\underline{M}_2$ is i-better than $\underline{M}_1$ then $Asf_{\sim}(\underline{M}_2) = 1$.

3.2.11 Lemma. (1) If q_2 is a-better than q_1 then q_2 is i-better than q_1.
(2) Consequently, each implicational quantifier is associational.

3.2.12 Remark. The reader easily sees that the quantifiers of implication and of founded p-implication (2.3.4) are implicational whereas the quantifier of simple association is not implicational. Cf. 4.5.1, 4.5.4.

3.2.13 Definition. A 2-card $\underline{v}$ i-improves a 2-card $\underline{u}$ (notation: $\underline{v} \geq_i \underline{u}$) if for each model $\underline{M}$ of type $\langle 1,1 \rangle$ and each $o \in M$ we have the following: If the card of o is $\underline{u}$ then $\underline{M}(o:\underline{v})$ is i-better than $\underline{M}$.

3.2.14 Theorem. (1) $\underline{v} \geq_a \underline{u}$ umplies $\underline{v} \geq_i \underline{u}$. (2) $\geq_i$ is a quasiordering completely described by the following conditions:

$$\langle 1,0 \rangle <_i \langle 0,0 \rangle \equiv_i \langle 0,1 \rangle <_i \langle 1,1 \rangle .$$

$$\begin{array}{l} \langle 1, 1 \rangle \\ \uparrow \\ \langle 0, 0 \rangle, \langle 0, 1 \rangle \\ \uparrow \\ \langle 1, 0 \rangle \end{array}$$

Proof: (1) is evident. By (1) , it remains to show the following:

$\langle 1,1 \rangle \not\leq_i \langle 0,0 \rangle$, $\langle 0,0 \rangle \leq_i \langle 0,1 \rangle$ and $\langle 0,0 \rangle \not\leq_i \langle 1,0 \rangle$.

We show two of the above relations; the remaining one being proved analogously. $\langle 1,1\rangle \not\leq_i \langle 0,0\rangle$: Let $\underline{M}$ be a model with a $o \in M$, $C_{\underline{M}}(o) = \langle 1,1\rangle$: put $\underline{N} = \underline{M}(o : \langle 0,0\rangle)$. Then $q_{\underline{N}} = \langle a_{\underline{M}}-1, b_{\underline{M}}, c_{\underline{M}}, d_{\underline{M}}\rangle$ so that $\underline{M}$ is i-better than $\underline{N}$.

$\langle 0,0\rangle \leq_i \langle 0,1\rangle$: Similarly, if $C_{\underline{M}}(o) = \langle 0,0\rangle$ in a model $\underline{M}$ and if we put
$\underline{N} = \underline{M}(o: \langle 0,1\rangle)$ then $\underline{N}$ is i-better than $\underline{M}$.

3.2.15 Remark and Definition. Our next aim is to offer some transparent deduction rules sound for each implicational quantifier. For the sake of simplicity, call a formula φ designated if its only free variable is the designated variable x. Obviously, if φ, ψ are designated and $\sim$ is a quantifier of type $\langle 1,1\rangle$ then $\varphi\sim\psi$ is a sentence; pedantically, it should be written $(\sim x)(\varphi,\psi)$.

3.2.16 Lemma. Let a MOPC be given. If $\sim$ is an implicational quantifier then the following rules are sound:

(a) $$\left\{ \frac{\varphi\sim\psi}{\varphi\sim(\psi\vee\chi)} ;\ \varphi,\psi,\chi \text{ designated}\right\} ;$$

(b) $$\left\{ \frac{(\varphi \,\&\, \neg\chi)\sim\psi}{\varphi\sim(\psi\vee\chi)} ;\ \varphi,\psi,\chi \text{ designated}\right\} .$$

Proof. (a) Let $\underline{M}$ be a model. Put $\underline{M}_1 = \langle M, \|\varphi\|_{\underline{M}}, \|\psi\|_{\underline{M}}\rangle$ and $\underline{M}_2 = \langle M, \|\varphi\|_{\underline{M}}, \|\psi\vee\chi\|_{\underline{M}}\rangle$. Denote by m_{ijk} the number of objects such that $\|\varphi\|_{\underline{M}}[o] = i$, $\|\psi\|_{\underline{M}}[o] = j$ and $\|\chi\|_{\underline{M}}[o] = k$. Then

$$a_{\underline{M}_1} = m_{110} + m_{111},\ a_{\underline{M}_2} = m_{111} + m_{110} + m_{101},\ b_{\underline{M}_1} = m_{101} + m_{100}$$

and $b_{\underline{M}_2} = m_{100}$. Hence, $\underline{M}_2$ is i-better than $\underline{M}_1$.

(b) Similarly, put $\underline{M}_1 = \langle M, \|\varphi \& \neg\chi\|_{\underline{M}}, \|\psi\|_{\underline{M}} \rangle$ and let $\underline{M}_2$ and m_{ijk} be as above. Then $a_{\underline{M}_1} = m_{110}$, $a_{\underline{M}_2} = m_{110} + m_{101} + m_{111}$

and $b_{\underline{M}_1} = b_{\underline{M}_2} = m_{100}$. Hence, $\underline{M}_2$ is i-better than $\underline{M}_1$.

3.2.17 We present two simple deduction rules sound for certain reasonable associational quantifiers,

(a) The rule of symmetry is

$$\mathrm{SYM} = \left\{ \frac{\varphi \sim \psi}{\psi \sim \varphi} \; ; \varphi, \psi \text{ designated} \right\} \quad .$$

(b) The rule of simultaneous negation is

$$\mathrm{NEG} = \left\{ \frac{\varphi \sim \psi}{\neg\varphi \sim \neg\psi} \; ; \varphi, \psi \text{ designated} \right\} \quad .$$

3.2.18 Remark. Observe that the above two rules are sound for the simple associational quantifier but neither the implication nor the founded p-implication.
Cf. also 4.5.2.

3.2.19 In the remainder of this section we restrict ourselves to monadic OPC's ; an MOPC is supposed to be fixed in the sequel. We shall investigate designated elementary conjunctions and disjunctions, i. e. open formulae of the form

$$\varepsilon_1 P_{i_1}(x) \& \ldots \& \varepsilon_k P_{i_k}(x) \qquad \text{or}$$

$$\varepsilon_1 P_{i_1}(x) \vee \ldots \vee \varepsilon_k P_{i_k}(x)$$

respectively, where each ε_γ is either the negation symbol $\neg$ or the empty symbol (cf. 2.2.12) and x is the designated variable. Such formulae can be viewed as the simplest open formulae: sentences of the form $\varphi\sim\psi$ where φ,ψ are designated elementary conjunctions and/or disjunctions and $\sim$ is an associational quantifier (in particular, an implicational quantifier), will play an important role in the method described in Chapter VII.

Our first aim is to modify the rules of 3.2.16 (sound for each implicational quantifier) as follows: First, we shall be more specific as regards the formulae φ,ψ,χ. We are interested in formulae with $\sim$ having a designated elementary conjunction on the left hand side and a designated elementary disjunction on the right-hand side. Secondly, we shall join both rules into a single rule allowing the transfer of a part of the left-hand side onto the right-hand side (the transferred part being negated) and possibly to extension of the right - hand side by new disjuncts. Some definitions will be useful: EC abbreviates "elementary conjunction" , ED abbreviates "elementary disjunction".

3.2.20 Definition. An elementary association is a designated sentence of the form $\kappa\sim\delta$, where κ is a designated EC or the empty conjunction, δ is a designated ED and κ,δ have no predicates in common. In the sequel, $\kappa,\kappa_1,\kappa_2,\dots$ denote a designated EC or $\underline{1}$, $\delta,\delta_1,\delta_2,\dots$ denote a designated ED. For each EC κ, let $\mathrm{neg}(\kappa)$ be the ED logically equivalent to $\neg\kappa$; put $\mathrm{neg}(\underline{1}) = \underline{0}$. Similarly for $\mathrm{neg}(\delta)$. Two formulae are disjoint if they have no predicates in common.

Let $\kappa_1\sim\delta_1,\kappa_2\sim\delta_2$ be elementary associations. One says that $\kappa_1\sim\delta_1$ results from $\kappa_2\sim\delta_2$ by specification if either $\kappa_1\sim\delta_1$ is identical with $\kappa_2\sim\delta_2$ or there is an ED δ_0 disjoint from δ_1 such that δ_2 is logically equivalent to $\delta_1\vee\delta_0$ and κ_1 is logically equivalent to $\kappa_2\ \&\ \mathrm{neg}(\delta_0)$. (E.g. $P_1\ \&\ P_3\ \&\ \neg P_5\sim P_2\vee P_4$ results from $P_1\ \&\ \neg P_5\sim P_2\vee\neg P_3\vee P_4$ by specification.) We also say that $\kappa_1\sim\delta_1$ despecifies to $\kappa_2\sim\delta_2$. One says that $\kappa_1\sim\delta_1$ results from $\kappa_2\sim\delta_2$ by reduction or $\kappa_1\sim\delta_1$ dereduces to $\kappa_2\sim\delta_2$ if κ_1 is identical with κ_2 and δ_1 is a subdisjunction of δ_2.

(E.g., $P_1\&\neg P_5\sim P_2\vee\neg P_3\vee P_4$ results from $P_1\&\neg P_5\sim P_2\vee\neg P_3\vee P_4\vee\neg P_6\vee P_7$ by reduction.)

We denote by SpRd the inference rule on the set of all elementary associations defined as follows (the despecifying -dereducing rule):

$$\frac{\kappa_1 \sim \delta_1}{\kappa_2 \sim \delta_2} \in \mathrm{Sp\,Rd}$$

iff $\kappa_1 \sim \delta_1$ results from $\kappa_2 \sim \delta_2$ by successive reduction and specification. (In other words, if there is a $\delta_3 \subseteq \delta_2$ such that $\kappa_1 \sim \delta_1$ despecifies to $\kappa_2 \sim \delta_3$ and $\kappa_2 \sim \delta_3$ dereduces to $\kappa_2 \sim \delta_2$. For example $P_1 \& \neg P_5 \sim P_2 \vee \neg P_3 \vee P_4 \vee \neg P_6 \vee P_7$ is inferred from $P_1 \& P_3 \& \neg P_5 \sim P_2 \vee P_4$.

3.2.21 Theorem. If $\sim$ is an implicational quantifier then SpRd is a sound deduction rule, i.e., if $\kappa_2 \sim \delta_2$ is SpRd-inferred from $\kappa_1 \sim \delta_1$ and if $\|\kappa_1 \sim \delta_1\|_{\underline{M}} = 1$ then $\|\kappa_2 \sim \delta_2\|_{\underline{M}} = 1$

Proof. Evident from 3.2.16.

3.2.22 Theorem. The rule SpRd is transitive.

Proof. It suffices to prove that if $\frac{\kappa_1 \sim \delta_1}{\kappa_2 \sim \delta_2} \in \mathrm{Sp\,Rd}$ and if $\frac{\kappa_2 \sim \delta_2}{\kappa_3 \sim \delta_3} \in \mathrm{Sp\,Rd}$ then $\frac{\kappa_1 \sim \delta_1}{\kappa_3 \sim \delta_3} \in \mathrm{Sp\,Rd}$. Since, evidently, the composition of two dereductions (despecifications) is a dereduction (despecification) it suffices to show that successive dereduction and despecification can be replaced by successive despecification and dereduction. More precisely, let $\kappa_1 \& \kappa_2 \sim \delta_1$ dereduce to $\kappa_1 \& \kappa_2 \sim \delta_1 \vee \delta_2$ and let the last formula despecify to $\kappa_1 \sim \delta_1 \vee \delta_2 \vee \mathrm{neg}(\kappa_2)$. Then $\kappa_1 \& \kappa_2 \sim \delta_1$ despecifies to $\kappa_1 \sim \delta_1 \vee \mathrm{neg}(\kappa_2)$, which dereduces to $\kappa_1 \sim \delta_1 \vee \mathrm{neg}(\kappa_2) \vee \delta_2$ which is in turn identical with $\kappa_1 \sim \delta_1 \vee \delta_2 \vee \mathrm{neg}(\kappa_2)$:

$$\begin{array}{ccc} (\kappa_1 \& \kappa_2 \sim \delta_1) & \longrightarrow & (\kappa_1 \& \kappa_2 \sim \delta_1 \vee \delta_2) \\ \downarrow & & \downarrow \\ (\kappa_1 \sim \delta_1 \vee \mathrm{neg}(\kappa_2)) & \longrightarrow & (\kappa_1 \sim \delta_1 \vee \delta_2 \vee \mathrm{neg}(\kappa_2)) \end{array}.$$

The despecifying - dereducing rule is certainly not sound for each associational quantifier. We shall isolate a simple property of quantifiers that causes that rule to be "as invalid as possible".

3.2.23 Definition. Let $\sim$ be an associational quantifier in a MOPC $\mathcal{F}$. The quantifier $\sim$ is called saturable if the following holds:

(1) For each quadruple $\langle a,b,c,d\rangle$ with $d \neq 0$ there is an $a' \geq a$ such that $\mathrm{Asf}_{\sim}(\underline{M}) = 1$ whenever $q_{\underline{M}} = \langle a',b,c,d\rangle$.

(2) For each quadruple $\langle a,b,c,d\rangle$ with $a \neq 0$ there is a $d' \geq d$ such that $\mathrm{Asf}_{\sim}(\underline{M}) = 1$ whenever $q_{\underline{M}} = \langle a,b,c,d'\rangle$.

(3) For each model $\underline{M}$ there is a model $\underline{M}'$ containing $\underline{M}$ and such that $\mathrm{Asf}_{\sim}(\underline{M}') = 0$.

Note that the simple associational quantifier is saturable (and cf. 4.5.3).

3.2.24 Theorem. Let $\sim$ be a saturable associational quantifier. Let $\kappa_1 \sim \delta_1$ and $\kappa_2 \sim \delta_2$ be two elementary associations (κ_1, κ_2 not $\underline{0}$) such that

$$\frac{\kappa_1 \sim \delta_1}{\kappa_2 \sim \delta_2} \in \mathrm{SpRd}\,.$$

If $\kappa_2 \sim \delta_2$ logically follows from $\kappa_1 \sim \delta_1$ (i.e., $\|\kappa_1 \sim \delta_1\|_{\underline{M}} = 1$ implies $\|\kappa_2 \sim \delta_2\|_{\underline{M}} = 1$ for each $\underline{M}$) then $\kappa_1 = \kappa_2$ and $\delta_1 = \delta_2$.

Proof. (a) First, assume that $\kappa_1 \sim \delta_1$ dereduces to $\kappa_1 \sim \delta_1 \vee \delta_2$; let $\underline{M}$ be such that $\|\kappa_1 \sim \delta_1 \vee \delta_2\|_{\underline{M}} = 0$ and let $\underline{u}$ be a card such that $\|\kappa_1\|[\underline{u}] = 0$, $\|\delta_1\|[\underline{u}] = 0$ and $\|\delta_2\|[\underline{u}] = 1$. Extend $\underline{M}$ to a model M' by adding so many objects with the card $\underline{u}$ that $\|\kappa_1 \sim \delta_1\|_{\underline{M}'} = 1$ (this is possible by 3.2.23 (2)). Note that $\|\kappa_1 \sim \delta_1 \vee \delta_2\|_{\underline{M}'} = 0$ since extending

$$\langle M, \|\kappa_1\|_{\underline{M}}, \|\delta_1 \vee \delta_2\|_{\underline{M}}\rangle \text{ to } \langle M', \|\kappa_1\|_{\underline{M}'}, \|\delta_1 \vee \delta_2\|_{\underline{M}'}\rangle \text{ we only add}$$

many times the card $\langle 0,1\rangle$, hence, we make the model a-worse.

(b) Assume now that we have $\kappa_1 \& \kappa_2 \sim \delta_1$ and $\kappa_1 \sim \delta_1 \vee \mathrm{neg}(\kappa_2) \vee \delta_2$ (where δ_2 can be $\underline{1}$). Take a model $\underline{M}$ with $\|\kappa_1 \sim \delta_1 \vee \mathrm{neg}(\kappa_2) \vee \delta_2\|_{\underline{M}} = 0$ and let $\underline{u}$ be a card such that $\|\kappa_1\|[\underline{u}] = \|\delta_1\|[\underline{u}] = \|\kappa_2\|[\underline{u}] = 0$. Extend $\underline{M}$ to a model $\underline{M}'$ by adding so many objects with the card $\underline{u}$ that $\|\kappa_1 \& \kappa_2 \sim \delta_1\|_{\underline{M}} = 1$ (this is possible by 3.2.23 (2)).

Note that $\|\kappa_1 \sim \delta_1 \vee \text{neg}(\kappa_2) \vee \delta_2\|_{M'} = 0$ since $\|\kappa_1\|[u] = 0$ but $\|\delta_1 \vee \text{neg}(\kappa_2) \vee \delta_2\|[u] = 1$. Thus $\kappa_1 \& \kappa_2 \sim \delta \not\models \kappa_1 \sim \delta \vee \text{neg}(\kappa_2) \vee \delta_2$.

3.2.25 Corollary: If $I \subseteq$ SpRd and I is sound for each associational quantifier then I is the identity, i.e.,

$$\frac{\kappa_1 \sim \delta_1}{\kappa_2 \sim \delta_2} \in I \qquad \text{implies that } \kappa_2 \sim \delta_2 \text{ is the same as } \kappa_1 \sim \delta_1 .$$

3.2.21 Remark and Definition. Theorem 3.3.21 can be reformulated in the following way:

Let L be any language containing a quantifier $\sim$ of type $\langle 1,1 \rangle$.

Whenever $\dfrac{\kappa_1 \sim \delta_1}{\kappa_2 \sim \delta_2} \in$ SpRd then the sentence

$$(*) \qquad (\kappa_1 \sim \delta_1) \longrightarrow (\kappa_2 \sim \delta_2)$$

is a scheme of implicational tautologies, i.e., (*) is a tautology of each MOPC with the language L in which $\sim$ is an implicational quantifier.

We ask what the situation is for the (broader) class of associational quantifiers. First, we define: Let L be a language whose only quantifier is a quantifier $\sim$ of type $\langle 1,1 \rangle$ and let Φ be a sentence of L. Φ is a scheme of associational tautologies in L if Φ is a tautology in each MOPC with the language L in which $\sim$ is an associational quantifier.

Corollary 3.2.25 can be interpreted as saying that there is no non-trivial schema of associational tautologies of the form (*) . On the other hand, there are various schemes of associational tautologies, e.g.

$$(\varphi \sim \psi) \longrightarrow (\varphi \sim (\varphi \,\&\, \psi)) ,$$

as the reader can easily verify. We shall prove that, for each language L, the set of all schemes of associational tautologies is recursive.

We begin with some preliminary considerations.

3.2.27 Definitions. (1) Let Q be the set of all quadruples in the sense of 3.2.1 (2): write $q_2 \succcurlyeq q_1$ if q_2 is a-better than q_1. A set $S \subseteq Q$ is a cut on Q if $q_2 \succcurlyeq q_1$, $q_1 \in S$ implies $q_2 \in S$.

(2) Let n be a fixed natural number (think of the number of predicates in a language). Let K be the set of all n-cards. By "a partition" we mean a sequence $\langle A,B,C,D\rangle$ of four disjoint subsets of K whose union is K (one could say : "a 4-partition on K"). $\mathcal{R}$ denotes the set of all partitions. Recall the notion of a genus (a certain mapping of K into natural numbers, see 3.1.18). If g is a genus then g determines a natural-valued measure on $\mathcal{P}(K)$ (the field of all subsets of K) defined, for each $A \subseteq K$, by the equation $\mu_g(A) = \sum_{u\in A} g(u)$: μ_g is called the measure induced by g. If $R = \langle A,B,C,D\rangle \in \mathcal{R}$ then we put

$$\mu_g(R) = \langle \mu_g(A), \mu_g(B), \mu_g(C), \mu_g(D)\rangle .$$

(3) Let $T \subseteq \mathcal{R}$ (T is a set of partitions). T is satisfiable if there is a genus g and a cut S on $\mathcal{Q}$ such that, for each $R\in\mathcal{R}$, $R \in T$ iff $\mu_g(R) \in S$.

3.2.28 Remark. The definitions above have the following meaning: Let L be a language with n predicates and a quantifier $\sim$ (of type $\langle 1,1\rangle$).

(1) Cuts on $\mathcal{Q}$ correspond uniquely to MOPC's with the language L in which $\sim$ is associational; if $\mathcal{F}$ is such an MOPC then the set $S = \{q ;$ for any $\underline{M}$ with $q_{\underline{M}} = q$, $Asf_{\sim}(\underline{M}) = 1\}$ is a cut; conversely, if S is a cut and we put $Asf_{\sim}(\underline{M}) = 1$ iff $q_{\underline{M}} \in S$ then we obtain an associational quantifier. Note that $q_{\underline{M}_1} = q_{\underline{M}_2}$ implies $Asf_{\sim}(\underline{M}_1) = Asf_{\sim}(\underline{M}_2)$.

Consequently, if S corresponds to $Asf_{\sim}$ then $Asf_{\sim}$ is recursive iff S is recursive.

(2) With any open designated formulas φ, ψ we associate a partition $r(\varphi,\psi) = \langle A,B,C,D\rangle$ such that

A consists of all n-cards satisfying $\varphi \& \psi$,

B consists of all n-cards satisfying $\varphi \& \neg\psi$,

C consists of all n-cards satisfying $\neg\varphi \& \psi$,

D consists of all n-cards satisfying $\neg\varphi \& \neg\psi$.

Evidently, each partition can be obtained in this way: we have ($\varphi \Leftrightarrow \varphi'$ and $\psi \Leftrightarrow \psi'$) iff $r(\varphi,\psi) = r(\varphi',\psi')$ ($\Leftrightarrow$ stands for logical equivalence).

Genera correspond to models of type n; $g_{\underline{M}} = g_{\underline{N}}$ iff $\underline{M}$ and $\underline{N}$ are isomorphic. If $g_{\underline{M}} = g$ and $A \subseteq K$ then $\mu_g(A)$ is the cardinality of the set of all $o \in M$ such that the $\underline{M}$-card of o is in A.

(3) Note that the set $\mathcal{R}$ of all partitions is finite since K is finite. Observe that T is satisfiable iff there is a genus g and a recursive cut S on $\mathcal{Q}$ such that, for each $R \in \mathcal{R}$, $R \in T$ iff $\mu_g(R) \in S$. Indeed, let S_0 be an arbitrary cut satisfying the last condition and put $g \in S$ iff there is a $R \in T$ with $q \gtrless \mu_g(R)$. Evidently, S is a cut, S is recursive and $R \in T$ implies $\mu_g(R) \in S$; furthermore, $S \subseteq S_o$. Hence if $R \notin T$ then $\mu_g(R) \notin S$ and, a fortiori, $\mu_g(R) \notin S_o$.

Consequently, $T \subseteq \mathcal{R}$ is satisfiable iff there is a MOPC $\mathcal{F}$ with n predicates and one associational quantifier $\sim$ of type $\langle 1,1 \rangle$ and a model $\underline{M}$ of type $\langle 1^n \rangle$ such that, for each pair φ, ψ of designated open formulae, $\| \varphi \sim \psi \|_{\underline{M}} = 1$ iff $r(\varphi,\psi) \in T$.

3.2.29 Theorem. Let L be a language with n predicates and one quantifier of type $\langle 1,1 \rangle$. The set of all sentences of L that are schemes of associational tautologies is recursive.

Proof. Let $\mathcal{T}_{sat}$ be the (finite) set of all satisfiable sets of partitions (of K). Let Φ be a sentence. Bring $\neg\Phi$ into prenex normal form (a disjunction of elementary conjunctions of pure prenex formulae) using 3.1.30. Note that the procedure is uniform and yields a sentence Ψ such that $\neg\Phi$ is equivalent to Ψ in each MOPC with the language L.

. Call Ψ satisfiable iff there is a MOPC $\mathcal{F}$ with the language L in which $\sim$ is an associational quantifier and such that there is an $\underline{M}$ such that $\|\Psi\|_{\underline{M}} = 1$ (in $\mathcal{F}$). Evidently, Φ is a scheme of associational tautologies iff Ψ is not satisfiable.

Hence, we ask whether Ψ is satisfiable. Ψ is a disjunction, hence Ψ is satisfiable iff one disjunct of Ψ is aatisfiable. Hence, suppose that Ψ_1

is a elementary conjunction of pure prenex formulas. By 3.2.28 , Ψ_1 is satisfiable iff there is a $T \subseteq \mathcal{R}$ which is satisfiable (in the sense of 3.2.27 (3)) and coherent with Ψ_1, i.e., for each pair φ,ψ of open formulae, if $\varphi\sim\psi$ is a conjunct of Ψ_1 then $r(\varphi,\psi) \in T$ and if $\neg(\varphi\sim\psi)$ is a conjunct of Ψ_1 then $r(\varphi,\psi) \notin T$. To verify whether Ψ_1 is satisfiable, go through the finite set $\mathcal{T}_{sat}$ and ask whether it contains a set of partitions coherent with Ψ_1.

3.2.30 Remark. Compare this theorem with 3.1.34 ; our result is unsatisfactory since the recursiveness argument is based on the finiteness of the set $\mathcal{T}_{sat}$. But we shall show in Problem 5 that the assumption of finitely many predicates is inessential in the present case.

On the other hand, the question of the complexity of the decision problem for schemes of associational tautologies is open.

3.2.31 Key words: quadruples, a-better, i-better, associational and implicational quantifiers, elementary associations, the despecifying-dereducing rule, saturable quantifiers, schemes of associational tautologies.

III.3 Calculi with incomplete information

In the present section, we are going to investigate some observational function calculi that are not predicate calculi since they have more than two-element sets of abstract values. There are at least two reasons for generalizing truth values to abstract values: First, since we imagine observational structures to be results of observations, we must recognize that one can observe not only properties of objects but more general attributes as well, not two-valued but - in most cases - natural number-valued or rational-valued. Second, we shall consider the possibility that our information on observed objects may be incomplete, i.e., there can be an object and an attribute such that the value of the attribute for the object is unknown or the information is missing. This may have various causes, e.g., technically, the object was destroyed . It is reasonable to introduce a special value for missing information; then we necessarily have more than two values. There is a natural notion of calculi with incomplete information: it will be studied in the present section. We describe the way in which each function calculus, and in particular each predicate calculus, may be extended to a calculus with incomplete information.

3.3.1 Definitions and Discussion. Let $\mathcal{F}_1$ be a function calculus. A function calculus $\mathcal{F}_2$ extends $\mathcal{F}_1$ (is an extension of $\mathcal{F}_1$) if $V_1 \subseteq V_2$, $Fm_1 \subseteq Fm_2$, $\mathfrak{M}_1 \subseteq \mathfrak{M}_2$ (V_i is the set of abstract values of $\mathcal{F}_i$, etc), and for each $\varphi \in Fm_1$ and $\underline{M} \in \mathfrak{M}_1$ we have $Val_1(\varphi, \underline{M}) = Val_2(\varphi, \underline{M})$. The common value can be denoted by $\|\varphi\|_{\underline{M}}$ without any misunderstanding.

We are interested now in particular extensions by adding exactly one new abstract value for missing information. Thus, if $\mathcal{F}$ is an arbitrary function calculus and if V is its set of abstract values, call the elements of V regular values; take a value $\times$ not in V and call it the singular value. Let $V^\times = V \cup \{\times\}$, and consider $V^\times$-structures. A $V^\times$-structure $\underline{M}$ is regular (or a structure with complete information) if $\underline{M}$ is a V-structure. A V-structure $\underline{N} = \langle N, g_1, \ldots, g_n \rangle$ is a (regular) completion of a $V^\times$-structure $\underline{M} = \langle M, f_1, \ldots, f_n \rangle$ if $\underline{M}$ and $\underline{N}$ have the same field $M = N$ and the same type and if, for each i and each $o_1, \ldots \in M$, $f_i(o_1, \ldots) \neq \times$

implies $g_i(o_1,\dots) = f_i(o_1,\dots)$ (i.e., all crosses in $\underline{M}$ are converted into some regular values; nothing else is changed). Similarly, a V-card $\langle u_1, \dots, u_n\rangle$ is a regular <u>completion</u> of a $V^{\times}$-card $\langle v_1,\dots,v_n\rangle$ if, for each i, $v_i \neq \times$ implies $v_i = u_i$.

3.3.2 <u>Remark</u>. A <u>partial</u> V-<u>structure</u> is a tuple $\langle M, h_1, \dots\rangle$ where each h_i is a mapping whose domain is a subset of M and whose range is included in V. There is a natural one-to-one correspondence between $V^{\times}$-structures and partial V-structures: $\langle M, f_1,\dots,f_n\rangle$ corresponds to $\langle M, h_1,\dots,h_n\rangle$ iff, for each i, f_i extends h_i and(f_i takes the value $\times$ iff h_i is undefined).

3.3.3 <u>Definition.</u> Let $\mathcal{F}$, $\times$ and V be as above. The <u>secured</u> $\times$-extension of $\mathcal{F}$ is the calculus $\mathcal{F}^{\times}$ defined as follows:

(a) The set of abstract values is $V^{\times} = V \cup \{\times\}$

(b) The set $\mathcal{M}^{\times}$ of all models in the sense of $\mathcal{F}^{\times}$ consists of all $V^{\times}$-structures $\underline{M}$ such that a completion of $\underline{M}$ is in $\mathcal{M}$ (where $\mathcal{M}$ is the set of all models in the sense of $\mathcal{F}$).

(c) $\mathcal{F}$ and $\mathcal{F}^{\times}$ have the same formulae.

(d) Associated functions of junctors and quantifiers in $\mathcal{F}^{\times}$ are <u>secured</u> extensions of the corresponding associated functions in $\mathcal{F}$, i.e., if ι is an n-ary junctor and if $\underline{u} \in (V^{\times})^n$ then

$$\iota(\underline{u}) = \begin{cases} i \in V & \text{iff for each completion } \underline{v} \text{ of } \underline{u},\ \mathrm{Asf}_{\iota}(\underline{v}) = i, \\ \times & \text{otherwise;} \end{cases}$$

if q is a quantifier of the type t and if $\mathrm{Asf}_q(\underline{M})$ is defined then

$$\mathrm{Asf}_q(\underline{M}) = \begin{cases} i \in V & \text{iff for each completion } \underline{N} \text{ of } \underline{M},\ \mathrm{Asf}_q(\underline{N}) = i, \\ \times & \text{otherwise.} \end{cases}$$

3.3.4 <u>Lemma</u>. Let $\mathcal{F}^{\times}$ be the secured extension of $\mathcal{F}$, let φ be a formula, let $\underline{M}$ be a model, and let e be an $\underline{M}$-sequence for φ. Then $\|\varphi\|_{\underline{M}}[e] = i \in V$ implies that for each completion $\underline{N}$ of $\underline{M}$ we have $\|\varphi\|_{\underline{N}}[e] = i$.

Proof. The assertion is obvious for atomic formulas and for nullary junctors. We proceed by induction on the complexity of formulae. Let, e.g., ι be a binary junctor, take a formula $\iota(\varphi_1, \varphi_2)$, and let the assertion hold for φ_1 and φ_2. Let $i \in V$ and let $\|\iota(\varphi_1, \varphi_2)\|_{\underline{M}}[e] = i = \mathrm{Asf}_\iota(\|\varphi_1\|_{\underline{M}}[e], \|\varphi_2\|_{\underline{M}}[e])$; put

$u_i = \|\varphi_i\|_{\underline{M}}[e]$. We know that $\mathrm{Asf}_\iota(v_1, v_2) = i$ for each completion $\langle v_1, v_2\rangle$ of $\langle u_1, u_2\rangle$. In particular, if $\underline{N}$ is a completion of $\underline{M}$ and if $v_i = \|\varphi_i\|_{\underline{N}}[e]$, then $\langle v_1, v_2\rangle$ is a completion of $\langle u_1, u_2\rangle$ by the induction hypothesis: hence $\|\iota(\varphi, \varphi)\|_{\underline{N}}[e] = i$; **similarly for a quantifier.**

3.3.5 Remark. The above implication cannot be reversed: It is possible that $\|\varphi\|_{\underline{N}}[e] = i \in V$ for each completion $\underline{N}$ of $\underline{M}$ and **that** $\|\varphi\|_{\underline{M}}[e]$ **still** $= \times$, **for** the following reason: If $\langle u_1, u_2\rangle$ is as above, then the set

$$\{\langle \|\varphi_1\|_{\underline{N}}[e], \|\varphi_2\|_{\underline{N}}[e]\rangle;\ \underline{N} \text{ a completion of } \underline{M}\}$$

can be a proper subset of the set of all completions of $\langle u_1, u_2\rangle$. For example, if φ_i are equal formulae, say, $\varphi_1 = \varphi_2 = \varphi$, then each pair $\langle \|\varphi_1\|_{\underline{N}}[e], \|\varphi_2\|_{\underline{N}}[e]\rangle$ consists of two equal elements: but if $\|\varphi\|_{\underline{M}}[e] = \times$ then evidently $\langle u_1 . u_2\rangle$ has completions $\langle v_1, v_2\rangle$ with $v_1 \neq v_2$. (Cf. below.)

3.3.6 Definition. A formula φ is secured if the following holds for each $\underline{M}$ and each $\underline{M}$-sequence e for φ:

$$\|\varphi\|_{\underline{M}}[e] = \begin{cases} i \in V & \text{iff } \|\varphi\|_{\underline{N}}[e] = i \text{ for each completion } \underline{N} \text{ of } \underline{M}, \\ \times & \text{otherwise.} \end{cases}$$

3.3.7 Lemma. If $\varphi_1, \ldots, \varphi_k$ are secured and for $i \neq j$ the formulas φ_i, φ_j have no function symbol in common, then $\iota(\varphi_1, \ldots, \varphi_k)$ is secured and $(qx)(\varphi_1, \ldots, \varphi_k)$ is secured (ι is a k-ary junctor; q is a quantifier of type $\langle 1,\dots,1\rangle$.

Proof: Exercise. Show that in the present case the two sets in 3.3.5 coincide.

3.3.8 Remark and Definition. In most cases we shall be interested in calculi of the following kind: One starts with a calculus $\mathcal{F}$ and forms the secured $\times$-extension $\mathcal{F}^{\times}$. Then one extends $\mathcal{F}^{\times}$ to a calculus $\hat{\mathcal{F}}$ having the same values and models as $\mathcal{F}^{\times}$ but having more formulae (e.g., more quantifiers). Any such calculus $\hat{\mathcal{F}}$ is called a $\times$-extension of $\mathcal{F}$.

Definition 3.3.6 also makes sense for $\hat{\mathcal{F}}$, but observe that Lemma 3.3.7 need not hold if $\hat{\mathcal{F}}$ is a proper extension of $\mathcal{F}^{\times}$.

3.3.9 Definition (1) Let $\hat{\mathcal{F}}$ be a $\times$-extension of $\mathcal{F}$ and let q be a quantifier of $\hat{\mathcal{F}}$. q is regular-valued if, for each $V^{\times}$-structure $\underline{M}$ such that $Asf_q(\underline{M})$ is defined, we have $Asf_q(\underline{M}) \in V$.

(2) An important example of a regular-valued quantifier is the quantifier of strong equivalence $\Leftrightarrow$ of type $\langle 1,1\rangle$ defined as follows: Assume $0,1 \in V$. Then $Asf_{\Leftrightarrow}(\langle M,f,g\rangle) = 1$ if $f = g$ and $= 0$ otherwise. (Thus, if, e.g., φ, ψ are designated open then $\|\varphi \Leftrightarrow \psi\|_{\underline{M}} = 1$ iff $\|\varphi\|_{\underline{M}} = \|\psi\|_{\underline{M}}$ and $= 0$ otherwise.)

(3) One defines regular-valued formulas in the obvious way.

3.3.10 Discussion. We think of a $V^{\times}$-structure $\underline{M}$ as incomplete information on a particular completion $\underline{N}_o$ of $\underline{M}$: $\underline{N}_o$ is the true complete information on our objects but $\underline{N}_o$ is not at our disposal; $\underline{N}_o$ is the "heavenly" model and $\underline{M}$ is the "earthly" model . If φ is a secured sentence, then $\|\varphi\|_{\underline{M}} = i \in V$ means that we know that $\|\varphi\|_{\underline{N}_o} = i$, $\|\varphi\|_{\underline{M}} = \times$ means that we do not know the value of φ in $\underline{N}_o$. On the other hand, $\varphi_1 \Leftrightarrow \varphi_2$ is an example of a non-secured regular-valued sentence and $\|\varphi_1 \Leftrightarrow \varphi_2\|_{\underline{M}} = 1$ means that we know the same about φ_1 as about φ_2; we cannot conclude anything about the $\underline{N}_o$-value of $\varphi_1 \Leftrightarrow \varphi_2$.

Note in passing that Körner [1966] obtains - mutatis mutandis - ×-extensions of classical predicate calculi from another notion, namely that of "inexact properties". The philosophical distinction between the two notions lies outside the scope of the present book.

3.3.11 Discussion and definitions. Now, we shall consider ×-extensions of predicate calculi (called ×-predicate calculi); hence, $V = \{0,1\}$ and $V^{\times} = \{0,1,\times\}$ here. It is reasonable to introduce a natural ordering of $V^{\times}$ putting $0<\times<1$. Associated functions of $\neg$, &, $\vee$ extend by the securing principle as follows:

	$\neg$
1	0
×	×
0	1

&	1	×	0
1	1	×	0
×	×	×	0
0	0	0	0

	1	×	0
1	1	1	1
×	1	×	×
0	1	×	0

The nullary junctors $\underline{0}$, $\underline{1}$ behave like sentences with constant values: $\|\underline{0}\|_{\underline{M}} = 0$ and $\|\underline{1}\|_{\underline{M}} = 1$ for each $V^{\times}$-model $\underline{M}$. We shall make a brief inspection of Chapter II, Sect. 2. Note that $\Leftrightarrow$ means logical equivalence, i.e., $\varphi \Leftrightarrow \psi$ means that for each $\underline{M}$ and each $e : (FV(\varphi) \cup FV(\psi)) \to M$, $\|\varphi\|_{\underline{M}}[e] = \|\psi\|_{\underline{M}}[e]$ (or, more pedantically, $\|\varphi\|_{\underline{M}}[e \restriction FV(\varphi)] = \|\psi\|_{\underline{M}}[e \restriction FV(\psi)]$).

3.3.12 Lemma. Let $\hat{\mathcal{F}}$ be an ×-predicate calculus and let φ, ψ, χ be formulae. Then the logical equivalences (1) - (14) from 2.2.10 (i.e., commutativity, idempotence, associativity, behaviour of $\underline{0}$ and $\underline{1}$ as members of disjunctions and conjunctions, double negation, distributivity, de Morgan laws) hold true for $\hat{\mathcal{F}}$.

Proof. Proofs of (1) - (10) are immediate: we verify (11), i.e., $\varphi \,\&(\psi \vee \chi) \Leftrightarrow (\varphi \,\&\, \psi) \vee (\varphi \,\&\, \chi)$. It suffices to verify that the left-hand side has the value 1 iff the right-hand side has; and the same holds true for 0. Put $\|\varphi\|_{\underline{M}}[e] = u$, $\|\psi\|_{\underline{M}}[e] = v$, $\|\chi\|_{\underline{M}}[e] = w$.

Now, $\|\varphi \,\&(\psi \vee \chi)\|_{\underline{M}}[e]$ is 1 iff $u = 1$ and $(v \vee w) = 1$, i.e., iff $u = 1$ and ($v = 1$ or $w = 1$). On the other hand, $\|(\varphi \,\&\, \psi) \vee (\varphi \,\&\, \chi)\|_{\underline{M}}[e] = 1$ iff $(u \& v) = 1$ or (u&w) = 1, hence iff u = v = 1 or u = w = 1, which is equivalent to u = 1 and (v = 1 or w = 1); similarly for 0. The cases (12)- (14) are treated similarly.

3.3.13 Remarks. (1) The equivalences (15), (16) of 2.2.11, namely $(\varphi \& \neg\varphi) \Leftrightarrow \underline{0}$, $(\varphi \vee \neg\varphi) \Leftrightarrow \underline{1}$, are not true for $\times$-predicate calculi (cf. 3.3.5). Indeed, if $\|\varphi\|_{\underline{M}}[e] = \times$, then $\|\varphi \& \neg\varphi\|_{\underline{M}}[e] = \|\varphi \vee \neg\varphi\|_{\underline{M}}[e] = \times$, but $\|\underline{0}\|_{\underline{M}} = 0$ and $\|\underline{1}\|_{\underline{M}} = 1$.

(2) Generalized conjunctions and disjunctions $\bigwedge B$, $\bigvee B$ are introduced as in 2.2.11 by 3.3.12, the equivalences (17) - (22) of 2.2.11 (generalized distributive and de Morgan laws) hold true for $\times$-predicate calculi.

(3) Open formulae, in particular: literals, elementary conjunctions and disjunctions, formulae in conjunctive (disjunctive) normal form, are defined exactly as in 2.2.12. The "normal form" lemma from 2.2.12 does not hold for $\times$-predicate calculi since we do not have the logical equivalences (15), (16); we shall obtain a reasonable normal form lemma in the next section.

3.3.14 Theorem. Let $\mathcal{P}$ be a predicate calculus whose only quantifiers are the classical quantifiers $\forall, \exists$ and let $\mathcal{P}^{\times}$ be the secured $\times$-extension of $\mathcal{P}$. There is a recursive function r associating with each formula φ a formula $r(\varphi)$ with the following properties:

(i) φ and $r(\varphi)$ have the same predicates, free and bound variables,

(ii) φ and $r(\varphi)$ are logically equivalent,

(iii) $r(\varphi)$ is either $\underline{0}$ or $\underline{1}$ or does not contain any nullary connective.

Proof. We construct $r(\varphi)$ by induction on the complexity of φ. For atomic formulae and for $\underline{0}, \underline{1}$ put $r(\varphi) = \varphi$. The induction step: If $r(\varphi)$ is $\underline{1}$, $\underline{0}$, without nullary junctors, then $r(\neg\varphi)$ is $\underline{0}$, $\underline{1}$, $\neg r(\varphi)$, respectively, $r((\forall x)\varphi)$ is $\underline{1}$, $\underline{0}$, $(\forall x)\, r(\varphi)$, respectively, and similarly for $\exists$. For $\varphi \& \psi$ we have the following possibilities:

r(φ) \ r(ψ)	1	0	w.n.j.
1	1	0	r(ψ)
0	0	0	0
w.n.j.	r(φ)	0	r(φ)&r(ψ)

similarly for $\varphi \vee \psi$. It is evident that the procedure is effective.

3.3.15 Corollary. No formula without **nullary** junctors is a {1}-tautology.

Proof. In each model $\underline{M}$ consisting only of crosses (i.e., $\langle M, f_1, \ldots \rangle$ where each f_i constantly takes the value $\times$) the value of each formula without nullary junctors is $\times$ (for each $\underline{M}$-sequence for φ - proof by induction).

3.3.16 Discussion. Secured $\times$-extensions of classical predicate calculi with both finite and infinite models were investigated by Cleave [1975]. He defines that φ logically implies ψ iff, for each $\underline{M}$ and e, $\|\varphi\|_{\underline{M}}[e] \leq \|\psi\|_{\underline{M}}[e]$ (for the natural ordering of $\{1, \times, 0\}$). In our terminology, this means that $\varphi \models_{\{1\}} \psi$ and $\varphi \models_{\{1,\times\}} \psi$ simultaneously hold. (Equivalently, $\varphi \models_{\{1\}} \psi$ and $\neg\psi \models_{\{1\}} \neg\varphi$.) Cleave shows that the relation $LI = \{\langle\varphi, \psi\rangle ; \varphi$ logically implies $\psi\}$ is recursively enumerable by axiomatizing this relation (he has a rule I which is both {1}-sound and {1,×}-sound and shows that provability from assumptions coincides with logical implication. It is easy to see that LI is not recursive.)

3.3.17 Remark. Naturally, we are interested in observational $\times$-predicate calculi. Observe that the secured $\times$-extension of any observational predicate calculus is observational (since each three-valued model has only finitely many regular completions; if $R(\underline{M}, \ldots)$ is a recursive relation, then the relation $(\forall \underline{N}$ completion of $\underline{M}) R(\underline{N}, \ldots)$ is recursive).

Trachtenbrot's theorem 2.2.16 generalizes for observational $\times$-predicate calculi with classical quantifiers as follows:

3.3.18 Theorem. There is an observational predicate calculus whose only quantifiers are $\forall, \exists$ such that Cleave's logical implication LI defined in the secured $\times$-extension $\mathcal{P}^{\times}$ of $\mathcal{P}$ is not recursively enumerable.

Proof. Let $\mathcal{P}$ be a calculus satisfying Trachtenbrot's theorem, let $P_1, \ldots, P_n$ be its predicates, P_i of arity k_i. Let Φ be the sentence

$$\bigwedge_{i=1}^{n} (\forall \underline{x}_i)(P_i(\underline{x}_i) \vee \neg P_i(\underline{x}_i)) ,$$

where $\underline{x}_i$ is the sequence of the first k_i variables. Then, for each sentence φ, φ is a tautology of $\mathcal{P}$ iff Φ logically implies φ in $\mathcal{P}^{\times}$.

3.3.19 Remark. Remember the definition 3.2.2 of associational quantifiers in observational predicate calculi. The definition extends to observational $\times$-predicate calculi by the principle of **secureness**. Thus a quantifier $\sim$ of type $\langle 1,1 \rangle$ is associational if the following holds:

(1) If $\underline{M}_1, \underline{M}_2$ are two-valued models of type $\langle 1,1 \rangle$, then

(i) $Asf_{\sim}(\underline{M}_i) \in \{0,1\}$ ($i = 1,2$) and

(ii) $Asf_{\sim}(\underline{M}_1) = 1$ and $(a_{\underline{M}_1} \leq a_{\underline{M}_2}, b_{\underline{M}_1} \geq b_{\underline{M}_2}, c_{\underline{M}_1} \geq c_{\underline{M}_2}, d_{\underline{M}_1} \leq d_{\underline{M}_2})$

implies $Asf_{\sim}(\underline{M}_2) = 1$.

(2) If $\underline{M}$ is an $\{0,\times,1\}$-model of type $\langle 1,1 \rangle$, then

$Asf_{\sim}(\underline{M}) = 1$ if $Asf_{\sim}(\underline{N}) = 1$ for each completion $\underline{N}$ of $\underline{M}$,

$Asf_{\sim}(\underline{M}) = 0$ if $Asf_{\sim}(\underline{N}) = 0$ for each completion $\underline{N}$ of $\underline{M}$,

$Asf_{\sim}(\underline{M}) = \times$ otherwise.

3.3.20 Definition. We extend Definition 3.2.2 (2) (a model $\underline{M}_2$ is a-better than $\underline{M}_1$) to three-valued models as follows: Let $\underline{M}_1, \underline{M}_2$ be three-valued (i.e., $\{0, \times, 1\}$-valued models). $\underline{M}_2$ is said to be a-better than $\underline{M}_1$ if for each completion $\underline{N}_2$ of $\underline{M}_2$ there is a completion $\underline{N}_1$ of $\underline{M}_1$ such that $\underline{N}_2$ is a-better than $\underline{N}_1$.

3.3.21 Lemma. (1) The "a-better" relation is a quasiordering of the set of all three-valued models. (2) If $\sim$ is an associational quantifier, then, for arbitrary three-valued models $\underline{M}_1, \underline{M}_2$ such that $\underline{M}_2$ is a-better than $\underline{M}_1$, $Asf_\sim(\underline{M}_1) = 1$ implies $Asf_\sim(\underline{M}_2) = 1$. (3) If $\underline{M}_2$ is not a-better than $\underline{M}_1$, then one can define an associational quantifier $\sim$ such that $Asf_\sim(\underline{M}_1) = 1$ but $Asf_\sim(\underline{M}_2) = 1$.

Proof. (1) is elementary. (2) Let $\sim$ be associational and let $\underline{M}_2$ be a-better than $\underline{M}_1$. If $Asf_\sim(\underline{M}_1) = 1$, then $Asf_\sim(\underline{M}_2)$ also must be 1 since, otherwise, there would exist a completion $\underline{N}_2$ of $\underline{M}_2$ with $Asf_\sim(\underline{N}_2) \neq 1$; there is a completion $\underline{N}_1$ of $\underline{M}_1$ such that $\underline{N}_2$ is a-better than $\underline{N}_1$ but $Asf_\sim(\underline{N}_1)$ is 1 - a contradiction. (3) Let $\underline{N}_2$ be a completion of $\underline{M}_2$ such that, for no completion $\underline{N}_1$ of $\underline{M}_1$, $\underline{N}_2$ is a-better than $\underline{N}_1$. Put, for each two-valued model $\underline{N}$, $Asf_\sim(\underline{N}) = 1$ iff $\underline{N}$ is a-better than a completion $\underline{N}_1$ of $\underline{M}_1$. Extend $\sim$ to all three-valued models by the principle of **secureness: then** $Asf_\sim(\underline{M}_1) = 1$ but $Asf_\sim(\underline{N}_2) = 0$, hence $Asf_\sim(\underline{M}_2) \neq 1$. (Note that $Asf_\sim$ is a recursive function.)

3.3.22 Remark. (1) In analogy to 3.3.19, we extend the definition of an implicational quantifier to x-predicate calculi. Thus, a quantifier $\sim$ of type $\langle 1,1 \rangle$ is implicational if it satisfies (1)(i), (1)(ii$'$) and (2), where (1)(i) and (2) are as in 3.3.19 and (ii$'$) is as follows:

(ii$'$) ($Asf_\sim(\underline{M}_1) = 1$ and $a_{\underline{M}_1} \leq a_{\underline{M}_2}$, $b_{\underline{M}_1} \geq b_{\underline{M}_2}$ implies $Asf_\sim(\underline{M}_2) = 1$.

(2) One extends the "i-better" relation 3.2.10 to three-valued models in analogy to 3.3.20, then one easily proves the obvious analogue of 3.3.21.

(3) We introduced the notation $\underline{M}(o:\underline{u})$ in 3.2.5. Our next aim is to analyse the relations "$\underline{v}$ a-improves $\underline{u}$" and "$\underline{v}$ i-improves $\underline{u}$" for $\underline{u}, \underline{v} \in \{0, \times, 1\}^2$. The definition is identical with 3.2.7, 3.2.13 (with the new meaning of a-better and i-better):

3.2.23 Definition. Let $\underline{u}, \underline{v} \in \{0, \times, 1\}^2$. $\underline{v}$ a-improves $\underline{u}$ (notation: $\underline{u} \leq_a \underline{v}$) if for each (three-valued) model $\underline{M}$ of type $\langle 1,1 \rangle$ and each $o \in M$ we have: If the card of o is $\underline{v}$, then $\underline{M}$ is a-better than

$\underline{M}$ (o:$\underline{u}$); similary for "i-better".

The following theorem is a generalization of 3.2.9 and 3.2.14.

3.3.24 Theorem. The relations of a-improvement and i-improvement on $\{0, \times, 1\}^2$ are completely described by the following graphs (where, of course, each vertex corresponds to a set of elements mutually equivalent w.r.t. the quasiodering in question).

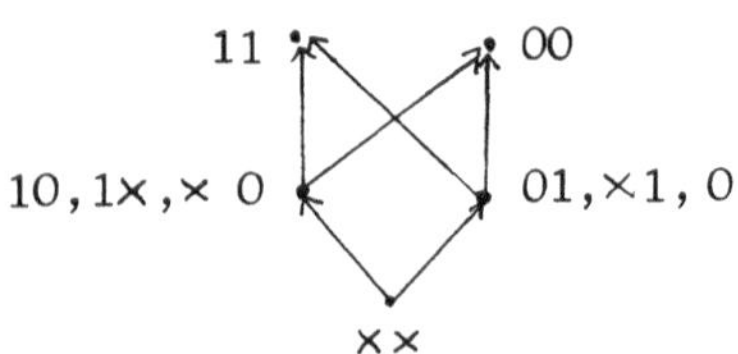

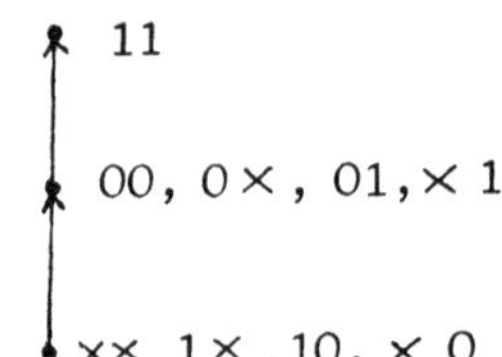

Proof. One can easily see that for pairs not containing $\times$ the result follows directly from 3.2.9 and 3.2.14. Hence, for a-improvement it suffices to show $\langle \times, \times \rangle <_a \langle 1,0 \rangle$, $\langle \times, \times \rangle <_a \langle 0,1 \rangle$, $\langle 1,0 \rangle \equiv_a \langle 1, \times \rangle \equiv_a \langle \times, 0 \rangle$ and $\langle 0,1 \rangle \equiv_a \langle \times, 1 \rangle \equiv_a \langle 0, \times \rangle$.

Let us show that $\langle \times, \times \rangle <_a \langle 1, 0 \rangle$. First, $\langle \times, \times \rangle \leq_a \langle 1,0 \rangle$ is obvious since if the card of o in $\underline{M}$ is $\langle 1,0 \rangle$, then each completion of $\underline{M}$ is a completion of $\underline{M}(o: \langle \times, \times \rangle)$. Further, show that $\langle 1,0 \rangle \not\leq_a \langle \times, \times \rangle$. Let $\underline{M}$ be two valued and let the card of o in $\underline{M}$ be $\langle 1,0 \rangle$. Then $\underline{M}(o: \langle 0,1 \rangle)$ is a completion of $\underline{M}(o; \langle \times, \times \rangle)$ and $\underline{M}(o: \langle 0,1 \rangle)$ is not a-better than $\underline{M}$. Since $\underline{M}$ is two-valued we see that $\underline{M}(o: \langle \times, \times \rangle)$ is not a-better than $\underline{M} = \underline{M}(o: \langle \times, \times \rangle)(o : \langle 1,0 \rangle)$.

Next, we show that $\langle 1,0 \rangle \equiv_a \langle 1, \times \rangle$. First, $\langle 1, \times \rangle \leq_a \langle 1,0 \rangle$ is obvious (as above). Conversely, if the card of o in $\underline{M}$ is $\langle 1, \times \rangle$ and if $\underline{M}' = \underline{M}(o: \langle 1,0 \rangle)$, then we have the following possibilities for a completion $\underline{N}$ of $\underline{M}$: Either the card of o in $\underline{N}$ is $\langle 1,0 \rangle$ and then $\underline{N}$ itself is a completion of $\underline{M}'$, or the card is $\langle 1,1 \rangle$ and then $\underline{N}(o: \langle 1,0 \rangle)$ is a completion of $\underline{M}'$ and $\underline{N}$ is a-better than $\underline{N}(o : \langle 1,0 \rangle)$.

All other cases concerning $\leq_a$ are treated similarly.

As regards $\leq_i$ we have to prove $\langle\times,\times\rangle \equiv_a \langle 1,\times\rangle \equiv_a \langle 1,0\rangle \equiv_a \langle\times,0\rangle$ and $\langle 0,0\rangle \equiv_a \langle 0,\times\rangle \equiv_a \langle 0,1\rangle \equiv_a \langle\times,1\rangle$. Let us verify the first claim. Since $\leq_a$ implies $\leq_i$ we have $\langle 1,0\rangle \equiv_i \langle 1,\times\rangle \equiv_i \langle\times,0\rangle$ and $\langle\times,\times\rangle \leq_i \langle 1,\times\rangle$ we verify $\langle 1,\times\rangle \leq_i \langle\times,\times\rangle$. Let the card of o in $\underline{M}$ be $\langle\times,\times\rangle$ and let

$\underline{M}' = \underline{M}(o: \langle 1,\times\rangle)$ let $\underline{N}$ be a completion of $\underline{M}$. If the card of o in $\underline{N}$ is $\langle 1,v\rangle$, then $\underline{N}$ is a completion of $\underline{M}'$: if it is $\langle 0,v\rangle$, then put $\underline{N}' = \underline{N}(o: \langle 1,0\rangle)$. Then $a_{\underline{N}} = a_{\underline{N}'}$ and $b_{\underline{N}'} = b_{\underline{N}} + 1$ so that $\underline{N}$ is i-better than $\underline{N}$.

For the second claim it suffices to show that $\langle 0,0\rangle \leq_i \langle 0,\times\rangle$. Let the $\underline{M}$-card of o be $\langle 0,\times\rangle$ and let $\underline{M}'$ be $\underline{M}(o: \langle 0,0\rangle)$; let $\underline{N}$ be a completion of $\underline{M}$. If the $\underline{N}$-card of o is $\langle 0,0\rangle$, then $\underline{N}$ is a completion of $\underline{M}'$; if the card is $\langle 0,1\rangle$, then put $\underline{N}' = (o: \langle 0,0\rangle)$. Then $\underline{N}'$ is a completion of $\underline{M}'$ and $\underline{N}$, $\underline{N}'$ are i-equivalent.

3.3.25 Remark. We visualize $\leq_a$ and $\leq_i$ in another way, representing $\{0, \times, 1\}^2$ as a 3×3 matrix, where a dotted line means equivalence and a heavy line together with a dotted line means that the transition from the side of the dotted line to the side of the heavy line signify a strict improvement (drivers understand).

a-improvement

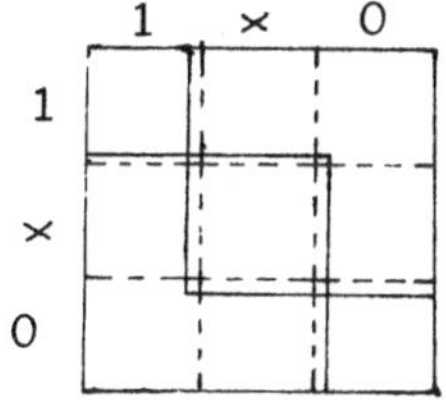

i-improvement

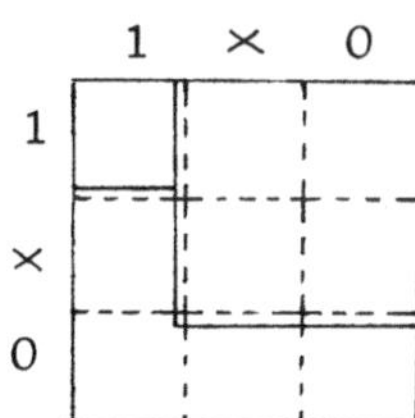

3.3.26 Remark. Associational quantifiers in calculi with incomplete information will play an important role in Part B; we shall use the last considerations in Chapter VI, Section 2.

3.3.27 Key words: extension of a function calculus, regular values and the singular value, completion of a structure, secured formulae, regular valued formulae, the quantifier of strong equivalence, associational and implicational quantifiers in x-predicate calculi, a-improvement, i-improvement.

III.4 Calculi with qualitative values

3.4.1 Discussion and Definition. As already mentioned in the introduction to the previous section, we must pay attention to the fact that observed attributes need not be two-valued. We now make a mild generalization by assuming that we have finite sets $V_1, \dots, V_n$ of abstract values (each V_i having at least two elements) and we consider all (observational) structures of the form $M = \langle M, f_1, \dots, f_n \rangle$ where each f_i is a V_i-valued function. Setting $\mathbb{V} = \langle V_1, \dots, V_n \rangle$ we can call M a $\mathbb{V}$-valued structure. Of course, we can replace each V_i having h_i elements by the segment $\{0, 1, \dots, (h_i - 1)\}$ of natural numbers, so that our structures become natural number-valued. This "normalization" does not depend on the character of the assumed observational attributes (qualitative, comparative, quantitative) but only on the finiteness assumption above. If we assume that our structures correspond to the behaviour of some qualitative attributes, i.e., if we assume no preferred structure on the sets V_i, then this fact will be reflected not by the structures themselves but by the language we shall use to speak about them.

3.4.2 Definition. Let $M = \langle M, f_1, \dots, f_n \rangle$ be a natural number valued structure and let $V_i = \{0, 1, \dots, h_i - 1\}$ $(i = 1, \dots, n)$. By saying that M is $\langle h_1, \dots, h_n \rangle$-valued we mean the same as saying that M is $\langle V_1, \dots, V_n \rangle$-valued, i.e., for each i, the range of f_i is included in V_i.

3.4.3 Remark. Let $M = \langle M, f_1, \dots, f_n \rangle$ be a $\langle h_1, \dots, h_n \rangle$-valued structure of type $t = \langle k_1, \dots, k_n \rangle$ (i.e., f_i is k_i-ary). We can associate with $\underline{M}$ an $\{0,1\}$-valued structure

$$\pi(\underline{M}) = \langle M, p_1^0, \dots, p_1^{h_1 - 1}, \dots, p_n^0, \dots, p_n^{h_n - 1} \rangle$$

of the type $\pi(t) = \langle \underbrace{k_1, \dots, k_1}_{h_1 \text{ times}}, \dots, \underbrace{k_n, \dots, k_n}_{h_n \text{ times}} \rangle$

defined as follows: $p_i^j(o_1, \dots, o_{n_i}) = 1$ iff $f_i(o_1, \dots, o_{n_i}) = j$;

$p_i^j(o_1, \dots, o_{n_i}) = 0$ otherwise

$\pi(\underline{M})$ fully represents $\underline{M}$ (i.e. π is one-to-one) and has the following evident property:

(*) for each $i = 1, \ldots, n$, and each $\varrho \in M^{k_i}$, exactly one of the numbers $p_i^o(\varrho), \ldots, p_i^{k_{i-1}}(\varrho)$ is 1 (the remaining ones being 0).

Conversely, each $\{0,1\}$-valued structure of type $\pi(t)$ satisfying (*) is $\pi(\underline{M})$ for some $\langle h_1, \ldots, h_n \rangle$ - valued $\underline{M}$.

Thus $\langle h_1, \ldots, h_n \rangle$ -valued structures with a given $\langle h_1, \ldots, h_n \rangle$ are uniquely representable by some specific two-valued structures. Nevertheless, it is reasonable to develop some autonomous logic for the former structures since working with them as like machine inputs one saves the computer's memory. (To represent an h-valued function defined on m objects one needs at most $m([\log h] + 1)$ bits; to represent h two-valued functions one needs $h.m$ bits.) Moreover, sentences of the language we shall use to speak about those structures, even if translatable into sentences speaking about the two-valued representations, are more transparent and useful than their translations.

We shall now describe observational calculi with qualitative values. This will be done in two steps: First, we describe open formulae and prove some lemmas on them, then we give the full definition. The numbers 0,1 have two roles: They are treated as some qualitative values as well as truth values.

3.4.4 Definition. (1) With each finite set X of natural numbers we associate a unary junctor (X) (called a coefficient) putting $Asf_{(X)}(u) = 1$ iff $u \in X$ and $Asf_{(X)}(u) = 0$ iff $u \notin X$ (hence, $Asf_{(X)}$ is the characteristic function of X).

(2) With each function $\alpha : \{0,1\}^j \to \{0,1\}$ we associate its canonical extension to $\mathbb{N}^j$ putting $\bar{\alpha}(u_1, \ldots u_j) = \alpha(\bar{u}_1, \ldots \bar{u}_j)$ where $\bar{0} = 0$ and $\bar{u} = 1$ for $u \geq 1$. (Hence, we "identify" non-zero values.)

(3) We introduce junctors $\&, \vee, \rightarrow, \neg$ whose associated functions are canonical extensions of their associated functions over $\{0,1\}$. Thus, e.g., $Asf_{\&}(u,v) = 1$ iff $u \geq 1$ and $v \geq 1$.

(4) Let $t = \langle k_1, \ldots, k_n \rangle$ be a type and let $h = \langle h_1, \ldots, h_n \rangle$, $h_i \geq 2$. For the time being call any observational function calculus of type t whose junctors are $\&, \vee, \rightarrow, \neg$ and the coefficients (X) for $X \subseteq \{0,1, \ldots, \max_i(h_i - 1)\}$ and whose models are exactly all observational h-valued models an <u>h-valued openly qualitative OFC</u>. (Nothing is assumed on quantifiers.) In the present context, V_i means $\{0,1,\ldots,h_i - 1\}$.

(5) Let $\mathcal{F}$ be an h-valued openly qualitative OFC. Each formula of the form $(X)F_i(\underline{x})$ where $1 \leq i \leq n$, $\emptyset \neq X \subsetneqq V_i$ and $\underline{x}$ is a k_i-tuple of variables is called a <u>literal</u>. (We write $(X)F_i$ instead of $(X)F_i(\underline{x})$ if there is no danger of confusion.) An <u>elementary disjunction</u> (ED) is a non-empty disjunction of literals in which each atom $F_i(\underline{x})$ occurs at most once; similarly, we define elementary conjunctions (EC). For example, let $h = \langle 3,3 \rangle$ and $t = \langle 2,1 \rangle$; then

$(0,2)\ F_1(x,y) \vee (1)\ F_1(z,x) \vee (1,2)\ F_2(x)$ is an ED.

(6) A formula φ is <u>two-valued</u> (or $\{0,1\}$-valued) if, for each model $\underline{M}$ and each e, $\|\varphi\|_{\underline{M}}[e] \in \{0,1\}$.

3.4.5 <u>Lemma</u>. Let $\mathcal{F}$ be an $\langle h_1, \ldots, h_n \rangle$-valued openly qualitative OMFC, let F_i be a functor, and let $X, Y \subseteq V_i$. Then

(1) $\neg(X)F_i \Leftrightarrow (V_i - X)F_i$,

(2) $(X)F_i \Leftrightarrow \bigvee_{k \in X}(\{k\})F_i$,

(3) $(X)F_i \Leftrightarrow \bigwedge_{k \notin X}\neg(\{k\})F_i$,

(4) $(X)F_i \vee (Y)F_i \Leftrightarrow (X \cup Y)F_i$,

(5) $(X)F_i \,\&\, (Y)F_i \Leftrightarrow (X \cap Y)F_i$,

(6) $(\emptyset)F_i \Leftrightarrow \underline{0}$,

(7) $(V_i)F_i \Leftrightarrow \underline{1}$.

Elementary proofs are left to the reader.

3.4.6 <u>Lemma</u>. Let $\mathcal{F}$ be an openly qualitative OFC and let φ, ψ, χ be formulae. Then the logical equivalences (1) - (16) from 3.3.2 (in particular, distributivity and de Morgan laws) hold true for $\mathcal{F}$. This is evident.

3.4.7 Definition. Let $\mathcal{F}$ be an h-valued openly qualitative OFC. A pseudoliteral is a formula of the form $(X)\, F_i(\underline{x})$ where $X \subseteq V_i$ and $\underline{x}$ is a k_i-tuple of variables.

A formula is (pseudo) regular if it results from (pseudo) literals by iterating applications of & and $\vee$.

A pseudoliteral $(X)\, F_i(\underline{x})$ is full if $X = V_i$, it is empty if $X = 0$.

A pseudoelementary conjunction (psEC) is a non-empty conjunction of non-full pseudoliterals (empty pseudoliterals allowed); a pseudoelementary disjunction (psED) is a non-empty disjunction of non-empty pseudoliterals (full pseudoliterals allowed).

Let $\varphi_o, \varphi_1 \ldots$ be an enumeration of all atoms. If $\kappa_1 = \bigwedge_I (X_i)\varphi_i$ and $\kappa_2 = \bigwedge_J (Y_j)\varphi_j$ are psEC's (i.e., the φ_i's are atoms) then put $\underline{\text{con}}(\kappa_1, \kappa_2) =$

$$= \bigwedge_{I \cup J} (Z_i)\varphi_i, \text{ where } Z_i = X_i \cap Y_i \text{ for } i \in I \cap J,\ Z_i = X_i \text{ for}$$

$i \in I - J$ and $Z_i = Y_i$ for $i \in J - I$. If $\delta_1 = \bigvee_I (X_i)\varphi_i$ and $\delta_2 = \bigvee_J (Y_i)\varphi_i$ are psED's then put $\underline{\text{dis}}(\delta_1, \delta_2) = \bigvee_{I \cup J} (Z_i)\, \varphi_i$, where $Z_i = X_i \cup Y_i$ for $i \in I \cap J$, $Z_i = X_i$ for $i \in I - J$ and $Z_i = Y_i$ for $i \in J - I$.

If $(X)\, F_i(\underline{x})$ is a literal then put $\underline{\text{neg}}((X) F_i(\underline{x})) = (V_i - X)\, F_i(\underline{x})$;

if $\kappa = \bigwedge_I (X_i)\, \varphi_i$ and $\delta = \bigvee_J (Y_j)\, \varphi_j$ then put

$\underline{\text{neg}}(\kappa) = \bigvee_I \underline{\text{neg}}((X_i)\varphi_i)$ and $\underline{\text{neg}}(\delta) = \bigwedge_J \underline{\text{neg}}((Y_j)\,\varphi_j)$.

3.4.8 Lemma. (1). If φ is (pseudo) regular, then $\neg\varphi$ is logically equivalent to a (pseudo) regular formula.

(2) If κ_1, κ_2 are psEC's then $\underline{\text{con}}(\kappa_1, \kappa_2)$ is a psEC logically equivalent to $\kappa_1 \& \kappa_2$; similarly for psED's and $\underline{\text{dis}}$.

(3) If κ is a psEC, then neg(κ) is a psED logically equivalent to $\neg\kappa$; similarly for a psED.

3.4.9 Corollary. Each open designated formula is logically equivalent to a formula of one of the following forms: $\underline{0}$, $\underline{1}$, atomic, pseudoregular.

Proof. By induction on the complexity of formulae, using 3.4.5, 3.4.6, 3.4.8. Note that, e.g., if φ is pseudoregular, then $(X)\varphi \Leftrightarrow \varphi$ if $1 \in X$ and $(X)\varphi \Leftrightarrow \neg\varphi$ otherwise; using 3.4.8, $\neg\varphi$ is logically equivalent to a pseudoregular formula. Note that 3.4.5 (6),(7) and 3.4.6 (15), (16) are not used.

3.4.10 Theorem. (Normal form .) Each open formula is logically equivalent to a formula of one of the following types: $\underline{0}$, $\underline{1}$, atomic, non-empty disjunction of elementary conjunctions. (Consequently, each non-atomic open formula is two-valued.)

Proof. If φ is pseudoregular, then one can express φ as a non-empty (possibly one-element) disjunction of conjunctions of pseudoliterals; in each such conjunction one can reduce the occurrences of each atom to one, by using some of 3.4.5 (6), (7) and 3.4.6 (7)-(10); each conjunction of pseudoliterals changes either to $\underline{0}$ or to $\underline{1}$ or to an EC. A disjunction of pseudolementary conjunctions changes either to $\underline{1}$ or to $\underline{0}$ or to a non-empty disjunction of EC's.

3.4.11 Remark. (1) One can easily prove the "dual form" of the Normal form theorem interchanging "conjunction" and "disjunction".

(2) What should we assume about quantifiers in a calculus to call it "a qualitative OFC"? As far as open formulae are concerned, we are interested in (pseudo) regular formulae. They are two-valued; hence, if q is a quantifier we are interested in the values $\mathrm{Asf}_q M$ for two-valued models only. But Asf_q is to be defined for natural number valued models; thus we use the device of "canonical extension" as in 3.4.3. This leads us to the following definition:

3.4.12 Definition. Let $\mathcal{F}$ be an openly qualitative OMFC and let q be a quantifier of $\mathcal{F}$ of type $<1^k>$. We call q essentially two-valued if for each (natural number-valued) model $\underline{M}$ of type $<1^k>$ we have

$Asf_q(\underline{M}) = Asf_q(\hat{\underline{M}})$, where $\hat{\underline{M}}$ results from $\underline{M}$ by replacing all non-zero values by 1 (i.e., if $\underline{M} = \langle M, f_1, \ldots \rangle$, then $\hat{\underline{M}} = \langle M, \bar{f}_1, \ldots \rangle$, where $\bar{f}_i$ is as in 3.4.4 .

3.4.13 Definition. An openly qualitative OMFC is qualitative if all its quantifiers are essentially two-valued.

In what remains of the present section we shall consider qualitative OMFC's with incomplete information.

3.4.14 Remark. Consider an OFC $\mathcal{F}$ which is a x-extension of an $\langle h_1, \ldots, h_n \rangle$ -valued qualitative OFC $\mathcal{F}_o$.

(1) Thus, models are structures $\underline{M} = \langle M, f_1, \ldots \rangle$ (finite) such that f_i maps M^{k_i} into $\{0, 1, \ldots, h_i - 1, \times\}$.

(2) If $\mathcal{F}$ is the secured x-extension of $\mathcal{F}_o$, then: (i) The junctors of $\mathcal{F}$ are secured x-extensions of the junctors of $\mathcal{F}_o$, i.e.,

$Asf_{(X)}(u) = 1$ if $u \in X$, $Asf_{(X)}(u) = 0$ if $u \in N - X$, $Asf_{(X)}(\times) = \times$.

The associated function of & is given by the following table:

&	≥1	×	0
≥1	1	×	0
×	×	×	0
0	0	0	0

etc.

(ii) The quantifiers of $\mathcal{F}$ are secured x-extensions of the quantifiers of $\mathcal{F}_o$, i.e.,

$$\mathrm{Asf}_q(\underline{M}) = \begin{cases} 1 & \text{iff } \mathrm{Asf}_q(\underline{N}) = 1 \text{ for each two-valued modification of } \underline{M}, \\ 0 & \text{iff } \mathrm{Asf}_q(\underline{N}) = 0 \text{ for each two-valued modification of } \underline{M}, \\ \times & \text{otherwise.} \end{cases}$$

Here, $\underline{N} = \langle M, g_1, \ldots \rangle$ is a two-valued modification of $\underline{M} = \langle M, f_1, \ldots \rangle$ if, for each $o \in M$,

$f_i(o) \geq 1$ implies $g_i(o) = 1$,

$f_i(o) = 0$ implies $g_i(o) = 0$,

$f_i(o) = \times$ implies $g_i(o) \in \{0,1\}$.

(3) In general, we shall work with calculi richer than the secured extension, namely containing new quantifiers. (Helpful quantifiers studied in Chapter VI, Section 3 are typical examples of non-secured quantifiers.) However, we restrict ourselves to quantifiers satisfying the following natural generalization of the notion "essentially two-valued":

3.4.15 Definition. (1) Let $\mathcal{F}$ be a cross-extension of a qualitative OFC $\mathcal{F}_0$. A quantifier q of $\mathcal{F}$ is essentially three-valued if, for each $\underline{M}$,

$\mathrm{Asf}_q(\underline{M}) = \mathrm{Asf}_q(\hat{\underline{M}}) \in \{1, \times, 0\}$, where $\hat{\underline{M}}$ results from $\underline{M}$ by replacing each regular value ≥ 1 by 1 (leaving 0 and $\times$ untouched), (ii) if $\underline{M}$ does not contain any $\times$ then $\mathrm{Asf}_q(\underline{M}) \in \{0,1\}$.

(2) A x-extension $\mathcal{F}$ of a qualitative OFC is a x-qualitative OFC if the junctors of $\mathcal{F}$ are &, $\vee$, $\rightarrow$, $\neg$ and the coefficients, and if each quantifier of $\mathcal{F}$ is essentially three-valued.

3.4.16 Remark. If q is an essentially three-valued quantifier, then its associated function is uniquely determined by its behaviour on three-valued (i.e., $\{1, \times, 0\}$-valued) models of the appropriate type.

3.4.17 Remark. Literals, EC's, ED's, psEC's, psED's and regular open formulae are defined as in qualitative calculi. One easily checks that logical equivalences 3.4.5(1)-(5) are true for each $\times$-qualitative calculus provided we assume $X \neq \emptyset$ in (2) and $X \neq V_i$ in (3); the last restriction is necessary since 3.4.5 (6),(7) are not true: $\|(\emptyset) F_i\|[\times] = \times$ but $\|\underline{0}\| = 0$; similarly, $\|(V_i) F_i\|[\times] = \times$ but $\|\underline{1}\| = 1$. As far as 3.4.6 is concerned, we easily verify 2.2.10 (1) - (14) but 2.2.10 (15) , (16) are not true for $\times$-qualitative calculi (cf. 3.3.12). These equivalences, true for qualitative but not for $\times$-qualitative calculi, were not used in the proof of 3.4.8, 3.4.9; hence, we have the following :

3.4.18 Theorem. (Normal form.) In an $\times$-qualitative OFC, each open formula is logically equivalent to a formula of one of the following forms: $\underline{0}$, $\underline{1}$, atomic, a non-empty disjunction of pseudoelementary conjunctions. Consequently, each non-atomic open formula is three-valued ($\{1, \times, 0\}$-valued).

Proof. As in the first part of the proof of 3.4.10, we arrive at a non--empty disjunction of conjunctions of pseudoliterals, each conjunction having the form $\bigwedge_I (X_i) \varphi_i$. We successively eliminate full literals as follows: If, e.g., $X_{i_o} = V_{i_o}$, then divide V_{i_o} into two disjoint non-empty subsets $X^1_{i_o}$, $X^2_{i_o}$ and define $X^1_i = X^2_i = X_i$ for $i \neq i_o$. Then

$$\bigwedge_I (X_i) F_i \Leftrightarrow (\bigwedge_I (X^1_i) F_i) \vee (\bigwedge_I (X^2_i) F_i) .$$

3.4.19 Remark. (1) "Pseudoelementary" cannot be strengthened to "elementary" - consider $(\emptyset) F_i$. On the other hand, one could continue the process of dividing coefficients to obtain a disjunction of pseudoelementary conjunctions with each coefficient of cardinality at most 1.

(2) Recall 3.4.3: a <2,...,2>-valued $\varkappa$-qualitative OFC is in fact an x-predicate observational calculus. So we have here the promised normal form for open formulae in $\varkappa$-predicate calculi.

3.4.20 Remark. We conclude this section with some remarks and definitions concerning monadic $\varkappa$-qualitative calculi. Let x be the designated variable. In part B we shall pay attention to designated (ps) ED's, these are formulae of the form $\bigwedge_I (X_i) F_i(x)$ and $\bigwedge_J (Y_j) F_j(x)$ respectively. Note that operations con, dis, neg preserve designated formulae. We define some syntactic relations between designated pseudoelementary conjunctions and disjunctions to be used later.

3.4.21 Definition. Let $\kappa = \bigwedge_I (X_i) F_i$ and $\lambda = \bigwedge_J (Y_i) F_i$

be two designated ps EC's.

(a) κ is included in λ ($\kappa \subseteq \lambda$) if $I \subseteq J$ and $X_i = Y_i$ for each $i \in I$.

(b) κ is poorer than λ ($\kappa \sqsubseteq \lambda$) if $I = J$ and $X_i \subseteq Y_i$ for each $i \in I$.

(c) κ is hidden in λ ($\kappa \lhd \lambda$) if $I \subseteq J$ and $X_i \subseteq Y_i$ for each $i \in I$.

(d) κ hoops λ ($\kappa \lessdot \lambda$) if $I \subseteq J$ and $X_i \supseteq Y_i$ for each $i \in I$. The definition is the same for psED's.

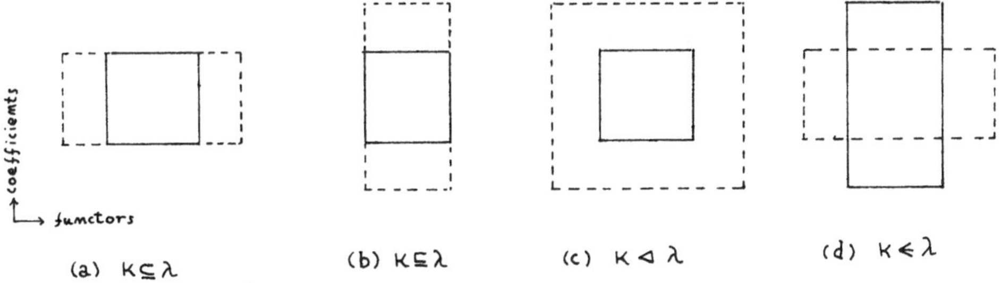

(a) $\kappa \subseteq \lambda$ (b) $\kappa \sqsubseteq \lambda$ (c) $\kappa \lhd \lambda$ (d) $\kappa \lessdot \lambda$

3.4.22 Lemma. Let κ, λ be designated psEC's.

(1) $\kappa \lhd \lambda$ iff there is a κ' such that $\kappa \subseteq \kappa'$ and $\kappa' \sqsubseteq \lambda$. $\kappa \lessdot \lambda$ iff there is a κ' such that $\kappa' \sqsubseteq \kappa$ and $\kappa' \subseteq \lambda$.

(2) If $\kappa \subseteq \lambda$, then λ logically implies κ, i.e., for each $\underline{M}$ and each $o \in M$, $\|\lambda\|_{\underline{M}}[o] = 1$ implies $\|\kappa\|_{\underline{M}}[o] = 1$. If $\kappa \sqsubseteq \lambda$, then κ logically implies λ. If $\kappa \lessdot \lambda$, then λ logically implies κ.

(3) Let γ, δ be designated psED's. If $\gamma \subseteq \delta$, then γ logically implies δ; if $\gamma \sqsubseteq \delta$, then γ logically implies δ; hence, if $\gamma \lhd \delta$, then γ logically implies δ.

(4) For κ, λ psEC's, $\underline{con}(\kappa, \lambda)$ is the $\lessdot$-supremum of κ and λ; for γ, δ psED's, $\underline{dis}(\gamma, \delta)$ is the $\lhd$-supremum of γ, δ. **This is** obvious from the definitions.

3.4.23 Remark. (1) The relation "is hidden in" can be thought of as a "syntactically simpler than"-relation; this is in accordance with the relation of logical implication for psED's but not for psEC's. This is why we study the "hoop"-relation for psEC's.

3.4.24 Key words: $\langle h_1, \ldots, h_n\rangle$ -valued structures, coefficients, (openly) qualitative OFC's, (pseudo) elementary conjunctions and disjunctions, essentially two-valued (three-valued) quantifiers, qualitative and x--qualitative OFC's, relations between psEC's (psED's): included in, poorer than, hidden in, hoops.

III.5 More on the logic of observational predicate calculi.

This is an additional section in which we collect some results of a logical and computational character concerning the observational predicate calculi but dependent on mathematical facts not presented in this book. We shall present definitions necessary for the understanding of theorems, but we refer to the literature for proofs of needed facts. Most proofs will only be briefly outlined; the results can be considered as possible starting points for further investigations.

3.5.1 Remember the observational predicate calculi - function calculi with truth values 0,1, with finite models and with recursive semantics. We shall compare OPC's with predicate calculi usually studied in Mathematical Logic, i. e., function calculi with truth values 0,1, with both finite and infinite models and with no restrictions on the associated functions of quantifiers. The latter calculi will be called usual predicate calculi-UPC. "Classical" is reserved to mean "with two quantifiers $\forall, \exists$ with their obvious semantics "; we speak of COPC's (classical observational predicate calculi) and CPC's (classical predicate calculi - more pedantically, but awkwardly, one could say classical usual predicate calculi: CUPC's). We shall first follow well-known notions and facts concerning UPC's and ask whether the notions are meaningful for OPC's and whether the facts remain valid if UPC's are replaced by OPC's ; then we shall investigate some particular OPC's with some remarkable properties. The reason for our concentration on observational predicate calculi is that their theory is more developed than the theory of other observational function calculi; similar investigations of other observational function calculi remain a task for the near future. We will make use of some facts on diophantine equations, weak monadic second order successor or arithmetics, and semisets.

3.5.2 We already know that COPC's differ from CPC's with respect to axiomatizability; whereas each non-monadic CPC is axiomatizable but undecidable (this follows from **Gödel's classical result), no non-monadic** COPC is axiomatizable. We stated the last fact in 2.1.17 as a (non-immediate !) consequence of Trachtenbrot's theorem 2.1.16. We shall prove Trachtenbrot's theorem later in this section.

A further well known property of CPC's is compactness:

For each set X of sentences which has no model there is a finite subset $A \subseteq X$ which has no model.

Now, almost no OPC is compact: For example, given an OPC containing the equality predicate and $\exists$, the set $X = \{(\exists^k x)(y=x);\ k$ a natural number$\}$ has no finite model but each finite subset of X has a model. (On the other hand, it follows from the Representation theorem 3.1.31 that each MOPC of finite dimension without equality is trivially compact. We shall consider various notions of classical definability of classes of models and apply them to observational calculi.

3.5.3 Definition. $\mathcal{K}$ is a variety of models if there is a type t such that $\mathcal{K}$ consists of some models of type t and $\mathcal{K}$ is closed under isomorphism, i.e., if $\underline{M} \in \mathcal{K}$ and $\underline{N}$ is isomorphic to $\underline{M}$, then $\underline{N} \in \mathcal{K}$. t is the type of $\mathcal{K}$.

3.5.4 Remark. (1) "Model" can mean either both finite and infinite $\{0,1\}$-structures or only $\{0,1\}$-structures that are **finite; in each** particular case the meaning will be clear from the context. (2) Varieties of models are in one-to-one correspondence with associated functions of quantifiers; in the observational case, a variety defines an observational quantifier iff it is a recursive class of models.

3.5.5 Definition. (1) Let $\mathcal{K}$ be a variety of type t. $\mathcal{K}$ is elementary if there is a classical sentence φ of type t such that $\mathcal{K}$ consists exactly of all models of φ.

(2) Let $t = \langle t_1, \ldots, t_n \rangle$ and $t' = \langle t_1, \ldots, t_n, t_{n+1}, \ldots, t_{n+m} \rangle$ be types; then t' is called an expansion of t. We call t' a 1-expansion of t if $t_{n+1} = \ldots = t_{n+m} = 1$. A structure $\underline{M}'$ of type t' is an expansion of $\underline{M} = \langle M, f_1, \ldots, f_n \rangle$ if M' has the form

$\langle M, f_1, \ldots, f_n, f_{n+1}, \ldots, f_{n+m} \rangle$ (it results from $\underline{M}$ by adding new $\{0,1\}$-functions). $\underline{M}'$ is a <u>1-expansion</u> of $\underline{M}$ if it results from $\underline{M}$ by adding unary functions to $\underline{M}$.

(3) A variety $\mathcal{K}$ of type t is <u>projective</u> (1-<u>projective</u>) if there is an expansion (1-expansion) t' of t and a classical formula ψ of type t' such that $\mathcal{K}$ consists of all structures $\underline{M}$ of type t that can be expanded (1-expanded) to a model of ψ, i.e., if

$$\mathcal{K} = \{\underline{M} \; ; \; (\exists \; \underline{M}' \text{expansion (1-expansion)of } \underline{M})(\|\psi\|_{\underline{M}'} = 1)\}.$$

3.5.6 <u>Remark</u>. The above definitions make sense both for CPC's and for COPC's. Note the following known facts <u>for CPC's</u>: (a) A variety $\mathcal{K}$ of type t is elementary iff both $\mathcal{K}$ and $-\mathcal{K}$ are projective ($-\mathcal{K}$ is the complement of $\mathcal{K}$ - it consists of all models of type t that are not in $\mathcal{K}$). This is a form of the so-called interpolation theorem. (b) There are projective non-elementary varieties (e.g., the variety of all so-called non-standard models of Peano arithmetic is 1-projective but not elementary). (c) Hence, there are projective (1-projective) varieties $\mathcal{K}$ such that $-\mathcal{K}$ is not projective.

3.5.7 We are going to investigate elementary and projective classes of <u>observational models</u>. First we show the close relationship of projective classes with languages recognizable in polynomial time. (The reader not interested in their relation may skip to 3.5.12 .)

We assume the following notions to be known (cf. Karp 1972): deterministic algorithm, indeterministic algorithm, operation in polynomial time, the class P of sets (languages) recognizable by a deterministic algorithm operating in polynomial time, the class NP of sets recognizable by a non-deterministic algorithm operating in polynomial time, the class π of functions defined by algorithms operating in polynomial time, polynomial reducibility, universal NP-problems. The famous P-NP problem is the problem whether P = NP, i.e., whether each set recognizable by a non-deterministic algorithm operating in polynomial time is recognizable by a deterministic algorithm operating in polynomial time.

3.5.8 It is obvious that observational $\{0,1\}$ - structures can be coded by words in a finite alphabet. For example, following Pudlák [1975 a,b], we associate with each $\{0,1\}$ -structure $\underline{M} = \langle M, r_1, \dots, r_n \rangle$ of type $t = \langle k_1, \dots, k_n \rangle$ its code cod($\underline{M}$). This is a word in the alphabet 2^{2+n} of length $m^{\max(t)}$ (where m is the cardinality of $\underline{M}$ and max(t) means $\max_i(k_i)$) defined as follows: let $M = \{u_1, \dots, u_m\}$ where $u_1 < \dots < u_m$ in the natural ordering of natural numbers. The code is an $\{0,1\}$ -matrix with $2+n$ rows and $m^{\max(t)}$ columns. The first row designates $m^{\max(t)}$, the second row designates m and the $(i+2)$-th row contains values of r_i for arguments ordered lexicographically, e.g., if $k_1 = 2$ and if $a_{ij} = r_1(u_i, u_j)$ then cod($\underline{M}$) looks like:

m	$m^{\max t}$
0 ...	 1
1	 0
$a_{11}, b_{12}, \dots$	$a_{21}, a_{22}, \dots$
...	...

If $\mathcal{K}$ is a variety then cod($\mathcal{K}$) = {cod($\underline{M}$); $\underline{M} \in \mathcal{K}$ }.

We have the following theorem:

3.5.9 Theorem. (Fagin 1973). A variety $\mathcal{K}$ is projective iff cod($\mathcal{K}$) is NP i.e. cod($\mathcal{K}$) is a language accepted by a nondeterministic Turing machine operating in polynomial time .

(Hint.) (1) Let $\mathcal{K}$ be projectively defined by a sentence $\varphi(\underline{R}, \underline{S})$. First prove that if $\mathcal{K}$ is elementary, i.e. there are no S-predicates, then cod($\mathcal{K}$) is deterministically polynomial. Then it is easy to see how to construct a nondeterministic machine for the general case: given the code cod M of a structure $\underline{M} = \langle M, r_1, \dots, r_n \rangle$, the machine first proceeds nondeterministically, guessing an expansion $\langle M, r_1, \dots, r_n, s_1, \dots, s_n \rangle$ and then continues deterministically, verifying $\varphi(\underline{r}, \underline{s})$. Hence cod($\mathcal{K}$) is in NP.

(2) Conversely. Let cod($\mathcal{K}$) be recognized by a non-deterministic Turing machine T operating in time m^k. Assume that T has q states, γ tape symbols (and one tape). Now, $\underline{M} \in \mathcal{K}$ iff there is an accepting computation of T on input cod($\underline{M}$) of the length $\leq c^k$ where c is the length of cod ($\underline{M}$); hence the length is polynomial in card(M), the type t being fixed . The computation is a certain sequence of configurations and can be represented as a matrix with c^{k+1} rows and $2c^k+1$ columns, whose elements are tape symbols, states and a marker showing the position of the head. Mutatis mutandis, the computation can be represented by $\{0,1\}$-matrix with $m^{\hat{k}}$ rows and columns, where $m = \text{card}(M)$ and $\hat{k}$ is larger than k but independent of m, i.e. as a $2\hat{k}$-ary relation on M. It is a tedious but straightforward exercise to show that there is a sentence $\varphi(S)$ where S is a $2\hat{k}$-ary relation such that $\langle \underline{M}, s \rangle$ satisfies $\varphi(S)$ iff s represents an accepting computation of T on input cod($\underline{M}$) as described above. Hence $\varphi(S)$ projectively defines $\mathcal{K}$.

3.5.10 Corollary. If there is a projective variety $\mathcal{K}$ such that $-\mathcal{K}$ is not projective then NP languages are not closed under complementation and hence $P \neq NP$.

3.5.11 Remark. Pudlák [1975a] also considers codes of languages by structures. Moreover, he defines the definitional complexity of a variety projectively defined by a sentence φ as the number of quantifiers in φ and shows close linear dependences between the definitional complexity of a projective class $\mathcal{K}$ and the computational complexity of the language cod($\mathcal{K}$) (i.e. the degree of the polynomial giving the time bound). Finally, he shows that for each non-monadic type t the hierarchy

$$Pr_k^t = \{ \mathcal{K} ; \mathcal{K} \text{ projectively defined by a formula whose definitional complexity is } k \}$$

is strictly increasing. We shall not go into more detail since these very important investigations are beyond the scope of the present book.

We now consider 1-projective classes. We show that 1-projective classes are not closed under complementation; unfortunately, this does not solve the P-NP problem.

3.5.12 Theorem. (Fagin 1975a, cf. Hájek 1975a) There is a 1-projective variety of type $\langle 2 \rangle$ whose complement is not 1-projective.

3.5.13 Remark. Each structure $\langle M, r \rangle$ of type $\langle 2 \rangle$ can be considered as a graph in the sense of graph theory : We call the elements of M vertices and pairs a,b such that $r(a,b) = 1$ edges of the graph. We assume the usual terminology concerning graphs.

We shall outline a proof of the fact that the class $\mathcal{K}_1$ of all (finite directed) disconnected graphs satisfies our theorem.

3.5.14 Extend the usual universe of sets by admitting the existence of proper semisets, i.e. nonsets that are subclasses of sets. (The notion of a semiset is due to Vopěnka, cf. Vopěnka and Hájek ; for a short survey of important facts about semisets see Hájek 1973 and/or Hájek 1972). In particular, assume that the semiset An of absolute natural numbers is a proper subsemiset of the set of all natural numbers. (Cf. [Čuda]; n is absolute if there is no semiset one-one mapping of n onto n + 1 .) $0 \in An$ and $(\forall n)(n \in An \rightarrow (n+1) \in An)$, hence if $An \neq \mathbb{N}$ then An is a proper semiset. Denote the theory of semisets with the above assumption by TSS. One has the following metamathematical result:

(Metatheorem) Let $\varphi(n)$ be a formula of set theory with one free variable ranging over natural numbers. If $(\forall n \in An)\, \varphi(n)$ is provable in TSS then $(\forall n)\, \varphi(n)$ is provable in set theory.

3.5.15 In particular, assume that formulae are coded by natural numbers in the usual manner. Call a variety $\mathcal{K}$ absolutely 1-projective if there is a $\varphi \in An$ 1-projectively defining $\mathcal{K}$ (i.e. $\mathcal{K}$ is 1-projectively defined by a short formula). By 3.5.14 , it suffices to prove in TSS that the **variety is not absolutely** 1-projective. Say that a semiset-mapping σ respects a set x if σ maps x onto a set (in general the image could be a proper semiset). Note that validity of short (i.e. absolute) sentences is preserved even by semiset isomophisms. Hence the following suffices for a variety $\mathcal{K}$ not to be absolutely 1-projective:

There is a structure $\underline{M} \in \mathcal{K}$ such that, for each short tuple $x_1, \ldots, x_k$ of subsets of $\underline{M}$, there is a σ mapping $\underline{M}$ isomorphically onto a set structure $\sigma(\underline{M}) \notin \mathcal{K}$ and respecting each $x_1, \ldots, x_k$. ($\underline{M}$ can be called critical for $\mathcal{K}$.)

3.5.16 We show that any elementary cycle of a long (non-absolute) length is critical for connected graphs. Let $\underline{G} = \langle G, R \rangle$ be such a cycle; let x be a subset of G; assume k = 1 for simplicity. For each $a \in G$ let a^+ be the immediate successor of a; let $a^{An} = \{a^+, a^{++}, \ldots \text{(An times)}\}$. There are a, b such that $a^{An} \cap b^{An} = \emptyset$ and for each $n \in An$, $a^{\overbrace{+\cdots\cdots+}^{}}_{n \text{ times}} \in x$ iff $b^{+\cdots\cdots+}_{n \text{ times}} \in x$.

Let $G' = G$ and let R' result from R by removing the edges $\langle a, a^+ \rangle$, $\langle b, b^+ \rangle$ and replacing them by $\langle a, b^+ \rangle$, $\langle b, a^+ \rangle$. Then $\underline{G}' = \langle G', R' \rangle$ decomposes into two elementary circuits and the mapping σ interchanging a^{An} and b^{An} and identical on the rest of G is a semiset isomorphism of $\underline{G}$, $\underline{G}'$ mapping x onto itself.

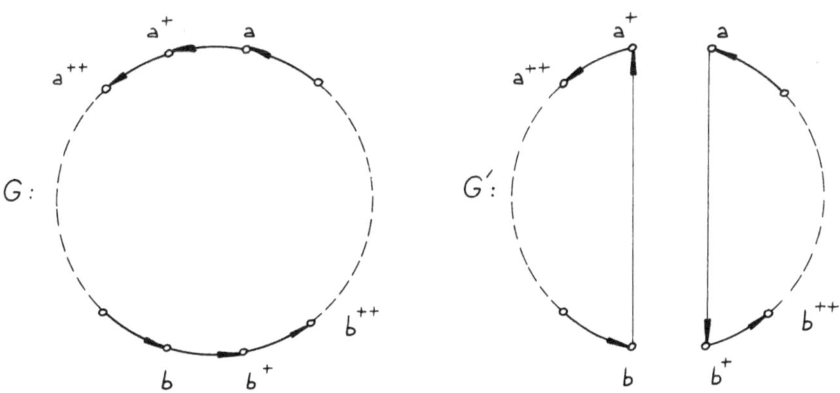

3.5.17 Remark. One can show that all the class $-\mathcal{K}_1$ is projective (even if not 1-projective). This also shows that, in the observational sense, there is a variety $\mathcal{K}$ such that both $\mathcal{K}$ and $-\mathcal{K}$ are projective but $\mathcal{K}$ is not elementary. (Cf. 3.5.6.) See also Problem (10).

The aim of the next ten paragraphs is to sketch a relatively rapid proof of Trachtenbrot's theorem:

3.5.18 Theorem. (Trachtenbrot 1950). There is a COPC which is not axiomatizable.

The proof we offer differs from the original proof and uses the notion of diophantine sets. We define necessary notions in a version useful for our purpose (Cf. Davis 1958).

3.5.19 Definition. (1) An n-ary polynomial is an arbitrary mapping $P(x_1, \ldots, x_n)$ of $\mathbb{N}^n$ into $\mathbb{N}$ having the following form:

$$\sum_{\substack{0 \le i_1 \le k_1 \\ \vdots \\ 0 \le i_n \le k_n}} a_{i_1} \cdots a_{i_n} x_1^{i_1} \cdots x_n^{i_n} \qquad (a_{i_1}, \ldots, a_{i_n} \in \mathbb{N})$$

(2) A set $A \subseteq \mathbb{N}$ is diophantine if there are polynomials $P(y, x_1, \ldots, x_n)$ and $Q(y, x_1, \ldots, x_n)$ such that

$$A = \{y; (\exists x_1, \ldots, x_n)\ P(y, x_1, \ldots, x_n) = Q(y, x_1, \ldots, x_n)\} .$$

(A is said to be a diophantine set corresponding to P, Q.)

3.5.20 Lemma. The set Pol_n of all n polynomials is the least set of mappings of $\mathbb{N}^n$ into $\mathbb{N}$ containing, for each $i = 1, \ldots, n$, the function $I_n^i(x_1, \ldots, x_n) = x_i$, for each $a \in \mathbb{N}$ the function $K_n^a(x_1, \ldots, x_n) = a$ and closed under sums and products of functions. **This is obvious.**

The following lemma presents the famous result due to Matiasevič and implies directly the unsolvability of Hilbert's 10th problem:

3.5.21 Lemma. A set $A \subseteq \mathbb{N}$ is recursively enumerable iff it is diophantine.

3.5.22 Corollary. There is a diophantine non-recursive set of integers.

3.5.23 Discussion. Now, we are going to describe a theory Ar whose finite models are exactly finite segments of natural numbers with the usual structure. The language will be as follows:

$=$ - equality predicate ,

$<$ - less-than predicate ,

Suc (-,-) (Suc(y,x) is read "y is the successor of x"),

Add (-,-,-) (Add (z,x,y) is read "z is the sum of x,y"),

Mult(-,-,-) (Mult (z,x,y) is read "z is the product of x,y").

It is easy to write axioms stating that Suc, Add, Mult describe partial functions, e.g., $(\mathrm{Suc}(y,x) \,\&\, \mathrm{Suc}(z,x)) \rightarrow y = z$ etc.: define $\mathrm{Least}(x) \Leftrightarrow \neg(\exists y)(y < x)$ and introduce formulae $\mathrm{num}_n(x)$ $(n \in \mathbb{N})$ by the following induction: $\mathrm{num}_o(x)$ is $\mathrm{Least}(x)$; $\mathrm{num}_{n+1}(x)$ is $(\exists y)(\mathrm{num}_n(y) \,\&\, \mathrm{Suc}(x,y))$.

Write finitely many axioms expressing the inductive behaviour of arithmetical operations, e.g.

$\mathrm{Suc}(y,x) \leftrightarrow x < y \,\&\, \neg(\exists z)(x < z \,\&\, z < y)$,

$\mathrm{Least}(y) \longrightarrow \mathrm{Add}(x,x,y)$;

$\mathrm{Add}(z,x,y) \,\&\, \mathrm{Suc}(\bar{y},y) \,\&\, \mathrm{Suc}(\bar{z},z) \longrightarrow \mathrm{Add}(\bar{z},x,\bar{y})$ etc.

The conjunction of the universal closures of these axioms is denoted by Ar. One proves the following lemma by induction using 3.5.20.

3.5.24 Lemma. For each polynomial $P(m_1, \ldots m_n)$ there is a formula $\pi(y,x_1, \ldots , x_n)$ such that the following are equivalent:

(i) $m = P(m_1, \ldots, m_n)$;

(ii) the formula

$$Ar \longrightarrow (\forall \underline{x})(\forall z)((num_{m_1}(x_1) \& \ldots \& num_{m_n}(x_n) \& num_m(y)) \longrightarrow \pi(y, x_1, \ldots, x_n)$$

is a tautology.

3.5.25 Definition and lemma. Let A be a non-recursive diophantine set, $A = \{n; (\exists \underline{m}) P(\underline{m}, n) = Q(\underline{m}, n)\}$; let $\pi(z, \underline{x}, y)$, $\varrho(z, \underline{x}, y)$ be the corresponding formulae satisfying 3.5.24. Then

(i) $n \in A$ iff $Ar \& (\exists \underline{x}, y, z)(num_n(y) \& \pi(z, \underline{x}, y) \& \varrho(z, \underline{x}, y))$

has a model, i.e.,

(ii) $n \notin A$ iff $Ar \rightarrow (\forall y)(num_n(y) \rightarrow \neg(\exists z, \underline{x})(\pi(z, \underline{x}, y) \& \varrho(z, \underline{x}, y)))$

is a tautology.

3.5.26 Corollary. The set of tautologies of the CPC with the present language is not recursively enumerable. (If it were recursively enumerable, then $\mathbb{N} - A$ would also be recursively enumerable which is not the case.) Hence, 3.5.18 follows.

To close the present section we shall consider a strengthening of classical MOPC's (with equality) by adding instead of equality a binary predicate $<$ (ordering predicate).

3.5.27 Definition. The classical monadic observational predicate calculus with ordering (CMOPC ($<$)) of type $\langle 1^k \rangle$ is the predicate calculus with k unary predicates $P_1, \ldots, P_k$, one binary predicate, the usual connectives, quantifiers $\forall$, $\exists$ and models $\underline{M} = \langle M, r, f_1, \ldots, f_k \rangle$ such that (i) M is finite,

(ii) the relation $<_r$ ($a <_r b$ iff $r(a,b) = 1$) is a linear ordering of M .

(This means: $\underline{M}$ is a model of CMOPC ($<$) in question if it satisfies (i), (ii).)

3.5.28 Remark. CMOPC's with ordering generalize CMPOC's with equality since = can be defined by:
$x = y$ iff $\neg(x < y) \& (y < x)$.

The definition is motivated by the idea of the model (data) as a sequence of objects with properties (a linearly ordered set rather than a set). Our aim is to show that each CMOPC (<) is decidable. We reduce the decision problem for a CMOPC (<) to the decision problem for the weak monadic second order successor arithmetic W2SA. This theory can be described as follows:

One has variables of two sets, x, y, ... (number variables) and X, Y, ... (set variables). One has two binary predicates Suc and $\in$; Suc may be followed by two number variables and $\in$ may be followed by one number variable and one set variable (one writes $x \in X$ instead of $\in(x,X)$). Formulas are built up using the usual connectives and the quantifiers $\forall$, $\exists$; the quantifiers may be applied both to number variables and to set variables.

The canonical model is as follows: Number variables vary over natural numbers, set variables vary over finite sets of natural numbers, Suc(y,x) means that y is the successor of x and $x \in X$ means that x is an element of X. Let True be the set of all sentences true in the canonical model. Büchi showed that True is a recursive set (cf. Siefkes).

One defines equality putting $x = y \Leftrightarrow (\forall X)(x \in X \leftrightarrow y \in X)$, segments putting $Seg(X) \Leftrightarrow (\forall x,y)((y \in X \,\&\, Suc(y,x)) \rightarrow x \in X)$, and ordering putting
$x < y \Leftrightarrow (\exists X)(Seg(X) \,\&\, x \in X \,\&\, y \notin X)$.

3.5.29 Construction. Let $\mathcal{P}_k$ be the CMOPC (<) of type $\langle 1^k \rangle$. With each sentence φ of $\mathcal{P}_k$ we associate effectively a formula $\varphi^*(z, X_1, \ldots, X_k)$ of W2SA such that φ is a tautology of $\mathcal{P}_k$ iff the formula

$(\forall z)(\forall X_1) \ldots (\forall X_k) \quad \varphi^*(z, X_1, \ldots, X_k)$ is in True. We define φ^* inductively for all formulae assuming that the variables of $\mathcal{P}_k$ coincide with the number variables of W2SA distinct from z . The idea is that if we fix a natural number $x = n$ and finite sets $A_1, \ldots, A_n$ of natural numbers, then the structure $\langle M, r, f_1, \ldots, f_n \rangle$ such that [(i) $M = \{o, \ldots, n\}$, (ii) for $a, b \leq n$, $r(a,b) = 1$ iff $a < b$, (iii) for $a \leq n$, $f_i(a) = 1$ iff $a \in A_i$] is a model of $\mathcal{P}_k$ and each model of $\mathcal{P}_k$ is isomorphic to a model of this kind.

$(x < y)^*$ is $x < y$;

$(P_i(x))^*$ is $x \in X_i$;

$(\varphi \,\&\, \psi)^*$ is $\varphi^* \,\&\, \psi^*$;

$(\neg\varphi)^*$ is $\neg\varphi^*$;

$((\forall x)\varphi)^*$ is $(\forall x)(x < z \rightarrow \varphi^*)$;

$((\exists x)\varphi)^*$ is $(\exists x)(x < z \,\&\, \varphi^*)$.

3.5.30 Lemma. If φ is a sentence of $\mathcal{P}_k$, then φ is a tautology of $\mathcal{P}_k$ iff the sentence

$$(\forall z)\ (\forall X_1) \ldots (\forall X_k)\ \varphi^*(z, X_1, \ldots, X_k)$$

is true in the canonical model of W2SA.

3.5.31 Corollary. $\mathcal{P}_k$ is decidable.

3.5.32 Remark. For results concerning the computational complexity of particular decidable theories see [Fisher and Rabin], [Meyer], [Rackoff].

3.5.33 Key words: varietes of models, elementary, projective and 1-projective varietes, NP classes of languages, CMOPC's with ordering.

PROBLEMS AND SUPPLEMENTS TO CHAPTER III.

(1) Let $Asf_{q_1}(\langle M, f_1, f_2\rangle) = 1$ iff there are at least three objects in M ; let $Asf_{q_2}(\langle M, f_1, f_2\rangle) = 1$ iff there are at least three distinct cards (i.e., there are $o_1, o_2, o_3 \in M$ such that $C_{\underline{M}}(o_1), C_{\underline{M}}(o_2), C_{\underline{M}}(o_3)$ are mutually distinct). Let $\mathcal{P}$ be a COMPC with at least two predicates, **without** equality, and let $\mathcal{P}_i$ be the extension of $\mathcal{P}$ by adding q_i. Show that q_1 is not definable in $\mathcal{P}_1$, but q_2 is definable in $\mathcal{P}_2$.

(2) Prove the following consequence of the Stability theorem 3.1.22: Under the assumption of 3.1.22, $\underline{M} \models \Phi$ iff for each $\underline{M}_o \subseteq \underline{M}$ with at most $p \cdot 2^n$ elements there is an $\underline{M}_1$ between $\underline{M}_o, \underline{M}$, $\underline{M}_1$ with at most $p.\ 2^{n+1}$ elements and such that $\underline{M}_1 \models \Phi$.

(3) Show that if the condition of Tharp's theorem is satisfied and the m constructed from the condition is given then one can effectively construct a sentence defining the quantifier q. Hint: Let $d_{\underline{M}_o}$ be a sentence such that, for each $\underline{M}$, $\underline{M} \models d_{\underline{M}_o}$ iff $\underline{M}_o$ can be isomorphically embedded into $\underline{M}$ (i.e., each card has frequency in $\underline{M}$ at least as large as in $\underline{M}_o$). Consider the sentence

$$\bigwedge d_{\underline{M}_o^i} \longrightarrow \bigvee d_{\underline{M}_o^{ij}}$$

where $\underline{M}_o^i$ varies over all models with at most m elements and $\underline{M}_o^{ij}$ varies over all submodels of M with at most 2m elements. (Use the considerations of (2).)

(4) Let $P_1, P_2, \ldots$ be a countably infinite sequence of unary predicates, let $\mathcal{R}$ be a quantifier of type <1> and $\exists^{!1}$ a quantifier of type <1>. Denote by $\mathcal{P}_n$ the MOPC with $P_1, \ldots, P_n$ and with $\mathcal{R}$, $\exists^{!1}$ where the quantifiers are interpreted as in the proof of 3.1.34. Let $Taut_n$ be the set of all tautologies of $\mathcal{P}_n$. Show that $\bigcup_n Taut_n$ is not recursive (and not

recursively enumerable). Hence in 3.1.34 finitely many predicates and infinitely many quantifiers may be replaced by infinitely many predicates and finitely many quantifiers.

Proceed analogously to the proof of 3.1.34 .

(5) Let $P_1, P_2, \ldots$ be as above and let $\sim$ be a quantifier of type $\langle 1,1 \rangle$. Let L_n be the predicate language with $P_1, \ldots, P_n$ and $\sim$. We shall show that

$$\mathrm{Sch} = \{\varphi ; \quad \varphi \text{ is a schema of associational tautologies in some } L_n\}$$

is recursive.

Remember 3.2.27 - 3.2.30. In the sequel, K_n denotes the set of all n-cards, $\mathcal{R}_n$ denotes the set of all 4-partitions of K_n, $\mathcal{T}_n$ is the power set of $\mathcal{R}_n$, etc.

(a) Let $T \in \mathcal{T}_n$. T is satisfiable iff there is a non-trivial natural number valued measure μ on the power set of K_n such that

(*) $R \in T$ and $R' \notin T$ implies $\mu(R) \not\leq \mu(R')$ for each $R, R' \in \mathcal{R}_n$,

(b) Corollary: The set of all satisfiable sets of partitions is recursively enumerable.

(c) The same as (a) with "natural number valued" replaced by "real valued strictly positive".

(d) A linear quasiordering $\preceq$ of $\mathcal{P}(K_n)$ is <u>realizable</u> if there is a real valued strictly positive measure μ on $\mathcal{P}(K_n)$ such that $X \preceq Y$ iff $\mu(X) \leq \mu(Y)$. A $T \in \mathcal{T}_n$ is satisfiable iff there is a realizable quasiordering $\preceq$ of K_n such that

(**) for each $R = \langle A,B,C,D \rangle \in T$ and $R' = \langle A', B', C', D' \rangle \in \mathcal{R}_n - T$ we have $(A \prec A'$ or $B \succ B'$ or $C \succ C'$ or $D \prec D')$.

(e) Let $\preceq$ be a linear quasiordering of $\mathcal{P}(K_n)$ and let $\underline{A} = \langle A_1, \ldots, A_k \rangle$, $\underline{B} = \langle B_1, \ldots, B_k \rangle$ be two k-tuples of subsets of K_n . Let $\Sigma\,\underline{A} = \Sigma\,\underline{B}$ mean that every element of K_n belongs to the same number of A_i' s as B_i' s. Call A, B an unwanted pair (of sequences) if $\Sigma \underline{A} = \Sigma \underline{B}$, $A_1 \prec B_1$ and $A_i \preceq B_i$ for $i = 2, \ldots, k$. Evidently, each pair $\underline{A}$, $\underline{B}$ is a finite object and we can effectively decide whether it is unwanted or not.

Lemma (Scott 1964). A linear quasiordering of $\mathcal{P}(K_n)$ is realizable by a real valued strictly positive measure iff there is no unwanted pair.

(f) $T \in \mathcal{T}_n$ is not satisfiable iff for each linear quasiordering of K_n, there is an unwanted sequence; hence the set of all unsatisfiable sets of partitions is recursively enumerable and the assertion follows.

(6) Theorem. Let $\sim$ be a saturable associational quantifier, let Π_1, Π_2 be two disjoint sets of unary predicates, let $\varphi, \varphi', \psi, \psi'$, be designated open formulae such that φ, φ' contain only predicates from Π_1 and ψ, ψ' only from Π_2. Suppose that each of the formulas is factual, i.e., it is satisfied by a card and its negation is also satisfied by a card. If $\varphi \sim \psi$ logically implies $\varphi' \sim \psi'$
then either ($\varphi \Leftrightarrow \varphi'$ and $\psi \Leftrightarrow \psi'$)
or ($\varphi \Leftrightarrow \neg\varphi'$ and $\psi \Leftrightarrow \neg\psi'$)
($\Leftrightarrow$ stands for logical equivalence).

Hint:(Lemma 1.) If $\|\varphi\|[u] = \|\psi\|[u]$ then $\|\varphi'\|[u] = \|\psi'\|[u]$.

(Lemma 2.) If $\varphi \Leftrightarrow \varphi'$ then $\psi \Leftrightarrow \psi'$.

(Lemma 3.) If $\varphi \Leftrightarrow \neg\varphi'$ then $\psi \Leftrightarrow \neg\psi'$.

(Lemma 4.) If there is a card $\underline{u}$ with $\|\varphi \,\&\, \varphi'\|[\underline{u}] = 1$ then φ logically implies φ'. If there is a card $\underline{u}$ with $\|\neg\varphi \,\&\, \neg\varphi'\|\ \underline{u} = 1$ then φ' logically implies φ.

(Represent cards as pairs $\langle \underline{u}_1, \underline{u}_2 \rangle$ where $\underline{u}_1$ evaluates predicates from Π_1 and $\underline{u}_2$ those from Π_2 and use the fact that $\|\varphi\|[\langle \underline{u}_1, \underline{u}_2 \rangle]$ depends only on $\underline{u}_1$ etc.)

(7) Let $\mathcal{F}$ be a monadic $\times$-predicate calculus with finitely many predicates and quantifiers (without equality). Then $\mathcal{F}$ is strongly decidable.

Hint (Bendová 1975): Extend $\mathcal{F}$ by adding a nullary junctor $\underline{\times}$ with the constant value $\times$. Imitate the proof of 3.1.30 to show that (in the extended calculus) each sentence is logically equivalent to a Boolean combination of pure prenex formulae. Conclude that there is a finite set S of sentences such that each sentence is logically equivalent to a sentence from S.

(8) There is an observational monadic $\times$-predicate calculus with infinitely many quantifiers and without equality which is $\{1\}$-undecidable.

(9) What is the best possible complexity of a decision procedure for schemes of associational tautologies? This is an open problem.

(10) Show that the variety of all planar graphs is not 1-projective (Hájek 1975a).

Hint: Remember that a graph containing a subgraph of the form:

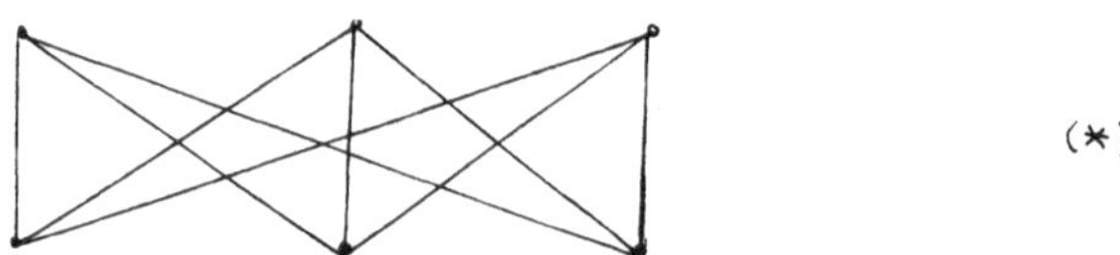

(*)

is not planar. For each n, let Z_n be the graph

where the vertices 0,1 are connected by three paths of length n: the left, middle and right one. Let K_n be the direct sum of n disjoint copies of Z_n. K_n is planar.

Claim: If n is large (non-absolute) then K_n is critical for planar graphs.

Proof. Let x be a subset of the field of K_n. Just one subset instead of several subsets for simplicity. If Z is a copy of Z_n in K_n and if r and ℓ are its right and left path then associate with each vertex a on r (on ℓ) which is far from 1 its x-type t(a) - the characteristic function of x on a^{An}. There are three distinct copies Z_n^1, Z_n^2, Z_n^3 of Z_n in K_n and elements $a_i \in \ell(Z_n^i)$,

$b_i \in r(Z_n^i)$ such that $t(a_1) = t(a_2) = t(a_3)$ and $t(b_1) = t(b_2) = t(b_3)$.

Remove edges $a_i \longrightarrow a_i^+$ and $b_i \longrightarrow b_i^+$; instead, add edges

$a_1 \longrightarrow a_3^+$, $b_1 \longrightarrow b_2^+$, $a_2 \longrightarrow a_1^+$, $b_2 \longrightarrow b_3^+$, $a_3 \longrightarrow a_2^+$, $b_3 \longrightarrow b_1^+$.

Then the modified graph K_n' has the form (*). One constructs a semiset isomorphism mapping K_n onto K_n' which is the identity outside $Z_n^1 \cup Z_n^2 \cup Z_n^3$ and maps x onto itself exactly as in 3.5.16.

Chapter IV. Logical Foundations of Computational Statistics

"Statistical data analysis and hypothesis testing does not involve logical deductive reasoning, as the words "inference" and "mathematical statistics" may suggest, but stochastic inductive reasoning. Especially **when done with the** computer, all problems inherent in inductive reasoning arise" (Van Reeken 1971).

Having this in mind, we shall use the term "computational statistics" for a theory of mechanized statistical inductive inference. To have clear and exact foundations of such a theory, one has to answer the following questions:

(1) What is the relation of probabilistic notions of Mathematical Statistics to the notions concerning computability?

(2) Can one formulate an exact logical framework for statistical procedures, in particular, for mechanized statistical procedures?

In sections 2 and 3 of the present chapter such logical foundations of computational statistics are elaborated. Some generalizations are presented in Chapter V, Section 1 . In Section 4 and 5 the results of our investigation are applied to predicate (two-valued) calculi. Definitions of some important particular statistical quantifiers (quantifiers based on statistical procedures) in observational predicate calculi are obtained. We also exhibit some useful logical properties of such quantifiers.

Statistical questions connected with the logic of suggestion will be considered in Chapter VIII.

IV.1 Preliminaries

4.1.1 We are interested in the exact logical and mathematical description of the properties of inference rules which bridge the gap between theoretical and observational sentences, speaking about theoretical and observational models respectively. Theoretical models that we have in mind now are random structures (cf. 2.4.5). Observational models can be viewed as "samples" from theoretical models. Hence each observational procedure, i.e. procedure operating with observational data has two sorts of properties:

(1) probabilistic

(2) logical and computational.

In Mathematical Statistics statistical procedures are treated purely analytically. But, in computational practice, one has "to learn what is computationally feasible as distinct from analytically possible" (Freiberger and Grenader).

The task is to study the interaction between the analytical and computational approach. Little, in fact, seems to have been done in this direction (c.f. Freiberger and Grenader). A practically orientated attempt has been made by the above cited authors; in their book they open a promising area of research.

We want to present a theoretical framework relating probabilistic and computational properties of statistical procedures. Therefore we must expound some probabilistic properties of those procedures. In the present section, we summarize notions of probability theory to the extent necessary for further investigations.

4.1.2 We assume that the reader is familiar with some basic ideas of measure and probability theory, but it seems to be useful to go over some basic definitions. The defined notions and basic facts about them will be freely used in the rest of the book.

We use the classical approach to probability theory found in the very readable introduction on the graduate level by Burril [1972]. Any other introduction to the Kolmogorov probability theory can also serve as a source of information.

4.1.3 The first basic notion is the notion of a field of sets, i.e. a class of subsets of a set, containing the empty set and closed under complement and finite union. A σ-field is required also to be closed under countable union.
The second notion is the notion of a measure; a mapping μ of a field $\mathcal{R}$ into $\mathbb{R}^* = \mathbb{R} \cup \{+\infty\}$ is called a measure if it is non-negative and additive (i.e. if $S_1, S_2 \in \mathcal{R}$, $S_1 \cap S_2 = \emptyset$ then $\mu(S_1 \cup S_2) = \mu(S_1) + \mu(S_2)$.

Suppose that $\mathcal{R}$ is the set of all subsets of a finite set K. Then a measure on $\mathcal{R}$ is uniquely determined by its values on the one-element subsets of K. In fact, each mapping $g: K \to \mathbb{R}$ determines a unique measure μ_g such that $\mu_g(\{u\}) = g(u)$. (For an $A \subseteq K$ we then have $\mu_g(A) = \sum_{u \in A} g(u)$ by additivity.) Such a measure is a generalized counting measure where different elements can have different weights.

4.1.4 A measure μ on a σ-field $\mathcal{R}$ is σ-additive if for each countable class $\mathcal{R}_o \subseteq \mathcal{R}$ of pairwise disjoint sets we have $\mu(\bigcup_{S \in \mathcal{R}_o} S) = \sum_{S \in \mathcal{R}_o} \mu(S)$.

The Borel field $\mathcal{B}$ is the minimal σ-field of sets of real numbers containing the class of all open subsets of $\mathbb{R}$. It can be equivalently characterized as the minimal σ-field of sets of real numbers containing all the half-open intervals.

Consider now an abstract set Σ and a σ-field $\mathcal{R} \subseteq \mathcal{P}(\Sigma)$. A real valued function f with domain Σ is called measurable if, for each $B \in \mathcal{B}$, $f^{-1}(B) \in \mathcal{R}$. (Naturally, $f^{-1}(B) = \{\sigma \in \Sigma \ ; \ f(\sigma) \in B\}$.)

4.1.5 We are now going to formalize the notion of probability. This can be done in many different more or less intuitive ways (see Fine 1973). We use the Kolomogorov axiomatic system, which makes use of measure theory and which is a rather useful frame for a mathematical theory of the statistical inference we are interested in.

Remember that probability is considered on a pair $\langle \Sigma, \mathcal{R} \rangle$ where Σ is a non-empty set and $\mathcal{R}$ is a σ-field, $\mathcal{R} \subseteq \mathcal{P}(\Sigma)$ (cf. 2.1.7).

Now, a σ-additive measure P on $\langle \Sigma, \mathcal{R} \rangle$ is called a <u>probability measure</u> if $P(\Sigma) = 1$. Elements of $\mathcal{R}$ are called <u>random events</u>; if $E \in \mathcal{R}$ then the number P(E) is called the probability of E. The triple $\langle \Sigma, \mathcal{R}, P \rangle$ is usually called a <u>probability space.</u>

4.1.6 Consider a probability space $\langle \Sigma, \mathcal{R}, P \rangle$; any measurable function from Σ to $\mathbb{R}$ is called a <u>random variate</u> (more frequently the term "random variable" is used; we use the term "variate" which was introduced by M.G.Kendall [1951]).

Now, if we have a random variate ϑ and a probability measure P on $\langle \Sigma, \mathcal{R} \rangle$ we obtain a probability measure on $\langle \mathbb{R}, \mathcal{B} \rangle$: For each $A \in \mathcal{B}$, $P_\vartheta(A) = P(\vartheta^{-1}(A))$. This measure is called the measure <u>induced</u> by a random variate; similarly the σ-field $\mathcal{R}_\vartheta = \{E \in \mathcal{R} ; \vartheta(E) \in \mathcal{B}\}$ is called the σ-field induced by ϑ.

(The notion of a random variate can be naturally generalized to a notion of an n -dimensional random variate, where n is a natural number. Consider functions from Σ to $\mathbb{R}^n$.)

The expectation and variance of a random variate are defined as integrals $E\vartheta = \int \vartheta \, dP$ and $VAR\,\vartheta = E(\vartheta - E\vartheta)^2$ respectively.

4.1.7 Probabilistic properties of a random variate are fully described by its <u>distribution function</u>. The distribution function of a random variate is defined as follows: for each $x \in \mathbb{R}$, $D(x) = P(\vartheta^{-1}(-\infty, x))$. Each distribution function has the following properties: it is non-negative, non-decreasing, continuous from the left, $\lim_{x \to -\infty} D(x) = 0$ and $\lim_{x \to +\infty} D(x) = 1$.

Each distribution function uniquely defines a probability measure on B. If we consider a variate ϑ then this measure is exactly P_ϑ. This is the reason why the distribution function fully describes the probabilistic properties of a random variate.

4.1.8 Consider a random variate ϑ which maps Σ into a finite set $\{x_1, \dots, x_n\}$, hence $P(\vartheta^{-1}(\{x_1, \dots, x_n\})) = 1$ (such a variate can be called <u>discrete</u>); suppose $p_i = P(\vartheta^{-1}(\{x_i\})) > 0$. The distribution function D_ϑ is a step function and can be described as follows: $D_\vartheta(x) = \sum_{x_i < x} p_i$.

4.1.9 The investigation of sequences of independent experiments leads to the following notion of stochastical independence:

A finite sequence $\vartheta_1, \ldots, \vartheta_n$ of random variates is called <u>stochastically independent</u> if, for each $E_1 \in \mathcal{R}_{\vartheta_1}, \ldots, E_n \in \mathcal{R}_{\vartheta_n}$,

$$P(E_1 \cap \ldots \cap E_n) = P(E_1) \ldots P(E_n)$$

(see 4.1.6 for the definition of $\mathcal{R}_{\vartheta_i}$; an equivalent condition reads: For each $B_1, \ldots, B_n$ Borel,

$$P(\vartheta_1 \in B_1 \,\&\, \ldots \,\&\, \vartheta_n \in B_n) = \prod_{i=1}^{n} P(\vartheta_i \in B_i)).$$

An infinite sequence $\vartheta_1, \vartheta_2, \ldots$

is called stochastically independent if each finite subsequence of $\vartheta_1, \vartheta_2, \ldots$ is stochastically independent.

We can define the <u>joint distribution function</u> of a sequence $\vartheta_1, \ldots, \vartheta_n$ (i.e., of an n-dimensional random variate) by the equation:

$$D_{\vartheta_1, \ldots, \vartheta_n}(x_1, \ldots, x_n) = P(\vartheta_1^{-1}((-\infty, x_1)) \ldots \vartheta_n^{-1}((-\infty, x_n)).$$

It is easy to prove that the sequence $\vartheta_1, \ldots, \vartheta_n$ is stochastically independent iff

$$D_{\vartheta_1, \ldots, \vartheta_n} = D_{\vartheta_1} D_{\vartheta_2} \ldots D_{\vartheta_n}.$$

4.1.10 <u>Key words</u>: measure, probability measure, random variate, distribution function, joint distribution function, stochastical independence.

IV.2 The concept of statistics

We now try to construct a framework for statistical inference as a particular case of inductive reasoning (cf. 1.3.1). We shall be more specific on theoretical sentences; our theoretical sentences will have semantics related to random structures (cf. 2.1.7). Inductive inference rules that will be studied will be called statistical inference rules since the argument for their reasonability will be based on their statistical properties. The theory we are going to develop is a metatheory of statistical reasoning.

4.2.1 Definition. Let $\underline{\Sigma} = \langle \Sigma, \mathcal{R}, P \rangle$ be a probability space and let V be a set of real numbers. Let $\underline{U} = \langle U, Q_1, \ldots, Q_n \rangle$ be a $\underline{\Sigma}$-random V-structure of type $\langle \underbrace{1, \ldots, 1}_{n\text{-times}} \rangle$ (cf. 2.4.5). $\underline{U}$ is regular if the following conditions hold:

(0) U is a recursive, possibly infinite, set of natural numbers.

(1) Each $Q(o, .)$, as a function from Σ to V, is a random variate.

(2) For any sequence $o_1, \ldots, o_m$ of elements from U the sequence of n-dimensional random variates

$$\{\langle Q_1(o_i .), \ldots, Q_n(o_i, .)\rangle\}_{i=1,\ldots,m}$$

is stochastically independent.

For each $o \in U$, the variate $Q_i(o, .)$ is denoted by $\vartheta_{i\,o}$.

4.2.2 Discussion. The notion of a regular random structure is our formalization of the informal notion "the theoretical universe of discourse". The condition 0 is technical; it is useful for considerations concerning computability. The second condition, (1) enables us to define probabilities concerning different outcomes of the experiments; the fact that the $Q_i(o, .)$ are random variates makes it possible to use induced probability measures (cf. 4.1.6). The last condition is adequate for many real situations in which the properties of one object are independent of the other (e.g., in series of independent experiments).

Note that there are situations in which this condition is not satisfied. We restrict ourselves to structures satisfying 2 for the sake of simplicity. Note that our formalism is related to the formalism sketched by Suppes [1962].

4.2.3 Definition. The distribution function on the n-dimensional random variate $\underline{\vartheta}_o = \langle \vartheta_{1o}, \ldots, \vartheta_{no} \rangle$ will be denoted by $D_{\vartheta_{1o}, \ldots, \vartheta_{no}}$ or $D_{\underline{\vartheta}_o}$. (It is usually called the joint distribution function of random variates $\vartheta_{1o}, \ldots, \vartheta_{no}$; by definition, it is a function from $\mathbb{R}^n$ to $[0,1]$.

4.2.4 Definition. A regular Σ-random V-structure $\underline{U}$ is d-homogeneous (distribution homogeneous) if the joint distribution function $D_{\vartheta_{1o}, \ldots, \vartheta_{no}}$ is independent of o (i.e., for any o_1, o_2, $D_{\underline{\vartheta}_{o_1}} = D_{\underline{\vartheta}_{o_2}}$). Then we denote $D_{\vartheta_{1o}, \ldots, \vartheta_{no}}$ as $D^{\underline{U}}$.

4.2.5 Remark. (1) If a regular Σ-random V-structure is d-homogeneous we can say that all objects in U are equivalent with respect to the probabilistic properties of random quantities. Hence, probabilistic conclusions based on finite subsets of U will be independent of the particular choice of a finite subset. But it then may be dependent on its cardinality.

(2) Consider a sequence $o_1, \ldots, o_m$ of objects. Under the condition of d-homogeneity and regularity, sequences of the form $\langle \vartheta_{o_1}, \ldots, \vartheta_{o_m} \rangle$ are usually called sequences of independent identically distributed (i.i.d.) random variates.

4.2.6 Remark. A sequence of objects $o = \langle o_1, \ldots, o_m \rangle$ generates an $n \times m$-dimensional random variate

$$\underline{\vartheta} = \langle \vartheta_{1o_1}, \ldots, \vartheta_{no_1}, \vartheta_{1o_1}, \vartheta_{1o_2}, \ldots, \vartheta_{no_m} \rangle .$$

Under the assumption of d-homogeneity, the joint distribution function of this random variate is determined by the n-dimensional joint distribution function of the random variate $\underline{\vartheta}_o = \langle \vartheta_{1o}, \ldots, \vartheta_{no} \rangle$, where o is an arbitrary object from U. (I.e.,

$$D_{\underline{\vartheta}}(x_{11}, \ldots, x_{n1}, x_{12}, \ldots, x_{nm}) = \prod_{i=1}^{m} D_{\underline{\vartheta}_o}(x_{1i}, \ldots, x_{ni}).$$

Thus, we see that the joint distribution function is independent of o (and thus of $M = \{o_1, \ldots, o_m\}$). In the remainder of Chapter IV we shall restrict ourselves to d-homogeneous structures (for the sake of convenience only).

4.2.7 Discussion and Definition. Let U be a regular random structure of type $t = \langle 1, \ldots, 1 \rangle$. Let M be a finite subset of U (a sample) and let σ be a random state. Remember the definition 2.1.6 of the structure $\underline{M}_\sigma$ determined by M and σ. Structures of the form $\underline{M}_\sigma$ are finite V-structures, (V-structures with finite domain).

Let $\underline{M} \in \mathfrak{M}^V$ iff $\underline{M}$ is a V-structure of type t and the domain of $\underline{M}$ is a set of natural numbers.(2.1.4) We shall pay much attention to mappings $f: \mathfrak{M}^V \to V$. In general it has no meaning to ask whether such an f is recursive since $\mathfrak{M}^V$ can be uncountable and therefore its elements cannot be coded by words. The situation is clear if $V = \mathbb{Q}$ (rationals); it is obvious how to encode $\mathfrak{M}^{\mathbb{Q}}$ and what we mean by saying that a function $f: \mathfrak{M}^{\mathbb{Q}} \to \mathbb{Q}$ is recursive. More generally, if V is a recursive set of rationals then we call $f: \mathfrak{M}^V \to V$ recursive if it is a restriction of a recursive mapping $\hat{f}: \mathfrak{M}^{\mathbb{Q}} \to \mathbb{Q}$.

In general (V is an arbitrary set of reals) we shall work with rational elements of $\mathfrak{M}^V$ (i.e. elements of $\mathfrak{M}^{V \cap \mathbb{Q}}$) as approximations of structures from $\mathfrak{M}^V$. This is justified by the fact that if an $\underline{M}$ is the result of some measurement then the numbers we are dealing with are rational.

We have two requirements: (a) each structure from $\mathfrak{M}^V$ should be approximable by structures from $\mathfrak{M}^{V \cap \mathbb{Q}}$; (b) $V \cap \mathbb{Q}$ should be a recursive set of rationals. (Then having a function $f: \mathfrak{M}^V$ we can ask whether its restriction to $\mathfrak{M}^{V \cap \mathbb{Q}}$ is recursive in the above sense .) This leads to the notion of a regular set of values.

4.2.8 Definition. (1) (Auxiliary.) Let $V \subseteq \mathbb{R}$ and $x \in \mathbb{R}$. We call x a boundary point of V if each open interval containing x intersects both V and $\mathbb{R} - V$. (2) A set V is a regular set of values if (a) all boundary points of V are rational and (b) the set $V \cap \mathbb{Q}$ is a recursive set of rationals.

4.2.9 Remark. Assumption (a) means that if x is irrational then either a whole open interval containing x is in V (hence, all sufficiently close rational approximations of x are in V) or such an interval is in $\mathbb{R} - V$.

Examples of regular sets: $\mathbb{N}$, $\mathbb{R}$, intervals of an arbitrary kind with rational end-points, finite unions of such intervals, etc. Examples of non-regular sets: $\mathbb{Q}$, intervals with irrational end-points, Cantor's discontinuum.

4.2.10 Theorem.

(1) Regular sets form a field of sets.

(2) If V is a regular set then $V \cap \mathbb{Q}$ is dense in V, i.e., if $x \in V$ then each open interval containing x contains some rational elements of V.

(3) If V_1, V_2 are regular then $V_1 \neq V_2$ implies $V_1 \cap \mathbb{Q} \neq V_2 \cap \mathbb{Q}$.

(4) Each regular set is Borel.

Proof. (1) Denote the system of all regular subsets of $\mathbb{R}$ by $\mathcal{A}$. Then $\mathbb{R} \in \mathcal{A}$. If $X \in \mathcal{A}$ then its complement X^c has only rational boundary points and $\mathbb{Q} \cap X^c = \mathbb{Q} - (\mathbb{Q} \cap X)$ is recursive. This is similar for the union. (2) Note that each irrational point of a regular set X is an interior point of X. (3) Easy from (2). (4) Each regular set X can be decomposed as follows: $X = (X - \mathbb{Q}) \cup (X \cap \mathbb{Q})$; $X - \mathbb{Q}$ is open and hence Borel, $X \cap \mathbb{Q}$ is clearly Borel.

4.2.11 Example and Discussion. In accordance with Section 4 of Chapter II we consider theoretical sentences, i.e. sentences of a theoretical function calculus. For simplicity, we restrict ourselves to two-valued theoretical sentences, i.e. theoretical sentences Φ such that, for each random structure $\underline{U}$ (of the appropriate type) either $\|\Phi\|_{\underline{U}} = 1$ or $\|\Phi\|_{\underline{U}} = 0$. As usual, we write $\underline{U} \models \Phi$ for $\|\Phi\|_{\underline{U}} = 1$.

We can now give an example of statistical inference. Roughly, the inference has the following form: We have two theoretical sentences Φ and Ψ; we have accepted Φ (and called Φ the frame assumptions) and we ask whether we should accept Ψ. To decide this question we first fix a set $V_0 \subseteq V$ of designated values and a function f associating with each structure $\underline{M}_\sigma$ ($M \in \mathcal{P}_{fin}(U), \sigma \in \Sigma$) a value $f(\underline{M}_\sigma) \in V$. Then we make observations (get a particular structure $\underline{M}_\sigma$) and compute $f(\underline{M}_\sigma)$; if $f(\underline{M}_\sigma) \in V_o$ we accept Ψ (and if $f(\underline{M}_\sigma) \notin V_o$ we do not claim anything as concerns Ψ).

This procedure is justified in statistics by choosing f and V_o such that the following holds: For each Σ-random V-structure $\underline{U}$, satisfying Φ, if $\underline{U} \not\models \Psi$ then the probability $P(\{\sigma ; f(\underline{M}_\sigma) \in V_o\})$ is small (say, less than 0.05 or whatever value we wish). Hence assuming $\underline{U} \models \Phi$ and verifying $f(\underline{M}_\sigma) \in V_o$ for our observed $\underline{M}_\sigma$, if Ψ were not true in $\underline{U}$ then our observation $\underline{M}_\sigma$ would be very improbable (since $f(\underline{M}_\sigma) \in V_o$ would be improbable). Hence, we accept Ψ.

Three very important questions arise: (a) Is the probability $P(\{\sigma ; f(\underline{M}_\sigma) \in V_o\})$ well-defined (at least under the condition $\underline{U} \models \Phi$ and $\underline{U} \models \Psi$) ?

(b) How is our reasoning affected by the fact that our observation is approximate, i.e., that we restrict ourselves to rational structures?

(c) Can we really compute $f(\underline{M}_\sigma)$, i.e., is f computable in some sense?

Note also the following paradox. Let f_1, f_2 be two functions from $\mathfrak{M}^V_M$ into V. Let $f_1 \upharpoonright \mathfrak{M}^{V \cap Q}_M = 1$, $f_2 \upharpoonright \mathfrak{M}^{V \cap Q}_M = 0$ and $f_1 \upharpoonright \mathfrak{M}^{V - Q}_M = f_2 \upharpoonright \mathfrak{M}^{V - Q}_M$. Moreover, let $V_o = \{1\}$. Under some conditions (continuity of distributions on $\mathfrak{M}^V_M$) both these functions can have the same probabilistic properties, i.e. they both can fulfill the above rationality criterion (and some optimality conditions; see below). But using f_1 we accept Ψ in every case and using f_2 we never accept Ψ.

How can we prevent this situation?

These questions lead us to some further assumptions concerning f and V_o formulated below. First, we need some auxiliary definitions and notations.

4.2.12 Definition. (1) Consider a finite set M of natural numbers and let $\mathfrak{M}^V_M$ be the set of all V-structures of type $\langle 1^n \rangle$ with field M. In particular, note that each $\underline{M}_\sigma$ ($\sigma \in \Sigma$) is in $\mathfrak{M}^V_M$. Convert $\mathfrak{M}^V_M$ into a metric space,putting, for $\underline{M}_1 = \langle M, f_1, \ldots, f_n \rangle$ and $\underline{M}_2 = \langle M, g_1, \ldots, g_n \rangle$,

$\varrho(\underline{M}_1, \underline{M}_2) = \max\{ |f_i(o) - g_i(o)|; \; o \in M, i = 1, \ldots, n \}$

It is routine to show that ϱ is a metric.

4.2.13 Definition. A mapping $f: \mathcal{M}^V \to V$ is a continuous computable statistic (cc-statistic; or, if no confusion can arise, only statistic) if the following conditions hold:

(a) f is invariant under isomorphism, i.e., if $\underline{M}_1, \underline{M}_2$ are isomorphic then $f(\underline{M}_1) = f(\underline{M}_2)$.

(b) For each M, the function $f \restriction \mathcal{M}^V_M$ is continuous.

(c) The function $f \restriction \mathcal{M}^{V \cap Q}$ is a recursive mapping of $\mathcal{M}^{V \cap Q}$ into $V \cap Q$.

4.2.14 Theorem. Let $\underline{U}$ be a regular d-homogeneous random structure. If f is a cc-statistic then:

(1) For each sample M, the function f_M defined by the equality $f_M(\sigma) = f(\underline{M}_\sigma)$ is a random variate.

(2) If M and N are two samples of the same cardinality then $D_{f_M} = D_{f_N}$.

Proof. Use the fact that continuous functions are measurable (see Problem 5) (2) is a consequence of 4.2.6 (see Problem 6).

4.2.15 Discussion. (1) Assumption (a) in Definition 4.2.13 is very natural: It guarantees that the value depends only on the structure but not on the particular samples. Assumption (b) in 4.2.13 corresponds to our questions 4.2.11 (a) and (b): First, it guarantees that small changes of values in a model $\underline{M}$ cause only a small shift of $f(\underline{M})$. Secondly, it follows from 4.2.14 that if $M \subseteq U$, where $\underline{U}$ is a $\underline{\Sigma}$-random V-structure (obviously regular), then for each V_o Borel (in particular, for each V_o regular; cf. 4.2.10) $\{\sigma \in \Sigma; f(\underline{M}_\sigma) \in V_o\}$ has a probability, i.e., $P(\{\sigma \in \Sigma; f(\underline{M}_\sigma) \in V_o\})$ is defined. Finally, assumption (c) of 4.2.13 answers our question 4.2.11 (c): whenever we have a rational-valued structure $\underline{M}$ (which is a finite structure) we can calculate $f(\underline{M})$ since f (restricted to such structures) is recursive. For notions of computable functions of real variables see [Pour - El 1975].

(2) Finally, it is easy to see that if f_1, f_2 are cc-statistics then the paradoxical situation described at the end of 4.2.11 cannot occur. Definition 4.2.13 can be generalized for k-dimensional statistics, i.e., mappings of $\mathfrak{M}^V$ into V^k .

(3) The notion of cc-statistics covers, in fact, almost all statistics used in classical **statistics** or they can be transformed to the multidimensional cc-statistics. For example, consider the usual Student test statistic T or the correlation coefficient r. The pairs $\langle T^2, \text{sign } T \rangle$ and $\langle r^2, \text{sign } r \rangle$ are cc-statistics.

Note that to cover rank statistics, considered in the next chapter, we shall have to generalize the notion of a cc-statistic to the notion of an almost continuous computable statistic.

4.2.16 Remark. We are now going to investigate the question what is the relation of the notion of a cc-statistic to observational languages. More precisely, we ask whether the notion of cc-statistic can be used for the construction of particular observational function calculi. First we answer this question under the assumption $V \subseteq \mathbb{Q}$ (V a regular set of values, e.g., $V = \mathbb{N}$ or $V = \{0,1\}$). This radically simplifies the situation since then the question of approximation (4.2.11 (b)) is superflous. Note that in Sections 4 and 5 of the present chapter we shall deal with various sets $V \subseteq \mathbb{Q}$.

4.2.17 Theorem. Let $V \subseteq \mathbb{Q}$ be a regular set of values and let $f: \mathfrak{M}^V \to V$ be a cc-statistic. Then there is an OFC with the abstract values V and with $\mathfrak{M}^V$ as the set of models in which f is nameable, i.e. there is a sentence φ such that $f(\underline{M}) = \|\varphi\|_{\underline{M}}$ for each $\underline{M} \in \mathfrak{M}^V$.

Proof. The simplest thing we can do is to take the calculus $\mathcal{F}$ with no junctors and one quantifier q whose associated function is f. The desired sentence is $(qx)(F_1 x, \ldots, F_n x)$. $\mathcal{F}$ is observational since f is recursive.

4.2.18 Remark. If $V \not\subseteq \mathbb{Q}$ we can construct the calculus $\mathcal{F}$ as described in the proof of 4.2.17 but we cannot claim that $\mathcal{F}$ is observational since it can have "too many" models. But the restriction of $\mathcal{F}$ to $V \cap \mathbb{Q}$ (in the obvious sense) is an observational calculus. Hence we give the following definition:

4.2.19 Definition. Let V be a regular set of values and let $\mathcal{F}$ be a function calculus with n unary function symbols whose set of values is V and whose set of models is $\mathfrak{M}^V$. $\mathcal{F}$ is pseudo-observational if the following holds:

(1) For each k-ary junctor ι of $\mathcal{F}$, Asf_ι maps $(V \cap \mathbb{Q})^k$ into $V \cap \mathbb{Q}$;

(2) For each quantifier q of type t, if $\mathrm{Asf}_q(\underline{M})$ is defined and $\underline{M}$ is rational-valued **then** $\mathrm{Asf}_q(\underline{M}) \in \mathbb{Q}$.

(3) Let $\mathcal{F}_o$ be the restriction of $\mathcal{F}$ to $V \cap \mathbb{Q}$ - i.e. values are restricted to $V \cap \mathbb{Q}$, models are restricted to elements of $\mathfrak{M}^{V \cap \mathbb{Q}}$ and associated functions are appropriately restricted. Then $\mathcal{F}_o$ is observational.

4.2.20 Theorem. Let V be a regular set of abstract values and let $f : \mathfrak{M}^V \to V$ be a cc-statistic. **Then** there is a pseudo-observational calculus $\mathcal{F}$ in which f is nameable (in the sense of 4.2.17).

The proof is obvious from the preceding.

4.2.21 Theorem. Let V be a regular set of values and let $\mathcal{F}_o$ be an OFC whose set of values is $V \cap \mathbb{Q}$ and whose set of models is $\mathfrak{M}^{V \cap \mathbb{Q}^o}$. Let φ be a sentence of $\mathcal{F}_o$. If there is a cc-statistic $f : \mathfrak{M}^V \to V$ such that $\|\varphi\|_{\underline{M}} = f(\underline{M})$ for each $\underline{M} \in \mathfrak{M}^{V \cap \mathbb{Q}}$ then f is determined uniquely by φ , i.e. for each statistic $g: \mathfrak{M}^V \to V$ satisfying $\|\varphi\|_{\underline{M}} = g(\underline{M})$ for each $\underline{M} \in \mathfrak{M}^V$ we have $f = g$.

Proof. Use the continuity condition and the fact that the regularity of V implies $V \cap \mathbb{Q}$ to be dense in V, so that $\mathfrak{M}_M^{V \cap \mathbb{Q}}$ is dense in $\mathfrak{M}_M^V$ w.r.t. the metric ϱ .

4.2.22 Discussion. (1) We shall deal with richer calculi in which statistics are nameable; in particular, the methods described in Part B make use of various junctors.

Remember 4.1.1 where we claimed that each observational procedure has two sorts of properties - (1) probabilistic and (2) logical and computational.

How is this claim related to our investigation of statistical inference? Assume that we have a theoretical calculus $\hat{\mathcal{F}}$ with the set $Sent_T$ of sentences and random V-structures as models and a pseudo-observational calculus $\mathcal{F}$ with the set $Sent_0$ of sentences and $\mathfrak{M}^V$ as the set of models; let $\mathcal{F}_0$ be the restriction of $\mathcal{F}$ to $V \cap Q$ ($\mathcal{F}_0$ is an OFC with the same sentences as $\mathcal{F}$).

The above considerations lead to statistical inference rules I consisting of pairs $\frac{\Phi, \varphi}{\Psi}$ where $\Phi, \Psi \in Sent_T$ and $\varphi \in Sent_0$. We have an observational procedure consisting of the evaluation of $\|\varphi\|_{\underline{M}}$ for various observational $\underline{M}$. Probabilistic properties of this procedure concern the relation of $\hat{\mathcal{F}}$ and $\mathcal{F}$. In particular, one has the rationality criterion as expressed in 4.2.11 : I is V_0-rational only if for each theoretical $\underline{U}$ such that $\underline{U} \vdash \Phi \& \neg \Psi$ and for each sample M the probability $P(\|\varphi\|_{\underline{M}_\sigma} \in V_0)$ is small. Logico-computational properties concern the calculus $\mathcal{F}_0$ (and its relation to $\mathcal{F}$).

(3) In fact, rules of the form described in (2) are used in the following way: One has accepted Φ and observed a sample M in a random state σ. $\underline{M}_\sigma$ need not be rational; but observing M we cannot distinguish $\underline{M}_\sigma$ from a rational approximation $\underline{M}'_\sigma$. We compute $f(\underline{M}'_\sigma)$ (which is possible since f is recursive on $\mathfrak{M}^{V \cap Q}$); if $f(\underline{M}'_\sigma \in V_0$ then we accept Ψ (cf. 4.2.11).

(4) To summarize, we ask what has been achieved by our considerations. We claim the following:

(a) We have a logical analysis of statistical inference. "Logical" means that statistical inference takes the form of inference rules containing theoretical and observational sentences. The semantics of both kinds of sentences have been clarified and some rationality criteria for the rules have been formulated.

(b) In particular, we have related cc-statistics with some quantifiers in observational function calculi. Thus various statistics used in practice lead to particular OFC's whose logical properties have been and will be further investigated (Chapter III, Section IV.5 and other places).

We close the present section with an example.

4.2.23 Example. Consider universes of the form $\langle U, Q \rangle$.(Think of the population of all marigold seeds which were influenced by gamma rays and a possible mutation.)

Let Φ (frame assumption) say the following: $\underline{U}$ is d-homogeneous and $\vartheta_o = Q(o, .)$ has an alternative distribution, i.e., it can attain only two values 0 and 1; 1 with probability p and 0 with probability 1-p, where $p \in [0.5, 1)$. $\underline{U} \models \Phi$ implies that $\underline{U} = \langle U, Q \rangle$ is a $\{0,1\}$-structure. Consider a theoretical sentence Ψ which means: ϑ_o has the alternative distribution with $p > 0.5$. $\neg\Psi$ then implies $p = 0.5$, i.e., probabilities of zeroes and ones are equal. Thus, if Q is a mutation, then $p = 0.5$ says that the probability that a marigold will possess the mutation is 0.5; the chance of having or not having the mutation is equal. If $p > 0.5$ then the chance of having the mutation is greater.

We suppose that $\underline{U}$ is d-homogeneous; then the joint distribution function of $\langle \vartheta_{o_1}, \ldots, \vartheta_{o_m} \rangle$, for any sequence $\underline{o}$, is independent of $\underline{o}$.

$\underline{M}_\sigma$ here has the form $\langle M; Q(., \sigma)\rangle$, where $Q(., \sigma)$ is a column of zeros and ones. The function f can be defined as the number of ones in this column. As we know from Problem (3e) for a given M with card(M)= = m, $\underline{U} \models \Phi$ implies that $f_M = \sum_{o \in M} \vartheta_o$ has the binomial distribution function, i.e., $P(f_M = k) = \binom{m}{k} p^k (1-p)^{m-k}$.

Now let the cardinality m be fixed; say $m = 5$. We have only five plants at our disposal. Take $V_o = \{5\}$ and an observational sentence $\varphi = \Sigma F$ (quantifier of the sum, i.e., if $\underline{M} = \langle M, g \rangle$ then $\|\varphi\|_{\underline{M}} = \sum_{o \in M} g(o)$).

Then the probability $P(\{\sigma ; \|\varphi\|_{\underline{M}_\sigma} \in V_o\}) = P(f_M = 5) = 0.5^5 = 0.031$.

Now if $\underline{M}_{\sigma_0}$ is our observation and $\|\varphi\|_{\underline{M}_{\sigma_0}} \in V_o$ then we infer Ψ .

4.2.24 Key words: regular random V-structures, d-homogeneity, regular sets of values, continuous computable statistics, pseudo-observational function calculi, statistical inference rules, frame assumptions, rationality criteria.

IV. 3 The form of theoretical sentences and inference rules

4.3.0 To explain the sense of statistical inference rules more thoroughly we have to be more specific as to the form of theoretical sentences. We shall do this and then make a review of some common statistical rules. It will clarify the rationality conditions used in statistics.

4.3.1 Definition and Discussion. For any regular d-homogeneous random V-structure $\underline{U}$ let $D^{\underline{U}}$ be the joint distribution function of

$\underline{\vartheta}_o = \langle \vartheta_{1o}, \ldots, \vartheta_{no} \rangle$ for an $o \in U$.

A theoretical sentence Φ is called distributional if $\underline{U} \models \Phi$ and $D^{\underline{U}} = D^{\underline{U}'}$ implies $\underline{U}' \models \Phi$ for any $\underline{U}$, $\underline{U}'$. Note that Φ is distributional iff there is a system $\mathcal{D}_T = \{D_t ; t \in T\}$ such that $\underline{U} \models \Phi$ iff $D^{\underline{U}} \in \mathcal{D}_T$ (i.e., there is a $t \in T$ such that $D^{\underline{U}} = D_t$). Iff $\mathcal{D}_T$ has this property we shall express Φ informally by $D \in \mathcal{D}_T$.

Statistical inference rules consist, in general, of some pairs of the form

$$\frac{D \in \mathcal{D}_T, A}{D \in \mathcal{D}_{T'}}$$

, where A is a finite set of observational sentences and T' is a proper subset of T.

4.3.2 Remark and Convention. Let f be a cc-statistic; consider, all Σ-random V-structures with a fixed domain U. Let $M \subseteq U$ be a sample and let $V_o \subseteq V$ be a regular set of values. Then the set $\{\sigma ; f(\underline{M}_\sigma) \in V_o\}$ depends on the random structure on U. We should write $\{\sigma ; f(\underline{M}_\sigma^{\underline{U}}) \in V_o\}$ where $\underline{U}$ varies over all random structures with the domain U. But, evidently, if $D^{\underline{U}_1} = D^{\underline{U}_2}$ then the probabilities $P(\{\sigma ; f(\underline{M}_\sigma^{\underline{U}_1}) \in V_o\})$ and $P(\{\sigma ; f(\underline{M}_\sigma^{\underline{U}_2}) \in V_o\})$ are equal and we denote the common value by $P(f_M \in V_o \mid D_o)$, where D_o denotes the distribution function $D_o = D^{\underline{U}_1} = D^{\underline{U}_2}$. If we let D_o vary over a system $\mathcal{D}_T = \{D_t; t \in T\}$

then $P(f_M \in V_o \mid D_t)$ becomes a function of t.

The same holds for the moments of the random variate f_M ; the random variate itself is determined by the choice of a particular $\underline{U}$ but **its moments** depend only on $D^{\underline{U}}$. Hence, if we let $D^{\underline{U}}$ vary over $\mathcal{D}_T$ the moments become functions of t and we write $E(f_M \mid D_t)$ and $VAR(f_M \mid D_t)$ for the expectation and variance of f_M respectively.

We shall write $E(\varphi)$, $VAR(\varphi)$ instead of E f, VAR f , where f is a statistic named by φ .

4.3.3 General survey of statistical inference rules.

First we must point out that even if some of the following inference rules seem to be identical, they differ in the metatheoretical criteria imposed on them.

For the sake of convenience, let us make the following assumptions: (1) The considered type is $\langle 1 \rangle$; our structures have the form $\langle U, Q_1 \rangle$, (2) $\mathcal{D}_T$ is a system of distribution functions with $T \subseteq \mathbb{R}$, (3) our observational language contains (at least): (a) the binary junctors = and $\leq$ whose associated functions are the characteristic functions of equality and the less-than-or-equal-to relation (such that if φ , ψ are sentences then $\|\varphi = \psi\|_{\underline{M}} = 1$ iff $\|\varphi\|_{\underline{M}} = \|\psi\|_{\underline{M}}$, etc.) and (b) for each $t \in V \cap \mathbb{Q}$ its name - a nullary junctor $\dot{t}$ such that $\|\dot{t}\|_{\underline{M}} = t$ for each $\underline{M}$.

We now briefly describe three types of statistical inference:

(1) Estimation (for a particular example see Problem (7)):

$$\frac{D \in \mathcal{D}_T , (\varphi = \dot{t}),}{D = D_t}$$

Here φ names a statistic, V_o is $\{1\}$. So we estimate the index of the distribution function D by the value $\|\varphi\|_{\underline{M}}$ (estimate) , i.e., we determine one particular distribution function (φ is called the estimator). The criteria here are, e.g., the following (for each $t \in T$):

(a) $E(\varphi \mid D_t) = t$,

(b) $VAR(\varphi \mid D_t)$ is as small as possible.

Remember that if f is a statistic such that $f(\underline{M}) = \|\varphi\|_{\underline{M}}$ for $\underline{M} \in \mathfrak{M}^{V \cap Q}$ then, by the strong law of large numbers, we see that if we consider a sequence of disjoint samples $M_1, M_2, \ldots$, we have (a.s.) $\lim_n \frac{1}{n} f_{M_n} = t$

(cf. Problem (5) of Chapter VIII).

Let us note that this is the reason why estimates obtained by an estimator fulfilling (a) can be pooled, i.e., we can use the number $\frac{1}{n} \|\varphi\|_{\underline{M}_n}$ as an estimate.

(2) Identification. Let $T = \{t_1, \ldots, t_k\}$. Consider inference rules having the form:

$$\frac{D \in \mathcal{D}_T, \varphi_1}{D = D_{t_1}}, \ldots, \frac{D \in \mathcal{D}_T, \varphi_k}{D = D_{t_k}}$$

The sentences $\varphi_1, \ldots, \varphi_k$ name some statistics $f_1, \ldots, f_k$ such that, for each $\underline{M} \in \mathfrak{M}^V$, $f_i(\underline{M}) \in V_o$ for exactly one i. If, for a $\underline{U}$, $D^{\underline{U}} = D_{t_i}$ and if for our observed $\underline{M}_\sigma$ we obtain $f_j(\underline{M}_\sigma) \in V_o$ for a $j \neq i$ then we make an erroneous inference. Thus our criterion is that the probabilities of the errors, i.e., $P(\{\sigma ; f_{i,M}(\sigma) \in V_o\} \mid D_{t_j})$ $(i \neq j)$ be as small as possible (even if perhaps unknown).

In particular, in the case $k = 2$, the probabilities $P(\{\sigma ; \|\varphi_2\|_{\underline{M}_\sigma} \in V_o\} \mid D_{t_1})$ and $P(\{\sigma ; \|\varphi_1\|_{\underline{M}_\sigma} \in V_o\} \mid D_{t_2})$ are to be as small as possible.

(3) Simple hypothesis testing. Consider a rule of the form

$$\frac{D \in \mathcal{D}_T, \varphi}{D = D_{t_2}} \qquad (\text{supposing } T = \{t_1, t_2\})$$

The situation is like (2) for $k = 2$ but the two possible errors are treated asymmetrically; one error is supposed to be substantial, namely $\|\varphi\|_{\underline{M}_\sigma} \in V_o$ under the assumption $D = D_{t_1}$. This error is called an error of the first kind (which error is substantial is the question of the actual meaning of the theoretical sentences); the error $\|\varphi\|_{\underline{M}_\sigma} \notin V_o$ under the assumption $D = D_{t_2}$ is called an error of the second kind (as opposed to (2), for $k = 2$, we do not make any inference if $\|\varphi\|_{\underline{M}} \notin V_o$; thus, this error signifies that no inference is made under the assumption that the conclusion, i.e. $D = D_{t_2}$, is true). An error of the first kind is substantial, so we require the probability

$$P(\{\sigma ; \|\varphi\|_{\underline{M}_\sigma} \in V_o\} \mid D_{t_1})$$

to be bounded from above by a small positive number α given in advance. Under the condition $P(\{\sigma ; \|\varphi\|_{\underline{M}_\sigma} \in V_o\} \mid D_{t_1}) \leq \alpha$ we require $P(\{\sigma ; \|\varphi\|_{\underline{M}} \notin V_o\} \mid D_{t_2})$ to be as small as possible.

4.3.4 More on hypothesis testing. Consider, for the sake of convenience, only a single quantity Q; then we have for each $o \in U$ a variate $\vartheta_o = Q(o, .)$ Now let T_1, T_2 be two disjoint subsets of T. The sentence $D \in \mathcal{D}_{T_1}$ will be called the null hypothesis, $D \in \mathcal{D}_{T_2}$ will be called the alternative hypothesis. We consider inference rules of the following form:

$$(*) \qquad \frac{D \in \mathcal{D}_{T}, \varphi}{D \in \mathcal{D}_{T_2}} .$$

If $D \in \mathcal{D}_T$ has been accepted and if $\|\varphi\|_{\underline{M}_\sigma} \in V_o$ for our observation $\underline{M}_\sigma$, we infer $D \in \mathcal{D}_{T_2}$, i.e., we accept the alternative hypothesis (we reject the null hypothesis). The observational sentence φ has to fulfil some conditions guaranteeing the reasonability of our inference.

4.3.5 Definition. Let $\alpha \in (0, 0.5]$. An observational sentence naming a cc-statistic f and used in an inference of the form (*) is called an observational test for the null hypothesis $D \in \mathcal{D}_{T_1}$ and alternative hypothesis $D \in \mathcal{D}_{T_2}$ on the significance level α if

$$P(\{\sigma ; f_M(\sigma) \in V_o\} \mid D_t) \leq \alpha$$

for each $t \in T_1$ (independently of the cardinality of the sample M).

If φ is an observational test and if M is a sample then the function $B_M(\ , t) = P(\{\sigma ; f_M \in V_o\} \mid D_t)$ is called the power function of φ (w.r.t. M). According to 4.3.2, this is a well-defined function. Note that (under our assumption of d-homogeneity) $B_M(\varphi, t)$ does not depend on the particular choice of M but only on the cardinality of M. Hence, we can write $B_m(\varphi, t)$ instead of $B_M(\varphi, t)$ for card M = m.

We want $B_m(\varphi, t)$ to be as large as possible for $t \in T_2$ under the condition $B_m(\varphi, t) \leq \alpha$ for $t \in T_1$.

(For $t \in T_2$, $B_m(\varphi, t)$ is the probability of inferring the alternative hypothesis $D \in \mathcal{D}_{T_2}$ under the assumption that $D = D_t$.)

Now let T_1, T_2 be given. Suppose that Φ_α is a class of observational tests (on the level α) for $D \in \mathcal{D}_{T_1}$ and $D \in \mathcal{D}_{T_2}$. A test φ_o is called uniformly most powerful w.r.t. Φ_α if $\varphi_o \in \Phi_\alpha$ and for each $\varphi \in \Phi_\alpha$ and each $t \in T_2$ we have $B_m(\varphi_o, t) \geq B_m(\varphi, t)$ independently of m.

4.3.6 Example. (Continuation of 4.2.23). The power function of the test $\varphi = \sum F$ (assuming m = 5 and $V_o = \{5\}$), is $B_5(\varphi, p) = p^5$; e.g., for p = 0.7 (i.e., we consider a single alternative hypothesis p = 0.7) $B(\varphi, 0.7) = 0.19$. We see that the probability of not inferring the alternative hypothesis under the condition that this alternative hypothesis holds is rather large ($1 - B(\varphi, 0.7) = 0.81$) and cannot be decreased for the given cardinality of samples (m = 5). Hence, if we want to have more powerful procedures, we have to use larger samples.

4.3.7 Remark.

(1) Note that we do not suppose $T_1 \cup T_2 = T$; so that there is a theoretical sentence $D \in \mathcal{D}_{T_3}$, where $T_3 = T - (T_1 \cup T_2)$. We construct our tests not taking into consideration the errors under the assumption $D \in \mathcal{D}_{T_3}$; if $T_3 \neq \emptyset$ then we should know that $D \in \mathcal{D}_{T_3}$ is rather unlikely and/or that the errors under the assumption $D \in \mathcal{D}_{T_3}$ have little importance.

(2) In mathematical statistics larger classes of **functions** are considered as tests; in general, only measurability is demanded. Our restriction concerning the notion of a statistic as introduced in 4.2.13 is due to our emphasis on observationality. Hence, if we use the results of mathematical statistics, observationality must be additionally verified.

(3) Note that, if a test (in the common statistical sense) is uniformly most powerful and if it is based on a cc-statistic in our sense and on a regular set of values V_o, we can conclude that this test is the uniformly most powerful observational test in the sense of 4.3.5.

(4) The construction of uniformly most powerful tests is in many cases impossible. Then the class of considered tests should be restricted. Probably, the most natural condition for all cases is the following:

$B_m(\varphi, t) \geq \alpha$ for each $t \in T_2$ and $m \in \mathbb{N}$, i.e., the probability of the acceptance of the alternative hypothesis under the assumption $D = D_t$ is larger than α. Such a test is called unbiased. One can define observational tests directly as unbiased (and so obtain - as opposed to our more general definition - a condition concerning the alternative hypothesis in the definition.)

4.3.8 Definition and Remark. (1) In many cases the weaker notion of an asymptotical observational test is useful:

Consider a statistic f. If there is a variate $\mathcal{W}$ such that, for any (strictly) increasing sequence $M_1 \subset M_2 \subset \ldots$ of samples, $D \in \mathcal{D}_{T_1}$ implies (d) $\lim_i f_{M_i} = \mathcal{W}$ and $P(\mathcal{W} \in V_o) \leq \alpha$, then **f is an asymptotical observational test.**

(2) Under our assumption of d-homogeneity it is sufficient to prove

(d) $\lim_i f_{M_i} = \mathcal{W}$ and $P(\mathcal{W} \in V_o) \leq \alpha$, under the assumption $D \in \mathcal{D}_{T_1}$

for <u>one particular</u> sequence of samples.

(3) For practical purposes, the speed of convergence has to be considered: only if the approximation is rather good can the asymptotical test be used. The investigation of these properties is a very interesting area of computer simulation, but it is beyond the scope of our book.

4.3.9 <u>Key words</u>: Distributional sentences and statistical theoretical sentences. Types of statistical inference: estimation, hypothesis testing, observational tests, error of the first kind, significance level, power function, asymptotical observational tests.

IV.4 Observational predicate calculi based on statistical procedures.

4.4.0 We are now interested in random variates taking only the values 0 and 1. Hence, consider regular Σ-random $\{0,1\}$ - structures. We restrict ourselves to d-homogeneous structures. If $\underline{U} = \langle U, Q_1, \ldots, Q_n \rangle$ is such a structure then each $\sigma \in \Sigma$ and each non-empty finite $M \subseteq U$ determines the $\{0,1\}$ -structure $\underline{M}_{\sigma}$. Our aim is to study some tests used in statistical hypotheses testing concerning $\{0,1\}$ - valued random quantities and build up various monadic observational predicate calculi in which these tests can be appropriately named.

Let $\langle 1^n \rangle$ be a fixed type. Each OPC of this type will contain n unary predicates $P_1, \ldots, P_n$ and, say, the classical connectives &, $\vee$, $\rightarrow$, $\neg$. Hence the notion of open formulae is fixed in advance. We shall be particularly interested in designated open formulae, i.e. open formulae containing no variable except the designated variable x. Designated open formulae can be abbreviated by omitting the variable x at all occurences, e.g. P_1x & $\neg P_2x$ is abbreviated as P_1 & $\neg P_2$.

On the other hand, if we want to speak of state dependent $\{0,1\}$ -structures we use state dependent predicate calculi. A state dependent predicate calculus of type $\langle 1^n \rangle$ has the same predicates $P_1, \ldots, P_n$; we define <u>designated</u> open formulae as open formulae containing no variable except x and the state variable s. Designated open formulae can be abbreviated by omitting x and s at all occurences; e.g. $P_1(x,s)$ & $\neg P_2(x,s)$ is abbreviated as P_1 & $\neg P_2$. In this way we identify designated open formulae of any OPC of type $\langle 1^n \rangle$ and designated open formulae of any state dependent predicate calculus of the same type. In particular, if φ is such a formula and $\underline{M}$ is an observational model then $\|\varphi\|_{\underline{M}}$ is the mapping of M into $\{0,1\}$ defined in accordance with 2.2.6 ; if $\underline{U}$ is a state dependent $\{0,1\}$ -structure then $\|\varphi\|_{\underline{U}}$ is the mapping of $U \times \Sigma$ into $\{0,1\}$ defined in accordance with 2.4.6. For each $o \in M$, $\|\varphi\|_{\underline{M}}[o] \in \{0,1\}$ and for each $o \in U$, $\|\varphi\|_{\underline{U}}[o]$ is a state dependent variate: $\|\varphi\|_{\underline{U}}[o]: \Sigma \rightarrow \{0,1\}$.

Let us make the convention that as models of our monadic calculus we have exactly all the $\{0,1\}$ - structures (of the given type) whose domain is a finite set of natural numbers (i.e. the set of all models in $\mathcal{M}^{\{0,1\}}$). What is not given in advance are quantifiers (and their associated functions) of our monadic observational predicate calculi. In fact, the considerations of this section will lead us to some definitions of particular quantifiers. (Logical properties of the classes of such quantifiers were considered in Chapter III, Sections 1,2.)

4.4.1 Notation. (1) Let $\underline{U} = \langle U, Q_1 \rangle$ be a d-homogeneous random $\{0,1\}$ - structure. Then the probability of success for $\underline{U}$ is the probability $p_{\underline{U}} = P(\{\sigma ; Q_1(o,\sigma) = 1\})$ where o is an arbitrary element of U (by d-homogeneity, $p_{\underline{U}}$ is independent of the choice of o and determines the alternative distribution of $Q_1(o,-)$).

(2) If $\underline{U} = \langle U, Q_1, \ldots, Q_n \rangle$ is an arbitrary d-homogeneous random $\{0,1\}$ - structure and if φ is a designated open formula then $\underline{U}_\varphi$ denotes the structure $\langle U, \|\varphi\|_{\underline{U}} \rangle$; evidently, this is a d-homogeneous random $\{0,1\}$ - structure and we write p_φ instead of $p_{\underline{U}_\varphi}$ if there is no danger of misunderstanding.

(3) More generally, if $\underline{U}$ is as in (2) and if $\varphi_1, \ldots, \varphi_n$ are designated open formulae then we put $\underline{U}_{\varphi_1, \ldots, \varphi_n} = \langle U, \|\varphi_1\|_{\underline{U}}, \ldots, \|\varphi_n\|_{\underline{U}} \rangle$; it is easy to show that $\underline{U}_{\varphi_1, \ldots, \varphi_n}$ is d-homogeneous.

(4) In what remains of the present section "random structure" stands for "d-homogeneous regular Σ-random $\{0,1\}$ - structure".

4.4.2 Lemma. Let $\underline{U} = \langle U, Q \rangle$ be a random structure and let $M \subseteq U$ be finite non-empty. Put $m = \mathrm{card}(M)$. Then the variate $\mathcal{W}_M$ defined by $\mathcal{W}_M(\sigma) = \mathrm{card}\{o \in M; Q(o,\sigma) = 1\}$ has the binomial distribution $P(\mathcal{W}^{-1}_M(k)) = \binom{m}{k} p^k_{\underline{U}} (1 - p^k_{\underline{U}})^{m-k}$ (independently of M).

Proof. See Problem (3e).

4.4.3 Remark. Without changing our present notion of observational models ({0,1} - structures with finite domains) it is sometimes useful to extend the set of abstract values $\{0,1\}$ to $\mathbb{N}$ (or even $\mathbb{Q}$) ; then we are able to introduce various useful quantifiers whose associated functions associate a natural rational number with each model. In this case, the associated functions of the junctors $\&, \vee, \neg$ are extended arbitrarily for arguments different from $\{0,1\}$. For example, consider a quantifier $\hat{m}$ (of type $\langle 1 \rangle$) with

$$\mathrm{Asf}_{\hat{m}}(\langle M,f\rangle) = \frac{\mathrm{card}\{o;\ f(o) = 1\}}{\mathrm{card}(M)}$$

(relative frequency). The sentences $\hat{\varphi} = \hat{m}(\varphi)$ then generate estimates for p_φ (in the sense of 4.3.3 (1)).

Given $\underline{U}$, φ and $M \subseteq U$, note that $\|\hat{\varphi}\|_{\underline{M}}$ is a variate and $E(\|\hat{\varphi}\|_{\underline{M}}) = p_\varphi$, $\mathrm{VAR}(\|\hat{\varphi}\|_{\underline{M}}) = \frac{1}{m} p_\varphi (1 - p_\varphi)$ where $m = \mathrm{card}(M)$.

We shall now follow the usual method of the treatment of alternative experiments (from the point of view of the construction of our appropriate monadic observational predicate calculi). Let us begin with a useful lemma.

4.4.4 Diagonal lemma. Let $\underline{U}$ be a random structure and let $M \subseteq U$ be a sample; put $m = \mathrm{card}(M)$. Let $g(\underline{M}_\sigma)$ be a statistic taking values from $\{0,1, \ldots, m\}$ and put

$$f(\underline{M}_\sigma) = P(\{\tau\ ;\ g(\underline{M}_\tau) \leq g(\underline{M}_\sigma)\})$$

(Imagine that it is desirable that $g(\underline{M}_\sigma)$ be large; then $f(M_\sigma)$ measures the probability that g is equal or worse then the observed value $g(\underline{M}_\sigma)$.) Then $P(f(\underline{M}_\sigma) \leq \alpha) \leq \alpha$.

Proof. We have $P(\{\sigma\ ;\ f(\underline{M}_\sigma) \leq \alpha\}) =$

$$= P(\{\sigma\ ;\ P(\{\tau\ ;\ g(\underline{M}_\tau) \leq g(\underline{M}_\sigma)\}) \leq \alpha\}) = P(\{\sigma\ ;\ g(\underline{M}_\sigma) \in A\}),$$

where $A = \{ k \in \{0, \ldots, m\}\ ;\ P(\{\tau\ ;\ g(\underline{M}_\tau) \leq k\}) \leq \alpha \}$

Note that $k' \leq k \in A$ implies $k' \in A$, hence either A is empty (and then there is nothing to prove) or, otherwise, putting $k_o = \max A$ we obtain $P(\{\sigma ; g(\underline{M}_\sigma) \in A\}) = P(\{\sigma ; g(\underline{M}_\sigma) \leq k_o\})$. But, by the definition of k_o, we have $P(\{\tau ; g(\underline{M}_\tau) \leq k_o\}) \leq \alpha$.

Hence, by renaming τ to σ, $P(\{\sigma ; g(\underline{M}_\sigma) \leq k_o\}) \leq \alpha$ and the lemma follows.

4.4.5 Definition. Consider $\{0,1\}$ - structures of type $\langle 1 \rangle$. Define, for a given structure $\underline{M} = \langle M, f \rangle$,

$k_{\underline{M}} = \mathrm{card}\ \{o;\ f(o) = 1\}$ and $m_{\underline{M}} = \mathrm{card}(M)$.

(If it does not lead to a misunderstanding we shall write k and m only.)

Let p be a real number, $p \in (0,1)$.
We define two functions $\underline{f}_p$ and $\overline{f}_p$ on $\{0,1\}$ - structures as follows:

(a) $\overline{f}_p(\underline{M}) = \sum_{i=0}^{k} \binom{m}{i} p^i (1-p)^{m-i}$,

(b) $\underline{f}_p(\underline{M}) = \sum_{i=k}^{m} \binom{m}{i} p^i (1-p)^{m-i}$.

4.4.6 Remark. (1) $\underline{f}$ and $\overline{f}$ are functions of k,m,p; we shall sometimes write $\overline{f}_p(m,k)$ and $\underline{f}_p(m,k)$.

(2) Observe that by 4.4.2, if $\underline{U}$ is a random structure such that $p_{\underline{U}} = p$ and $M \subseteq U$, $\mathrm{card}(M) = m$, then

$$\overline{f}_p(m,k) = P(\{\sigma ; k_{\underline{M}_\sigma} \leq k\})$$

and

$$\underline{f}_p(m,k) = P(\{\sigma ; k_{\underline{M}_\sigma} \geq k\}).$$

(3) Note the relation of $\overline{f}_p$ and $\underline{f}_p$ with binomial distribution function

$$D_m(p,x) = \sum_{o \leq i < x}^{m} \binom{m}{i} p^i (1-p)^{m-i}.$$

(4) $\overline{f}_p$ and $1-\underline{f}_p$ are (a) strictly decreasing in p,

(b) continuous in p and (c) for any k, m $\lim_{p\to 0} \overline{f}_p(k,m) = \lim_{p\to 0}(1-\underline{f}_p(k,m)) = 0$ and $\lim_{p\to 1} \overline{f}_p(k,m) = \lim_{p\to 1}(1-\underline{f}_p(k,m)) = 1$.

4.4.7 Lemma. Let $\underline{U} = \langle U, Q \rangle$ be a random structure, let $M \subseteq U$ be a sample and let α be a real number, $\alpha \in (0,1)$.

Then (a) for $p_{\underline{U}} \geq p$ we have

$$P(\{\sigma ; \overline{f}_p(\underline{M}_\sigma) \leq \alpha\}) \leq \alpha$$

and (b) for $p_{\underline{U}} \leq p$ we have

$$P(\{\sigma ; \underline{f}_p(\underline{M}_\sigma) \leq \alpha\}) \leq \alpha .$$

Proof. (a) note that if $p_{\underline{U}} \geq p$ then $\overline{f}_p(\underline{M}_\sigma) \leq \alpha$ implies $\overline{f}_{p_{\underline{U}}}(\underline{M}_\sigma) \leq \alpha$ hence

$$P(\{\sigma ; \overline{f}_p(\underline{M}_\sigma) \leq \alpha\}) \leq P(\{\sigma ; f_{p_{\underline{U}}}(\underline{M}_\sigma) \leq \alpha\})$$. Use lemma 4.4.4.

(b) can be proved similarly.

4.4.8 Remark. A dual form of the above lemma holds. Put $\alpha' = 1-\alpha$ and apply Lemma 4.4.4 to this α'. Then we obtain

$P(\{\sigma ; \overline{f}_p(\underline{M}_\sigma) > \alpha\}) \geq \alpha$ for $p_{\underline{U}} < p$ and

$P(\{\sigma ; \underline{f}_p(\underline{M}_\sigma) > \alpha\}) \geq \alpha$ for $p_{\underline{U}} > p$. Moreover, the dual form can be proved with strict inequality.

4.4.9 Definition. For each rational $\alpha \in (0, 0.5]$ and $p \in (0,1)$ define the quantifiers $!_{p,\alpha}$ and $?_{p,\alpha}$ of type $\langle 1 \rangle$ with the associated functions

$$\mathrm{Asf}_{?_{p,\alpha}}(\langle M,f\rangle) = 1 \text{ iff } \overline{f}_p(\langle M,f\rangle) > \alpha$$

and

$$\mathrm{Asf}_{!_{p,\alpha}}(\langle M,f\rangle) = 1 \text{ iff } \underline{f}_p(\langle M,f\rangle) \leq \alpha .$$

4.4.10 Theorem. Let $\underline{U}$ be a random structure. Consider a designated open formula φ. Under our assumptions, $\underline{U}_{\varphi}$ is a random structure with probability of success p_{φ}. Let α and $p \in (0,1)$ be given.

Then $!_{p,\alpha}(\varphi)$ is an observational test for the null hypothesis $p_{\varphi} \leq p$ (or, in more detail, $D_{\mathcal{V}} \in \mathcal{D}_{p_{\varphi} \leq p}$) and the alternative hypothesis $p_{\varphi} > p$. Similarly, $\neg\ ?_{p,\alpha}(\varphi)$ is an observational test of the null hypothesis $p_{\varphi} \geq p$ and the alternative hypothesis $p_{\varphi} < p$.

Proof. The observationality of the quantifiers defined above is clear.

Let $\underline{U}$ be a regular d-homogeneous random $\{0,1\}$ - structure and assume the null hypothesis, $p_{\varphi} \leq p$. We have to prove $P(\| !_{p,\alpha}(\varphi) \|_{\underline{M}_{\sigma}} = 1) \leq \alpha$. But $P(\| !_{p,\alpha}(\varphi) \|_{\underline{M}_{\sigma}} = 1) = P(\underline{f}_p(\underline{M}_{\sigma}^{\underline{U}_{\varphi}}) \leq \alpha) \leq \alpha$ by Lemma 4.4.7.

4.4.11 (1) By 4.4.8 we know that the above mentioned tests are unbiased.

(2) We shall now define some quantifiers of type $\langle 1,1\rangle$ which belong to the class of quantifiers studied in Chapter III (associational quantifiers) and will be used in the methods of Chapter VII. The associationality (and implicationality) of the defined quantifiers will be studied in Section 5.

4.4.12 Definition. Consider $\{0,1\}$ - structures of type $\langle 1,1\rangle$. For such a structure $\langle M,f_1,f_2\rangle$ we denote

$a_{\underline{M}} = \mathrm{card}\,\{o \in M;\ f_1(o) = 1 \text{ and } f_2(o) = 1\}$,

$b_{\underline{M}} = \mathrm{card}\,\{o \in M;\ f_1(o) = 1 \text{ and } f_2(o) = 0\}$,

$c_{\underline{M}} = \mathrm{card}\,\{o \in M;\ f_1(o) = 0 \text{ and } f_2(o) = 1\}$,

$d_{\underline{M}} = \mathrm{card}\,\{o \in M;\ f_1(o) = 0 \text{ and } f_2(o) = 0\}$,

$k_{\underline{M}} = \text{card}\{o \in M;\ f_1(o) = 1\}$,

$l_{\underline{M}} = \text{card}\{o \in M:\ f_1(o) = 0\}$,

$r_{\underline{M}} = \text{card}\{o \in M:\ f_2(o) = 1\}$,

$s_{\underline{M}} = \text{card}\{o \in M:\ f_2(o) = 0\}$,

$m_{\underline{M}} = \text{card}\{o \in M\}$.

If there is no danger of misunderstanding we shall only write a,b,c,d,k,l, r,s,m.

(1) The quantifier $\Rightarrow^{?}_{p,\alpha}$ of type $\langle 1,1\rangle$ with the associated function

$$\text{Asf}_{\Rightarrow^{?}_{p,\alpha}}(\langle M, f_1, f_2\rangle) = 1 \text{ iff } \bar{f}_p(a,k) > \alpha$$

is called the <u>suspicious p-implication quantifier</u> (on level α).

(2) The quantifier $\Rightarrow^{!}_{p,}$ of type $\langle 1,1\rangle$ with the associated function

$$\text{Asf}_{\Rightarrow^{!}_{p,\alpha}}(\langle M, f_1, f_2\rangle) = 1 \text{ iff } \underline{f}_p(a,k) \leq \alpha$$

is called the <u>likely p-implication</u> (on level α).

Before examining the statistical meaning of these quantifiers we have to introduce the notion of conditional probability. We shall restrict ourselves to a particular case sufficient for our purposes.

4.4.13 <u>Definition</u>. Let ϑ_1, ϑ_2 be two variates such that for their ranges we have $|\vartheta_1| \subseteq \{0, \ldots, n_1\}$ and $|\vartheta_2| \subseteq \{0, \ldots, n_2\}$. The joint distribution of ϑ_1, ϑ_2 is then given by the probabilities

$p_{ij} = P(\vartheta_1^{-1}(i) \cap \vartheta_2^{-1}(j))$ for $i = 0, \ldots, n_1$ and $j = 0, \ldots, n_2$. The distribution of ϑ_1 is given by the probabilities $p_{i\cdot} = P(\vartheta_1^{-1}(i))$ for $i = 0, \ldots, n_1$.

(Note that $p_{i\cdot} = \sum_{j=0}^{n_2} p_{ij}$.) The same holds for ϑ_2 $(p_{\cdot j})$.

Suppose that, for each $j = 0, \ldots, n_2$, $p_{\cdot j} > 0$. The conditional distribution of ϑ_1 relative to ϑ_2 is then given by the conditional probabilities defined as follows:

$$P(\vartheta_1^{-1}(i) / \vartheta_2^{-1}(j)) = p_{ij} / p_{\cdot j}$$

4.4.14 Lemma. Denote $A_i = \vartheta_1^{-1}(i)$ and $B_j = \vartheta_2^{-1}(j)$.

Then (1) $P(A_i / B_j) = P(A_i \cap B_j) / \sum_{i=1}^{n_1} P(A_i \cap B_j)$

and (2)

$$P(A_i) = \sum_{j=1}^{n_2} P(A_i / B_j) P(B_j) .$$

Proof. Note that $\sum_{i=1}^{n_i} P(A_i \cap B_j) = P(B_j)$.

4.4.15 Remark. $P(\vartheta_1^{-1}(i) / \vartheta_2^{-1}(j))$ is then the probability of the event $\vartheta_1^{-1}(i)$ (i.e., $\vartheta_1 = i$) under the assumption that $\vartheta_2 = j$. If $\underline{U}$ is a random structure, ϑ_1 is $\|\varphi_1\|_{\underline{U}}[o]$ and ϑ_2 is $\|\varphi_2\|_{\underline{U}}[o]$, then $p^{\underline{U}}_{\varphi_2/\varphi_1}$ denotes the conditional probability

$$P(\vartheta_2^{-1}(1) / \vartheta_1^{-1}(1)) = P(\|\varphi_2\|_{\underline{U}}[o] = 1 / \|\varphi_1\|_{\underline{U}}[o] = 1) .$$

It follows from the d-homogeneity of $\underline{U}$ that the above probability does not depend on o.

In the sequel, for each number p we denote by $p_{\varphi_2/\varphi_1} \geq p$ the theoretical sentence true in a random structure $\underline{U}$ iff $p^{\underline{U}}_{\varphi_2/\varphi_1} \geq p$, i.e., $\underline{U} \models p_{\varphi_2/\varphi_1} \geq p$, iff $p^{\underline{U}}_{\varphi_2/\varphi_1} \geq p$. The same holds for $<$ instead of $\leq$.

4.4.16 Theorem. Consider two open designated formulae φ_1, φ_2. Then the sentence $\varphi_1 \Rightarrow^{!}_{p,\alpha} \varphi_2$ is an observational test (on the level α) for the null hypothesis $p_{\varphi_2/\varphi_1} \leq p$ and the alternative hypothesis $p_{\varphi_2/\varphi_1} > p$.

Similarly $\neg(\varphi_1 \Rightarrow^{?}_{p,\alpha} \varphi_2)$ is a test for the null hypothesis $p_{\varphi_2/\varphi_1} \geq p$ and the alternative hypothesis $p_{\varphi_2/\varphi_1} < p$.

For the proof see Problem (8).

4.4.17 Discussion. The meaning of the theoretical sentence $p_{\varphi_2/\varphi_1} > p$ is (assuming that p is close to 1): φ_2 is "quasi-implied" by φ_1. Here the value $p^{\underline{U}}_{\varphi_2/\varphi_1}$, measures the quality of this "quasi-implication" in $\underline{U}$.

We have some inductive inference rules. Let Φ_o be a theoretical sentence summarizing our frame assumptions: $\underline{U} \models \Phi_o$ iff $\underline{U}$ is a d-homogeneous Σ-random $\{0,1\}$- structure.

First, consider a rule consisting of some pairs of the form

$$\frac{\Phi_o, \varphi_1 \Rightarrow^{!}_{p,\alpha} \varphi_2}{p_{\varphi_2/\varphi_1} > p}$$

The probability of an erroneous inference (i.e., the probability $P(\|\varphi_1 \Rightarrow^{!}_{p,\alpha} \varphi_2\|_{\underline{M}_\sigma} = 1)$ under the assumption $p^{\underline{U}}_{\varphi_2/\varphi_1} \leq p$; (cf. 4.4.16) is small ($\leq \alpha$). So if we have $\|\varphi_1 \Rightarrow^{!}_{p,\alpha} \varphi_2\|_{\underline{M}_\sigma} = 1$ for the observed $\underline{M}_\sigma$ then the assertion $p_{\varphi_2/\varphi_1} > p$ is relatively reliable. On the other hand, assume $p^{\underline{U}}_{\varphi_2/\varphi_1} > p$; then we cannot say anything about the probability $P(\|\varphi_1 \Rightarrow^{!}_{p,\alpha} \varphi_2\|_{\underline{M}_\sigma} = 1)$ i.e., about the probability that we shall actually assert $p_{\varphi_2/\varphi_1} > p$.

On the other hand, if we have a rule of the form

$$\frac{\Phi_o, \varphi_1 \Rightarrow^{?}_{p,\alpha} \varphi_2}{p_{\varphi_2/\varphi_1} \geq p}$$

then the situation is dual, i.e. this is not a case of hypothesis testing in the sense of 4.3.4 : We know that if $\underline{U} \models p_{\varphi_2/\varphi_1} \geq p$ then the probability $P(\| \varphi_1 \Rightarrow^{?}_{p,\alpha} \varphi_2 \|_{\underline{M}} = 1) \geq 1 - \alpha$; i.e. the probability that $p_{\varphi_2/\varphi_1} \geq p$ will be inferred (and asserted) is large ($\geq 1 - \alpha$) . Hence the probability do we reject the null hypothesis p $p_{\varphi_2/\varphi_1} \geq p$ if it is true is small but our conclusion is rather unreliable.

Observe the important property of the above two inference rules: they have a fixed frame assumption Φ_o, and the observational sentences in the antecedent ($\varphi_1 \Rightarrow^{!}_{p,\alpha} \varphi_2$ in the first rule and $\varphi_1 \Rightarrow^{?}_{p,\alpha} \varphi_2$ in the second rule) determines uniquely the hypothesis in the succedent ($p_{\varphi_2/\varphi_1} > p$ and $p_{\varphi_2/\varphi_1} \geq p$ respectively) and vice versa . Cf. 1.1.6 (L3).

We now turn our attention to some other observational quantifiers of a statistical nature of type $< 1, 1 >$.

4.4.18 Definition and Discussion. Let $\underline{U}$ be a given random structure. In practical considerations the following situation very often occurs. There are two designated open formulas φ_1, φ_2 and we do not know the probabilities

$$P(\| \varphi_1 \& \varphi_2 \|_{\underline{U}}[o] = 1) = p_{11}, \quad P(\| \varphi_1 \& \neg \varphi_2 \|_{\underline{U}}[o] = 1) = p_{10},$$

$$P(\| \neg \varphi_1 \& \varphi_2 \|_{\underline{U}}[o] = 1) = p_{01}, \quad P(\| \neg \varphi_1 \& \neg \varphi_2 \|_{\underline{U}}[o] = 1) = p_{00}$$

(note that these numbers are independent of o by homogeneity).

The question is whether φ_1 and φ_2 are independent or associated; i.e., whether the satisfaction of φ_1 affects the satisfaction of φ_2 and vice versa.

Put $p_{1.} = P(\| \varphi_1 \|_{\underline{U}}[o] = 1) = p_{10} + p_{11}$, and analogously for $p_{0.}, p_{.0}, p_{.1}$. Independence is statistically expressed by $p_{ij} = p_{i.} \cdot p_{.j}$ ($i, j \in \{0, 1\}$).

We shall suppose further that $P_{ij} > 0$ for each $i, j \in \{0, 1\}$.

Independence is then equivalent to $\dfrac{p_{11}\, p_{00}}{p_{10}\, p_{01}} = 1$.

Edwards has proved in that each reasonable measure of association is a strictly monotone function of ratio

$\Delta = \dfrac{p_{11}\, p_{00}}{p_{10}\, p_{01}}$ or, equivalently, of $\delta = \log \Delta$; Δ is called the interaction and δ the logarithmic interaction (sometimes we shall write $\delta(p_{11}, p_{1.}, p_{.1})$, since $p_{11}, p_{1.}, p_{.1}$ determine all the probabilities in question). If $\delta > 0$ we say that the properties are positively associated, if $\delta < 0$ we say that they are negatively associated.

4.4.19 Remark.

(1) Note that independence is equivalent to $\delta = 0$; formally, $D_{12} \in \mathcal{D}_{T_1}$,

where $T_1 = \{\langle p_{11}, p_{1.}, p_{.1}\rangle ;\ \delta(p_{11}, p_{1.}, p_{.1}) = 0\} \subseteq \{0,1\}^3$

and D_{12} is the joint distribution of $\|\varphi_1\|_{\underline{U}}[o]$, $\|\varphi_2\|_{\underline{U}}[o]$ independent of o.

(2) The alternative hypothesis of positive association is then $D_{12} \in \mathcal{D}_{T_2}$, where

$T_2 = \{\langle p_{11}, p_{1.}, p_{.1}\rangle ;\ \delta(p_{11}, p_{1.}, p_{.1}) > 0\}$.

(3) Negative association of φ_1, φ_2 is equivalent to positive association of $\neg\varphi_1, \varphi_2$; moreover, $\delta_{\varphi_1, \varphi_2} = -\delta_{\neg\varphi_1, \varphi_2}$

4.4.20 Definition. For a given number $\alpha \in (0, 0.5]$ consider a quantifier $\sim_\alpha$ of type $\langle 1,1 \rangle$ with the following associated function:

$\mathrm{Asf}_{\sim_\alpha}(\langle M, f_1, f_2\rangle) = 1$

iff, putting $a = a_{\underline{M}}$, $b = b_{\underline{M}}$, etc., we have

$ad > bc$ and $\sum_{i=a}^{\min(r,k)} \sigma(i,r,k,m) \leq \alpha$, where

$$\sigma(a,r,k,m) = \frac{r!\, s!\, k!\, l!}{m!\, a!\, b!\, c!\, d!}$$

(here, of course, $b = r - a$, $c = k - a$, $s = m - r$, $l = m - k$, $d = s - c =$ $= l - b = m + a - r - k$).

We put $\mathrm{Fish}(a,r,k,m) = \sum_{i=a}^{\min(r,k)} \sigma(i,r,k,m)$.

The quantifier $\sim_\alpha$ is called the Fisher quantifier (on the level α) .

4.4.21 Theorem. Let α be a rational number, $0 < \alpha \leq 0.5$.

The sentence $\varphi_1 \sim_\alpha \varphi_2$ is an observational test on the level α of the null hypothesis $\delta \leq 0$ and the alternative hypothesis $\delta > 0$.

Proof. Let $\underline{U}$ and $M \subseteq U$ be given; hence $m = \mathrm{card}(M)$ is fixed. Consider samples from $\underline{U}_{\varphi_1, \varphi_2}$. For each $\underline{M}_\sigma$ put

$\mathrm{Marg}(\underline{M}_\sigma) = \langle r_{\underline{M}_\sigma}, k_{\underline{M}_\sigma} \rangle$ (marginal sums) . Write a_σ, b_σ, ... etc., instead of $a_{\underline{M}_\sigma}$, $b_{\underline{M}_\sigma}$, ...

(1) First, let us calculate the joint distribution of a_σ and b_σ under the assumption $a_\sigma + c_\sigma = k_\sigma = k$ and, thus, $b_\sigma + d_\sigma = m - k = l$. Under this assumption, a_σ and b_σ are two stochastically independent binomial variates with probability of success

$$p_{\varphi_1/\varphi_2} = \frac{p_{11}}{p_{.1}} \quad \text{and} \quad p_{\varphi_1/\neg\varphi_2} = \frac{p_{10}}{p_{.0}}$$

respectively. Moreover,

$$1 - p_{\varphi_1/\varphi_2} = \frac{p_{01}}{p_{.1}} \quad \text{and} \quad 1 - p_{\varphi_1/\neg\varphi_2} = \frac{p_{00}}{p_{.0}} \quad .$$ Then the

conditional joint probabilities of a_σ , b_σ are the following:

$P(a_\sigma = a \,\&\, b_\sigma = j / k_\sigma = k) =$

$$= \binom{k}{a} \left(\frac{p_{11}}{p_{.1}}\right)^{a} \left(\frac{p_{01}}{p_{.1}}\right)^{k-a} \binom{1}{j} \left(\frac{p_{10}}{p_{.0}}\right)^{j} \left(\frac{p_{00}}{p_{.0}}\right)^{1-j} =$$

$$= \binom{k}{a}\binom{1}{j} \left(\frac{p_{11}}{p_{.1}}\right)^{a} \left(\frac{p_{01}}{p_{.1}}\right)^{c} \left(\frac{p_{10}}{p_{.0}}\right)^{j} \left(\frac{p_{00}}{p_{.0}}\right)^{1-j} .$$

(2) Now we calculate the conditional probability

$P(\{\sigma \,;\, a_\sigma = a\} / \mathrm{Marg}(\sigma) = \langle r,k \rangle)$.

Using 4.4.14 we express this probability as

$$(*) \qquad \frac{P(a_\sigma = a \,\&\, r_\sigma = r / k_\sigma = k)}{\sum_{i=\max(0,r+k-m)}^{\min(r,k)} P(a_\sigma = i \,\&\, r_\sigma = r / k_\sigma = k)} .$$

Now, (using $r = a + b$) we obtain:

$P(a_\sigma = a \,\&\, r_\sigma = r / k_\sigma = k) = P(a_\sigma = a \,\&\, b_\sigma = b) =$

$$\binom{k}{a}\binom{1}{b} \left(\frac{p_{11}}{p_{.1}}\right)^{a} \left(\frac{p_{01}}{p_{.1}}\right)^{c} \left(\frac{p_{10}}{p_{.0}}\right)^{b} \left(\frac{p_{10}}{p_{.0}}\right)^{d} = C(r,k) \binom{k}{a}\binom{1}{b} \frac{p_{11}\, p_{00}}{p_{10}\, p_{01}}$$

where $C(r,k)$ depends only on r,k . Remember that $\Delta = \dfrac{p_{11}\, p_{00}}{p_{01}\, p_{10}}$,

(*) equals

$$(**) \qquad \frac{\binom{k}{a}\binom{1}{b} \Delta^{a}}{\sum_{i=\max(0,r+k-m)}^{\min(r,k)} \binom{k}{i}\binom{1}{r-i} \Delta^{i}} ;$$

note that for $\Delta = 1$ (**) reduces to the hypergeometrical distribution

$$P(a_\sigma = a / \mathrm{Marg}(\sigma) = \langle r,k\rangle) = \frac{\binom{k}{a}\binom{l}{b}}{\binom{m}{r}} = \sigma(a,r,k,m).$$

(Use $\sum_{i=\max(0,r+k-m)}^{\min(r,k)} \binom{k}{i}\binom{l}{r-i} = \binom{m}{r}$; see Problem (9).)

(3) Assume the null hypothesis $\Delta \leq 1$ (i.e., $\delta \leq 0$); we want to estimate $P(a_\sigma \geq a / \mathrm{Marg}(\sigma) = \langle r,k\rangle)$. By (**) , this probability equals

$$\frac{\sum_{j=a}^{\min(r,k)} \binom{k}{j}\binom{l}{r-j} \Delta^j}{\sum_{i=\max(0,r+k-m)}^{\min(r,k)} \binom{k}{i}\binom{l}{r-i} \Delta^i}$$

Using $\Delta \leq 1$, it is a matter of elementary treatment of inequalities to show that the above expression is less than or equal to

$$\frac{\sum_{j=a}^{\min(r,k)} \binom{k}{j}\binom{l}{r-j}}{\sum_{i=\max(0,r+k-m)}^{\min(r,k)} \binom{k}{i}\binom{l}{r-j}}$$

which equals $\sum_{i=a}^{\min r,k} \sigma(i,r,k,m)$, hence to $\mathrm{Fish}(a,r,k,m)$.

Consequently, we have proved

$$P(a_\sigma \geq a / \mathrm{Marg}(\sigma) = \langle r,k\rangle) \leq \mathrm{Fish}(a,r,k,m).$$

(4) Hence, assuming $\Delta \leq 1$, we have

$$P(\{\sigma ; \mathrm{Fish}(\underline{M}_\sigma) \leq \alpha\} / \mathrm{Marg}(\sigma) = \langle r,k\rangle) \leq$$

$$\leq P(\{\sigma ; P(a_\tau \geq a / \mathrm{Marg}(\sigma) = \langle r,k\rangle) \leq \alpha\}) \leq \alpha ;$$

the last inequality follows from the Diagonal Lemma 4.4.4 applied to conditional probabilities . But then

$$P(\{\sigma \ ; \ \mathrm{Fish}(\underline{M}_\sigma) \le \alpha\}) =$$

$$= \sum_{\langle r,k\rangle} P(\ \mathrm{Fish}(\underline{M}_\sigma) \le \alpha \ / \mathrm{Marg}(\sigma) = \langle r,k\rangle) P(\mathrm{Marg}(\sigma) = \langle r,k\rangle) \le$$

$$\le \alpha \sum_{\langle r,k\rangle} P(\mathrm{Marg}(\sigma) = \langle r,k\rangle) = \alpha \quad ,$$

using Lemma 4.4.14 (2) .

(5) For α rational the function $\mathrm{Asf}_{\sim\alpha}$ is recursive. This completes the proof.

4.4.22 Remark. (1) Via facti, the Fisher test can be considered to be a test on the level

$$\alpha_{crit} = \sum_{i=a_{\underline{M}}}^{\min(r,k)} \sigma(i, r_{\underline{M}}, k_{\underline{M}}, m_{\underline{M}}) \le \alpha$$

(for a given $\underline{M}$) i.e., had we used $\sim_{\alpha_{crit}}$ we should have rejected the null hypothesis too (assuming $\alpha_{crit} \le \alpha$) . Remember Lemma 4.4.4,

$$\alpha_{crit} = f(\underline{M}) = P(\tau \ ; g(\underline{M}_\tau) \ge g(\underline{M})) \ .$$

(2) The Fisher test is an unbiased test of the null hypothesis $\delta \le 0$ and the alternative hypothesis $\delta > 0$ (see Problem (10)).

(3) Recall the notions from 4.3.5 - 4.3.7. As proved in [Lehmann 1959], the Fisher test is uniformly most powerful in the class of unbiased tests of the null hypothesis $\delta \le 0$ and the alternative hypothesis $\delta > 0$. Thus, the Fisher test is a uniformly most powerful observational test of the above hypothesis.

4.4.23 Discussion and Definition.

On the other hand, the computation of the values of the Fisher test for larger m is complicated, the complexity of computation increasing rapidly. For these practical reasons, another test (the χ^2 - test) is widely used.

This test is only asymptotical, but the approximation is rather good for reasonable cardinalities ($a,b,c,d \geq 5$, $m \geq 20$). As will be seen later (in Section 5), the two tests have similar logical properties. Before defining the new quantifier, we have to define the notion of quantiles which will be used in many ways in the sequel.

Let a continuous one-dimensional distribution function $D(x)$ be given. For each $\alpha \in [0,1]$, the value $D^{-1}(\alpha)$ is called the α-quantile of D. (If ϑ is a variate and $D = D_\vartheta$ then $P(\vartheta^{-1}([D^{-1}(1-\alpha), +\infty))) = \alpha$).

Consider now the quantifier $\sim^2_\alpha$ of type $\langle 1,1\rangle$ with the associated function $\mathrm{Asf}_{\sim^2_\alpha}(\langle M,f_1,f_2\rangle) = 1$ iff $ad > bc$ and $m(ad-bc)^2 \geq \chi^2_\alpha\, r\,s\,k\,l$, where χ^2_α is the $1-\alpha$ quantile of the χ^2 - distribution function (i. e. the first χ^2 - distribution function; see Problem (3)). This quantifier is called the χ^2-quantifier on the level α.

4.4.24 Remark. (1) The quantifier $\sim^2_\alpha$ was defined for all real numbers $\alpha \in (0, 0.5]$. On the other hand, if we want $\sim^2_\alpha$ to be an observational quantifier (i.e., if we want to use it in an MOPC) then we shall restrict ourselves to those numbers α for which χ^2_α is a rational number. Remember that χ^2_α is the solution of the equation

$$\int_0^x \frac{e^{-\frac{y}{2}}}{\sqrt{2y}\,\Gamma(1/2)}\, dy = 1-\alpha \quad .$$

Hence, on the one hand, χ^2_α can be irrational even for rational **numbers**. On the other hand, χ^2_α is continuous as a function of α and, hence, if one starts with an α then one can deal with rational $\chi^2_{\alpha_0}$ for α_0 arbitrary close to α.

(2) Note that if $ad > bc$ then $r,s,k,l > 0$ and the ratio

$$m\,\frac{(ad-bc)^2}{r\,s\,k\,l}$$ is well defined.

(3) The class of χ^2 - tests used in statistics is very wide; these tests are useful in many situations (see Rao).

4.4.25 Lemma. Let $\{M_m\}_m$ be a sequence of samples such that $\operatorname{card}(M_m) = m$. Suppose $M_1 \subseteq M_2 \subseteq \ldots$. Consider the variates a_m, b_m, c_m, d_m given by the number of objects in M_m with cards $\langle 1,1\rangle$, $\langle 1,0\rangle$, $\langle 0,1\rangle$, $\langle 0,0\rangle$, respectively.

Denote

$$\mathcal{W}_m = \begin{cases} \dfrac{\left(a_m - \frac{r_m k_m}{m}\right)^2}{r_m k_m / m} + \dfrac{\left(b_m - \frac{r_m l_m}{m}\right)^2}{r_m l_m / m} + \dfrac{\left(c_m - \frac{s_m k_m}{m}\right)^2}{s_m k_m / m} + \dfrac{\left(d_m - \frac{s_m l_m}{m}\right)^2}{s_m l_m / m}, & \text{if } r_m, s_m, k_m, l_m > 0, \\ 0 & \text{otherwise,} \end{cases}$$

where $r_m = a_m + b_m$ etc.

Then, under the hypothesis $\delta = 0$, $(d)\lim_m \mathcal{W}_m = \mathcal{V}$, where $\mathcal{V}$ has the first χ^2-distribution function (for (d) lim see Problem (4)).

The proof is not trivial, see [Rao]6.a.1 - 6.d.2.

4.4.26 Theorem. Under the assumptions of 4.4.18, $\varphi_1 \sim_\alpha^2 \varphi_2$ is an asymptotical test (on the level α) of the null hypothesis $\delta = 0$ and the alternative hypothesis $\delta > 0$.

Proof. Note that

$$\mathcal{W}_m = m \frac{(a_m b_m - b_m c_m)^2}{r_m s_m k_m l_m} \quad (\text{if } r_m, s_m, k_m, l_m > 0).$$

Then $P(\|\varphi_1 \sim_\alpha^2 \varphi_2\| = 1 \mid \delta = 0) \leq P(\mathcal{W}_m \geq \chi_\alpha^2 \mid \delta = 0) = 1 - D_{\mathcal{W}_m}(\chi_\alpha^2)$.

Moreover, $\lim_{m \to +\infty} D_{\mathcal{W}_m}(\chi_\alpha^2) = D_{\mathcal{V}}(\chi_\alpha^2) = 1 - \alpha$ applying the definition of the number χ_α^2.

4.4.27 Remark. (1) The non-asymptotical distribution of $\mathcal{W}_m$ depends on the probabilities $p_{11}, p_{1.}, p_{.1}$. Remember that the null hypothesis is $D \in \mathcal{D}_{T_1}$, where

$$T_1 = \{ \langle p_{11}, p_{.1}, p_{1.} \rangle \; ; \; \delta(p_{11}, p_{1.}, p_{.1}) = 0 \} \subseteq (0,1)^3 \; ;$$

so that the probabilities $p_{11}, p_{1.}, p_{.1}$ are not specified.

(2) The number ad/bc (cross ratio) is, if defined, an estimate of Δ. Thus, if ad > bc (and hence $\log \frac{ad}{bc} > 0$, provided that bc > 0) and if, in addition,

$$\frac{(ad - bc)^2}{r\,s\,k\,l}\, m \geq \chi^2_\alpha$$

then we infer the alternative hypothesis $\delta > 0$. (If $(ad - bc)^2 m \geq \chi^2_\alpha \, r\,s\,k\,l$ we may have either ad > bc or ad < bc, so that $\frac{(ad - bc)^2}{r\,s\,k\,l} m \geq \chi^2_\alpha$ with ad > bc is more improbable, under $\delta = 0$, than $\frac{(ad - bc)^2 m}{r\,s\,k\,l} \geq \chi^2_\alpha$ by itself; for probabilities we have then $P(\|\varphi_1 \sim^2_\alpha \varphi_2\| = 1 \mid \delta = 0) \leq P(\mathcal{W}_m \geq \chi^2_\alpha \mid \delta = 0)$; see Problem (11).)

(3) If we omit frame assumptions summarized into a theoretical sentence Φ_o, the inference rules for sentences expressing association are of the following form:

(i) $$\frac{\varphi_1 \sim_\alpha \varphi_2}{\delta_{\varphi_1,\varphi_2} > 0} \, ,$$

and

(ii) $$\frac{\varphi_1 \sim^2_\alpha \varphi_2}{\delta_{\varphi_1,\varphi_2} > 0} \, .$$

Both of these are constructed from the point of view of the probability of an error of the first kind, i.e., they are of the type 4.3.3(3). Cf. 4.4.17(2), 1.1.6(L3).

Inference rules based on point estimation can also be used. Define the quantifier of simple association $\sim$ as a quantifier of type $\langle 1,1 \rangle$ with $Asf_\sim(\langle M, f_1, f_2 \rangle) = 1$ iff ad > bc and the quantifier of p-implication $\Rightarrow_p$ as a quantifier with $Asf_{\Rightarrow_p}(\langle M, f_1, f_2 \rangle) = 1$ iff $a \geq p(a+b)$. (Cf. 3.2.4.)

Corresponding inference rules are reasonable in our statistical framework only in the case of very large samples.

Nevertheless, the quantifiers mentioned can be useful in many non-statistical situations. In particular, they serve as simpler representatives of certain classes of quantifiers (e.g., associational quantifiers, see Chapter III) including (as complicated representatives) our quantifiers $\Rightarrow^?_p$, $\Rightarrow^!_p$, $\sim_\alpha$ and $\sim^2_\alpha$.

4.4.28 <u>Key words.</u> Likely p-implication, suspicious p-implication, the Fisher and χ^2 quantifiers; their test properties; p-implication and the simple association quantifier.

IV.5 Some properties of statistically motivated observational predicate calculi

In the previous section we defined some particular statistical quantifiers. Our first aim now is to prove that they belong to the class of associational or implicational quantifiers defined in Chapter III.

Our second aim is to discuss some properties of quantifiers based on tests in cross-nominal calculi and related topics.

4.5.1 Theorem. (1) The Fisher quantifier is associational.
(2) The χ^2-quantifier is associational.

Proof: The associationality of a quantifier can be proved in four steps: Let $\underline{M}_o$, $\underline{M}_4$ be two models, $\underline{M}_4$ a-better than $\underline{M}_o$. Consider models $\underline{M}_1$, $\underline{M}_2$, $\underline{M}_3$ such that if $q_{\underline{M}_o} = \langle a,b,c,d \rangle$ then $q_{\underline{M}_1} = \langle a+\Delta_1, b,c,d \rangle$, $q_{\underline{M}_2} = \langle a+\Delta_1, b-\Delta_2, c,d \rangle$, $q_{\underline{M}_3} = \langle a+\Delta_1, b-\Delta_2, c-\Delta_3, d \rangle$ and $q_{\underline{M}_4} = \langle a+\Delta_1, b-\Delta_2, c-\Delta_3, d+\Delta_4 \rangle$, where $\Delta_1, \Delta_2, \Delta_3, \Delta_4 \geq 0$.

It suffices to prove: If $\mathrm{Asf}_\sim(\underline{M}_o) = 1$ then $\mathrm{Asf}_\sim(\underline{M}_1) = 1$, if $\mathrm{Asf}_\sim(\underline{M}_1) = 1$ then $\mathrm{Asf}_\sim(\underline{M}_2) = 1$, if $\mathrm{Asf}_\sim(\underline{M}_2) = 1$ then $\mathrm{Asf}_\sim(\underline{M}_3) = 1$ and if $\mathrm{Asf}(\underline{M}_3) = 1$ then $\mathrm{Asf}_\sim(\underline{M}_4) = 1$. Associated functions of both the Fisher and χ^2- quantifiers are invariant under interchanging b and c and under interchanging a and d. Thus we have to prove the first two steps only. It is easy to see that we have to prove the desired property for the models $\underline{N}_1$ with $q_{\underline{N}_1} = \langle a+1,b,c,d \rangle$ and $\underline{N}_2$ with $q_{\underline{N}_2} = \langle a,b-1,c,d \rangle$ only.

(1) The Fisher quantifier : Remember the notation from 4.4.20

$$\sigma(i,r,k,m) = \frac{r!\, s!\, k!\, l!}{m!\, i!\, b!\, c!\, d!} \quad \text{and} \quad \Delta(a,r,k,m) = \sum_{i=a}^{\min(r,k)} \sigma(i,r,k,m).$$

We have defined $\mathrm{Asf}\ \underline{M} = 1$ iff $\Delta(a_{\underline{M}}, r_{\underline{M}}, k_{\underline{M}}, m_{\underline{M}}) \leq \alpha$ and $a_{\underline{M}} d_{\underline{M}} > b_{\underline{M}} c_{\underline{M}}$ (for a given α).

(a) First we prove that $\mathrm{Asf}_{\sim}(\underline{M}_o) = 1$ implies $\mathrm{Asf}_{\sim}(\underline{N}_1) = 1$. This means proving the inequality $\Delta(a+1, r+1, k+1, m+1) \leq \Delta(a,r,k,m)$ ($(a+1)d > bc$ being obvious). Observe that

$$\sigma(i+1,r+1,k+1,m+1) = \frac{(r+1)(k+1)}{(m+1)(i+1)} \sigma(i,r,k,m).$$

We note that $(r+1)(k+1)/(m+1)(i+1) \leq 1$ (since $(r+1)(k+1) = i^2+ic+i+ib+bc+b+i+c+1$, $(m+1)(i+1) = i^2 + ic + i + bc + id + b + i + c + 1$ and $id \geq bc$ for $i \geq a$).
Moreover, $\Delta(a+1, r+1, k+1, m+1) =$

$$= \sum_{i=a}^{\min(r+1,k+1)-1} \sigma(i+1,r+1,k+1,m+1)$$

and the number of members in this sum is equal to that in $\Delta(a,r,k,m)$.
Consider that $\sigma(a+1,r+1,k+1,m+1) \leq \sigma(a,r,k,m)$,
$\sigma(a+2, r+1, k+1, m+1) \leq \sigma(a+1, r,k,m)$ etc.

(b) We prove now that $\mathrm{Asf}_{\sim}(\underline{M}_o) = 1$ implies $\mathrm{Asf}_{\sim}(\underline{N}_2) = 1$. It is easy to see that

$$\sigma(i,r-1,k,m-1) = \sigma(i,r,k,m)\frac{mb}{rl} \text{ and } \frac{mb}{rl} \leq 1$$

(apply $ad > bc$). Compare members of $\Delta(a,r-1,k,m-1)$ with members of $\Delta(a, r, k, m)$:

$\sigma(a,r-1,k,m-1) \leq \sigma(a,r,k,m)$,

$\sigma(a+1,r-1,k,m-1) \leq \sigma(a+1,r,k,m)$ ·

etc. If $\min(r-1,k) = k$ then the last inequality is $\sigma(k,r-1,k,m-1) \leq \sigma(k,r,k,m)$,

if $\min(r-1,k) = r-1$ then the last inequality is

$$\sigma(r-1,r-1,k,m-1) \leq \sigma(r-1,r,k,m).$$

The inequality $\Delta(a,r-1,k,m-1) \leq \Delta(a,r,k,m)$ holds in both cases.

(2) χ^2-quantifier:

(a) $\mathrm{Asf}_{\sim}(\underline{M}_o) = 1$ implies $\mathrm{Asf}_{\sim}(\underline{N}_1) = 1$:

Remember that in the present case

$$\mathrm{Asf}_{\sim}(\underline{M}) = 1 \text{ iff } \frac{(a_{\underline{M}}d_{\underline{M}} - b_{\underline{M}}c_{\underline{M}})^2}{r_{\underline{M}}\, s_{\underline{M}}\, k_{\underline{M}}\, l_{\underline{M}}} m_{\underline{M}} \geq \chi^2_{\alpha} \text{ and } a_{\underline{M}}d_{\underline{M}} > b_{\underline{M}}c_{\underline{M}}$$

Thus we have to prove the following inequality (if $ad > bc$ then $(a+1)d > bc$ is obvious):

$$\frac{(ad-bc)^2}{rskl}\, m \le \frac{((a+1)d-bc)^2}{(r+1)(k+1)ls}\,(m+1) .$$

We prove a slightly stronger result:

$$\frac{(ad-bc)^2}{rkls} \le \frac{((a+1)d-bc)^2}{rkls+(r+k+1)ls} = \frac{((ad-bc)+d)^2}{rkls+(r+k+1)ls} . \qquad (*)$$

Let A, B, x, y be some numbers greater than 0. In this case

$$\frac{A}{B} \le \frac{A+x}{B+y} \quad \text{is equivalent to} \quad \frac{A}{B} \le \frac{x}{y} .$$

We apply this fact to

$$\frac{(ad-bc)^2}{rk} \le \frac{(ad-bc)^2+2d(ad-bc)+d^2}{rk+(r+k+1)}$$

which is obtained from $(*)$, and we see that the inequality in question is

$$\frac{a^2d^2-2adbc+b^2d^2}{rk} \le \frac{2ad^2-2bcd+d^2}{r+k+1}$$

and so

$$a^2d^2(r+k+1)+2bcdrk+b^2c^2(r+k+1) \le 2ad^2rk+2adbc(r+k+1)+d^2rk .$$

From $ad > bc$ we have $b^2c^2(r+k+1) \le adbc(r+k+1)$

and it remains to prove that

$$a^2d^2r+a^2d^2k+a^2d^2+2bcdrk \le 2ad^2rk+d^2rk+adbc(r+k+1) .$$

We use $rk = a^2+ab+ac+bc$ and $r+k+1 = 2a+b+c+1$

and we obtain, after the omission of equal members on both sides of the equality,

$$abc^2d+ab^2cd+2b^2c^2d \le$$

$$\le a^2bd+a^2cd^2+2ad^2bc+abd^2+acd^2+bcd^2+abcd .$$

Now use the inequalities $abc^2d \le a^2d^2c$, $ab^2cd \le a^2d^2b$ and $b^2c^2d \le abcd^2$.

(b) $Asf_{\sim}(\underline{M}_o) = 1$ implies $Asf_{\sim}(\underline{N}_2) = 1$.

Since ad > bc obviously implies ad >(b-1)c, we have to prove the following inequality:

$$\frac{(ad - bc)^2}{r\,s\,k\,l}\, m \leq \frac{(ad - (b-1)\,c)^2}{r-1 \;\; sk \;\; l-1}\,(m-1) .$$

It is equivalent to

$$\frac{m}{m-1} \quad \frac{r-1}{r} \quad \frac{l-1}{l} \leq \frac{(ad - b + c)^2}{(ad - bc)^2} . \qquad (**)$$

The right-hand side of (**) is greater than or equal to 1. Remember that $\frac{1}{r} \geq \frac{1}{m}$ and then $\frac{1 - 1/r}{1 - 1/m} \leq 1$. We obtain : $\frac{m}{m-1} \; \frac{r-1}{r} \leq 1$. Moreover, $\frac{l-1}{l} \leq 1$ and it can immediately be seen that the left-hand side of (**) is less than or equal to 1.

4.5.2 Theorem. Consider the following two inference rules in observational calculi with associational quantifiers

$$\text{SYM} = \left\{ \frac{\varphi \sim \psi}{\psi \sim \varphi} \; ; \quad \varphi, \psi \quad \text{designated open} \right\}$$

and

$$\text{NEG} = \left\{ \frac{\varphi \sim \psi}{\neg\varphi \sim \neg\psi} \; ; \quad \varphi, \psi \quad \text{designated open} \right\}. \quad \text{Cf. 3.2.17.}$$

These are the rules {1} -sound for the Fisher quantifier and χ^2-quantifier.

The proof is left to the reader. (Hint: the associated functions have the following form: $Asf_{\sim_\alpha}(\underline{M}) = 1$ iff ad > bc and $f_1(\langle a,b,c,d \rangle) \leq \alpha$ or $Asf_{\sim_\alpha}(\underline{M}) = 1$ iff ad > bc and $f_2(\langle a,b,c,d \rangle) \geq \chi^2_\alpha$ respectively. Prove that, for i = 1,2

$$f_i(\langle a,b,c,d \rangle) = f_i(\langle a,c,b,d \rangle) \qquad (\text{SYM})$$

and

$$f_i(\langle a,b,c,d \rangle) = f_i(\langle d,c,b,a \rangle) \qquad (\text{NEG}).$$

4.5.3 Theorem. The Fisher quantifier and the χ^2-quantifier are saturable.

Proof. Having in mind the three conditions of Definition 3. 2. 23, by the previous theorem we immediately see that the first and second conditions are equivalent.

Keep the notation from the previous proof. Now, note that under $ad > bc$ $f_1(\langle a,b,c,d\rangle)$ is decreasing in a and $f_2(\langle a,b,c,d\rangle)$ is increasing in a. Hence the first condition is satisfied.

For the third condition, note that the associated functions of both the quantifiers depend on the inequality $ad > bc$. If a model M has genus $\langle a,b,c,d\rangle$ take a model $\underline{M}'$ containing $\underline{M}$ with genus $\langle a,([ad/bc]+1)b,c,d\rangle$.

4.5.4 Theorem.

(1) The quantifier of suspicious p-implication is implicational.

(2) The quantifier of probable p-implication is implicational.

Proof. We shall use some facts known from mathematical statistics. These facts are fomulated in two lemmas.

(Lemma 1).

(1) $\underline{f}_p(k,m) \leq \alpha$ iff

$$\underline{g}_\alpha(k,m-k) = \frac{k}{k+(m-k+1)\,F_\alpha(2(m-k+1),\,2k)} \geq p,$$

(2) $\overline{f}_p(k,m) > \alpha$ iff

$$\overline{g}_\alpha(k,m-k) = \frac{(k+1)\,F_\alpha(2(k+1),\,2(m-k))}{(m-k)+F_\alpha(2(k+1),\,2(m-k))} > p,$$

where F_α is the $(1-\alpha)$ -quantile of the Fisher distribution (cf. Problem (3).) The proof of the lemma is purely analytical; the relation between $\overline{f}_p(k,m)$ and $I(p,k,m) = \int_0^{1-p} x^{m-k-1}(1-x)^k\,dx$

is used, namely $\bar{f}_p(k,m) = C(k,m)I(p,k,m)$.

(Lemma 2.)

(1) If $n_1 \leq n_2$ then $n_1 F_\alpha(n_1,n) \leq n_2 F_\alpha(n_2,n)$.

(2) If $n_1 \leq n_2$ then $n_2 F_\alpha(n,n_1) \geq n_1 F_\alpha(n,n_2)$.

Proof. (1) Let $\vartheta_{n_1}, \vartheta_{n_2-n_1}, \vartheta_{n_2}$ and ϑ_n be variates with the χ^2-distributions ; let ϑ_n be stochastically independent of $\vartheta_{n_2-n_1}$, ϑ_{n_2} and ϑ_{n_1}. Let $\mathcal{W}_i$ be a variate with the (n_i,n)-th F-distribution. Then (by definition) $n_i \mathcal{W}_i = \dfrac{\vartheta_{n_i}}{\vartheta_n/n}$. Using the properties of χ^2-distributions we can write $\vartheta_{n_2} = \vartheta_{n_1} + \vartheta_{n_2-n_1}$ provided that ϑ_{n_1} and $\vartheta_{n_2-n_1}$ are stochastically independent.

Thus $n_2 \mathcal{W}_2 = (\vartheta_{n_1} + \vartheta_{n_2-n_1})/(\vartheta_n/n) = n_1 \vartheta_1 + \vartheta_{n_2-n_1}/(\vartheta_n/n)$.

We have $P(\vartheta_{n_2-n_1}(\vartheta_n/n)^{-1} > 0) = 1$;

Hence $P(n_2 \mathcal{W}_2 > n_1 \mathcal{W}_1) = 1$. Thus, $P(n_1 \mathcal{W}_1 \geq x) \leq P(n_2 \mathcal{W}_2 \geq x)$, which is equivalent to $D_{n_1 \mathcal{W}_1}(x) \geq D_{n_2 \mathcal{W}_2}(x)$; applying the definition of the $(1-\alpha)$-quantile we obtain (1).

(2) Consider the variates $\vartheta_{n_1}, \vartheta_{n_2-n_1}, \vartheta_{n_2}$ and ϑ_n as above. Then $\mathcal{W}_i = \dfrac{n_i \vartheta_n}{n \vartheta_{n_i}}$ has the (n,n_i)-th F-distribution. Then $n_2 \mathcal{W}_1 \geq n_1 \mathcal{W}_2$

is equivalent (with probability 1) to $\frac{1}{\vartheta_{n_1}} \geq \frac{1}{\vartheta_{n_2}}$ and, hence, to

$\vartheta_{n_1} \leq \vartheta_{n_2}$; $P(\vartheta_{n_2} > \vartheta_{n_1}) = 1$ and so we obtain $P(n_2 \vartheta_{n_1} > n_1 \vartheta_{n_2}) = 1$.

Now we can prove the theorem. Recall the definition of

$\Rightarrow^{?}_{p,\alpha}$ and $\Rightarrow^{!}_{p,\alpha}$ 4.4.12

(1) $\Rightarrow^{?}_{p,\alpha}$: It is clear that, using Lemma 1, it suffices to prove

(a) $\bar{g}_\alpha(a,b) \leq \bar{g}_\alpha(a+1,b)$ and (b) $\bar{g}_\alpha(a,b) \leq \bar{g}_\alpha(a,b-1)$;

(a) is equivalent to

$$(a+1)\, F_\alpha(2(a+1),\ 2(n-a)) \leq (a+2)\, F_\alpha(2(a+2),\ 2(n-a))$$

which holds by Lemma 2, point (1) ;

(b) is equivalent to

$$(r-a-1)\, F_\alpha(2(a+1),\ 2(r-a)) \leq$$

$$\leq (r-a)\, F_\alpha(2(a+1),\ 2(r-a-1)).$$

Use Lemma 2, point (2) .

(2) $\Rightarrow^{!}_{p,\alpha}$: By Lemma 1 we have .

$\mathrm{Asf}_{\Rightarrow^{!}_{p,\alpha}}(\underline{M}) = 1$ iff $\underline{g}_\alpha(a_{\underline{M}}, b_{\underline{M}}) \geq p$.

Hence we have to prove the following:

(a) $\underline{g}_\alpha(a+1,b) \geq \underline{g}_\alpha(a,b)$ which is equivalent to

$$(a+1)\, 2(r-a+1)\, F_\alpha(2(r-a+1),\ 2a) \geq$$

$$a(2(r-a+1))\, F_\alpha(2(r-a+1),\ 2(a+1))$$

(use lemma 2, (2)) .

(b) $\underline{g}_\alpha(a, b-1) \geq \underline{g}_\alpha(a,b)$; here we consider an equivalent inequality

$$(r - a)\, F_{\alpha}(2(r - a), 2a) \leq (r - a + 1)\, F_{\alpha}(2(r - a + 1), 2a)$$

and Lemma 2,(1).

It remains to discuss some topics related to calculi with incomplete information on qualitative values.

4.5.5 <u>Discussion</u>. Now let $\underline{U} = \langle U, Q_1, \ldots, Q_n \rangle$ be a regular d-homogeneous random structure on a given Σ. For a given sample M, i.e., for a finite set of finite objects from U, we obtain a set of V-structures $\mathfrak{M}_M^V = \{\underline{M}_\sigma ; \sigma \in \Sigma\}$. Suppose now that $V \subseteq \mathbb{Q}$. Then the elements of $\mathfrak{M}_M^V$ are observational structures and we consider a situation as in 3.3.10. From observation we obtain a $V^\times$-structure $\underline{N} = \langle M, f_1, \ldots, f_n \rangle$ as incomplete information about a sample structure $\underline{M}_\sigma$.
Consider now random structures satisfying some frame assumption Φ.
We have two mutually incompatible distributional sentences Ψ_o and Ψ_1, and we will decide whether Ψ_1 is to be accepted on the basis of $\underline{N}$. Thus we are looking for an observational sentence φ such that **if** $\|\varphi\|_{\underline{N}} \in V_o$ we accept Ψ_1 and, for each given sample M, the probability of accepting Ψ_1, under the assumption of Ψ_o, is less than or equal to a number α given in advance. Then we have a <u>test based on incomplete information.</u> Now, the question is: (i) for which sentences is the above probability well defined, and (ii) how to construct such a test. The following theorem shows a way of solving these problems.

4.5.6 <u>Theorem.</u> Consider function calculi $\mathcal{F}$ and $\mathcal{F}^\times$. If a sentence φ of $\mathcal{F}$ is a test of a hypothesis Ψ_o against an alternative hypothesis Ψ_1 under a frame assumption Φ (on the significance level α) in the sense of models with complete information and if φ is secured in $\mathcal{F}^\times$, then φ is a test of Ψ_o and Ψ_1 under Φ (on the significance level α) based on incomplete information.

Proof. If $\underline{N}$ is a V-structure then $\|\varphi\|_{\underline{N}} = i$, $i \in V$, iff for each completion $\underline{M}$ of $\underline{N}$ $\|\varphi\|_{\underline{M}} = i$. If now $\underline{N}$ is some incomplete information about $\underline{M}$, then $\|\varphi\|_{\underline{N}} = i$ implies $\|\varphi\|_{\underline{M}_\sigma} = i$. Hence there is no σ such that

$\|\varphi\|_{\underline{M}_\sigma} \notin V_o$ and $\|\varphi\|_{\underline{N}} \in V_o$ for some incomplete information $\underline{N}$ about $\underline{M}$. Thus

$$\{\sigma \,;\, (\exists\, \underline{N})(\underline{N} \text{ incompl. inf. about } \underline{M}_\sigma \text{ and } \|\varphi\|_{\underline{N}} \in V_o\} \subseteq \qquad (*)$$
$$\subseteq \{\sigma \,;\, \|\varphi\|_{\underline{M}_\sigma} \in V_o\} .$$

4.5.7 Remark. (1) Note that Theorem 4.5.4 is independent of the particular deterministic or indeterministic way of obtaining the incomplete information $\underline{N}$ about the sample structure $\underline{M}_\sigma$. If this way is known, the theorem can be improved (we use now the most "pessimistic" way). See also 5.2.11 and Problem (6) of Chapter VII.

(2) Note that, in general, the inclusion in (*) from the proof of 4.5.6 is strict.

(3) The generalization for situations in which $V \nsubseteq \mathbb{Q}$ is straightforward.

We now turn our attention to tests related to multinomial distributions. These distributions describe probabilistic properties of theoretical models related to observational quantitative models as studied in Chapter III.

4.5.8 Definition and discussion. Consider a random variate ϑ ($\vartheta : \Sigma \to \mathbb{R}$). We say that ϑ has a multinomial distribution (h - valued) if there are numbers p_j , $j = 0,\ldots,h-1$, such that $p_j = P(\{\sigma \,;\, \vartheta(\sigma) = j\})$ and $\sum_{j=0}^{h-1} p_j = 1$. Analogously, we can say that $\langle \vartheta_1,\ldots,\vartheta_n\rangle$ has an n-dimensional multinomial distribution ($\langle h_1,\ldots,h_n\rangle$ -valued) if each ϑ_i has a multinomial distribution (h_i -valued). Now let $V_i = \{0,\ldots,h_i-1\}$ as above, and consider d-homogeneous regular $\langle V_1,\ldots,V_n\rangle$ -structures. Cf. 5.1.1 and 3.4.1.

Consider a distributional statement $\Phi(h_1,\ldots,h_n)$ such that

$\underline{U} = \langle U, Q_1, \ldots, Q_n\rangle \models \Phi(h_1,\ldots,h_n)$ iff, for each $o \in U$,

$< Q_1(o,-), \ldots, Q_n(o,-)>$ has $<h_1,\ldots,h_n>$ -valued n-dimensional multinomial distribution. It is clear that for each of the above mentioned structures we have $\underline{U} \models \Phi(h_1,\ldots,h_n)$.

4.5.9 Lemma. If $\underline{U} = < U, Q_1,\ldots,Q_n >$ is a d-homogeneous regular random $<V_1,\ldots,V_n>$ -structure and φ_1, φ_2 are two regular formulas of an MOFC with $<h_1,\ldots,h_n>$ -valued models, then $\underline{U}_{\varphi_1, \varphi_2}$ is d-homogeneous and regular.

Proof. Obvious. Remember that regular formulae are $\{0,1\}$ -valued.

4.5.10 Discussion. Under our homogeneity conditions, for each object $o \in U$, $\|\varphi_1\|_{\underline{U}}[o], \|\varphi_2\|_{\underline{U}}[o]$ has two dimensional alternative distribution. So we can apply our results obtained in Section 4 of Chapter IV to $\underline{U}_{\varphi_1, \varphi_2}$, and use the test quantifiers introduced there.

4.5.11 Key words: Associationality of Fisher and χ^2-quantifiers, implicationality of $\overset{?}{\Rightarrow}_{p,\alpha}$ and $\overset{!}{\Rightarrow}_{p,\alpha}$ quantifiers, tests based on incomplete information.

PROBLEMS AND SUPPLEMENTS TO CHAPTER IV

(1) A distribution function is absolutely continuous iff $D(x) = \int_{-\infty}^{x} f\,d\lambda$, where f is a non-negative measurable function and λ is the Lebesgue measure; one can write $D(x) = \int_{-\infty}^{x} f(y)\,dy$.

The function f is called the density of D.

More generally, a probability measure P is absolutely continuous w.r.t. a measure μ iff for each $E \in \mathcal{R}$ $P(E) = \int_E f\,d\mu$, where f is a measurable function from Σ to $\mathbb{R}$.

(2) Each discrete distribution function is absolutely continuous in the generalized sense. (Put

$$f(x) = \begin{cases} p_i & \text{if } x = x_i, \\ 0 & \text{otherwise} \end{cases}, \qquad \mu(A) = \begin{cases} 1 & \text{if there is an } x_i, x_i \in A \\ 0 & \text{otherwise} . \end{cases}$$

These definitions can be generalized for the case that ϑ maps Σ into a countable set $\{x_1, x_2, \dots\}$ and all conditions remain unchanged . Then μ is not a finite measure.)

(3) We collect some particular cases of distribution functions.

(a) The function $N_{\mu,\sigma}(x)$ with density

$$(2\pi)^{-\frac{1}{2}}\,\sigma^{-1}\,\exp(-(y-\mu)^2/2\sigma^2)$$

is called the normal distribution function (if a variate ϑ has this distribution function we say that ϑ has normal distribution) with parameters $\mu \in \mathbb{R}$ and $\sigma > 0$. Note that $E\vartheta = \mu$ and $VAR\,\vartheta = \sigma^2$. (Of course, exp x means the same as e^x.)

(Lemma) If a variate ϑ has normal distribution, then the variate $\dfrac{\vartheta - E\vartheta}{\sqrt{VAR\,\vartheta}}$ has normal distribution on with parameters 0 and 1 (normalized normal distribution with distribution function $N_{0,1}$).

(b) Consider a random variate ϑ such that $P(\vartheta^{-1}(\{0,1\}))=1$, i.e., ϑ can give with probability 1 only the values 0 or 1. Put $P(\vartheta^{-1}(\{1\}))=p$.

Then

$$D_{\vartheta}(x)=\begin{cases}0 & \text{for } x\leq 0\\ 1-p & \text{for } 0<x\leq 1\\ 1 & \text{for } x>1\end{cases};$$

this distribution is called the alternative distribution (with probability p of success). Our considerations of the statistical aspects of observational predicate calculi are naturally based on such distributions.

(c) Consider a sequence of s, independent random variates with the normalized normal distribution. Then the distribution function of the variate

$$\mathcal{W}_n=\sum_{i=1}^{n}\vartheta_i^2$$

is called the n-th χ^2-distribution function. It is absolutely continuous; its density is the following:

$$f(y)=\begin{cases}0 \text{ for } y<0,\\ c(n)\exp\left(-\frac{y}{2}\right)y^{\frac{n-2}{2}}, \text{ for } y\geq 0,\end{cases}$$

where

$$c(n)=\left[\int_0^{+\infty}\exp\left(-\frac{y}{2}\right)y^{\frac{n-2}{2}}\,dy\right]^{-1}.$$

(d) Consider two stochastically independent random variates ϑ_1, ϑ_2 with n_1-th and n_2-th χ^2-distributions respectively. Then

$\mathcal{W}=\frac{n_2}{n_1}\frac{\vartheta_1}{\vartheta_2}$ has the (n_1,n_2)-th F-distribution. (The (n_1,n_2)-th F-distribution function has the following density:

$$g(y) = \begin{cases} 0 & \text{for } y < 0, \\ c(n_1,n_2)\, y^{\frac{n_1}{2}-1}\left(1+\frac{n_1}{n_2}y\right)^{-\frac{n_1+n_2}{2}} & \text{for } y \geq 0, \end{cases}$$

where

$$c(n_1,n_2) = \left[\int_0^{+} y^{\frac{n}{2}-1}\left(1+\frac{n_1}{n_2}y\right)^{\frac{n_1+n_2}{2}} dy\right]^{-1}.)$$

(e) Consider an s.independent sequence $\vartheta_1,\ldots,\vartheta_n$ of random variates with an alternative distribution with equal probability of success p. The random variate $\mathcal{W} = \sum_{i=1}^{n} \vartheta_i$ (the number of successes in n independent alternative trials) has the so-called (n-th) binomial distribution:

$$P(\mathcal{W}^{-1}(k)) = \begin{cases} \binom{n}{k} p^k (1-p)^{n-k} & \text{for } k=0,\ldots,n \\ 0 & \text{otherwise}. \end{cases}$$

(The distribution function is then

$$D_{\mathcal{W}} = \sum_{k<x} \binom{n}{k} p^k (1-p)^{n-k} \quad .)$$

(4) We say that a sequence $\vartheta_1, \vartheta_2,\ldots$ of random variates converges in distribution to the random variate ($(d)\lim_i \vartheta_i = \vartheta$), if $\lim_i D_{\vartheta_i} = D_{\vartheta}$ at each continuity point of D_{ϑ}.

(5) Prove (1) of Theorem 4.2.14 (Hint: Consider a Borel σ-field $\mathcal{B}_{M,\rho}$ generated by the open subsets of $\mathcal{M}^V_M$ (w.r.t. the metric ρ). $f \restriction \mathcal{M}^V_M$ is then measurable in the following sense: If $A \in \mathcal{B}$ ($\mathcal{B}$ is the Borel σ-field of the subsets of $\mathbb{R}$) then for the inverse image $(f \restriction \mathcal{M}^V_M)^{-1}(A) \in \mathcal{B}_{M,\rho}$.

Investigate whether there are any $B \in \mathcal{B}_{M,\varrho}$ whose inverse image is an element of $\mathcal{R}$; use the measurability of the random variates $\vartheta_{1o}, \ldots, \vartheta_{no}$ for $o \in M$.)

(6) Prove (2) of Theorem 4.2.14 (Hint: By the definition of D_{f_M}, we have $D_{f_M}(x) = P(\{\sigma; f_M(\sigma) < x\}) = P(\{\sigma; \underline{M}_\sigma \in A\})$, where $A = \{\underline{M}_\tau \in \mathfrak{M}_M^V ; f(\underline{M}_\tau) < x\}$ is an element of $\mathcal{B}_{M,\varrho}$, so that we know its induced probability $P'(A) = P($ inverse image of $A)$; $P'(A) = \int_A d D_{\underline{\vartheta}}$ where $D_{\underline{\vartheta}}$ is an $n \times \text{card}(M)$ dimensional distribution function (Remark 4.2.5). Use d-homogeneity.)

(7) Consider the sentence $\hat{\varphi} = \hat{m}(\varphi)$ from 4.4.3. Prove

$E(\| \varphi \|_{\underline{M}_\sigma}) = p_\varphi$ (- an estimator with this property is called unbiased).

(8) Prove 4.4.16. (Hint: Let $\underline{U}$ and a sample $M \subseteq U$, card$(M) = m$, be given. Consider the conditional distribution of a_m under the condition $k_m = k$. Considerations of 4.4.5 - 4.4.9 can now be performed for such conditional distributions due to the fact that the conditional distribution of a_m under $k_m = k$ is the binomial distribution with the probability of success $p^{\underline{U}}_{\varphi_2/\varphi_1}$, i.e., $P(a_m = a / k_m = k) = \binom{k}{a} p^a_{\varphi_2/\varphi_1} (1 - p_{\varphi_2/\varphi_1})^{k-a}$ (under the assumption $\underline{U} \models p_{\varphi_2/\varphi_1}$). Apply theorem 4.4.10, using $\overset{!}{\Rightarrow}_{p,\alpha}, \overset{?}{\Rightarrow}_{p,\alpha}$ inetead of $!_{p,\alpha}, ?_{p,\alpha}$. Thus,

$$P(\| \varphi_1 \overset{!}{\Rightarrow}_{p,\alpha} \varphi_2 \|_{\underline{M}_\sigma} = 1 / k_{\underline{M}_\sigma} = k) \leq \alpha \text{ under the assumption}$$

$p_{\varphi_2/\varphi_1} \leq p$. Now, apply Lemma 4.4.14

$$(2): \quad P(\| \varphi_1 \overset{!}{\Rightarrow}_{p,\alpha} \varphi_2 \| = 1) = \sum_{k=0}^{m} P(\| \varphi_1 \overset{!}{\Rightarrow}_{p,\alpha} \varphi_2 \| = 1 / k_{\underline{M}} = k) P(k_{\underline{M}} = k)$$

$$\leq \alpha \sum_{k=0}^{m} P(k_m = k) = \alpha \quad .$$

(9) Prove that $\sum_{i=\max(0,r+k-m)}^{\min(r,k)} \binom{k}{i}\binom{m-k}{r-i} = \binom{m}{r}$.

(Hint: Note that $\frac{\binom{k}{i}\binom{m-k}{r-i}}{\binom{m}{r}}$ are hypergeometrical probabilities, so that

$$\sum_{i=\max(0,r+k-m)}^{\min(r,k)} \binom{k}{i}\binom{m-k}{r-i} \Big/ \binom{m}{r} = 1 \text{ . }$$)

(10) Prove the unbiasedness of the Fisher test. (Hint: Let m and α be given. For given r,k we can find a_o such that $\text{Fish}(a_o,r,k) \leq \alpha$ and $\text{Fish}(a_o-1,r,k) > \alpha$. Consider, further, $\alpha_o(r,k) = \text{Fish}(a_o,r,k)$.

Now let $\text{Marg}(\sigma) = \langle r,k \rangle$. Then $\| \varphi_1 \sim_\alpha \varphi_2 \|_{\underline{M}_\sigma} = 1$ iff $a_{\underline{M}_\sigma} = a_o$ and for $\delta = 0$

$$P(\| \varphi_1 \sim_\alpha \varphi_2 \|_{\underline{M}_\sigma} = 1 / \text{Marg}(\sigma) = \langle r,k \rangle) =$$

$$\frac{\sum_{j=a_o}^{\min(r,k)} \binom{k}{j} \binom{l}{r-j}}{\binom{m}{r}} = \alpha_o(r,k) \text{ .}$$

Remember (**) from the proof of 4.4.21 and then, for $\delta > 0$,

$$B_m(\varphi,\delta / \text{Marg}(\sigma) = \langle r,k \rangle) = P(\| \varphi_1 \sim_\alpha \varphi_2 \|_{\underline{M}_\sigma} = 1 / \text{Marg}(\sigma) = \langle r,k \rangle =$$

$$\frac{\sum_{j=a_o}^{\min(r,k)} \binom{k}{j}\binom{l}{r-j} \Delta^j}{\sum_{j=\max(0,r+k-m)}^{\min(r,k)} \binom{k}{j}\binom{l}{r-j}} > \text{Fish}(a_o,r,k) = \alpha_o(r,k) \text{ .}$$

Consider now $\alpha_o = \min_{\langle r,k \rangle} \alpha_o(r,k)$ and obtain thus the unbiasedness for given cardinality m (i.e., $B_m(\varphi, \delta) > \alpha_o$ for $\delta > 0$ and $B_m(\varphi, \delta) \leq \alpha_o$ for $\delta \leq 0$). But α_o depends on m. To prove general unbiasedness (and 4.4.22 (3)) it is necessary to use randomized tests (see[Lehmann]).)

(11) The quantifier $\sim^3_\alpha$ of type $\langle 1,1 \rangle$ with the associated function $Asf_{\sim^3_\alpha}(\langle M, f_1, f_2 \rangle) = 1$

$$\text{iff} \quad \frac{|\log ad/bc|}{\sqrt{\frac{1}{a}+\frac{1}{b}+\frac{1}{c}+\frac{1}{d}}} \geq \eta_{\frac{\alpha}{2}}$$

(if one of the frequences is zero, we replace it by 0.5), where $\eta_{\alpha/2}$ is the $(1 - \frac{\alpha}{2})$-quantile of the normalized normal distribution, is called the interaction quantifier (on the level $\frac{\alpha}{2}$) (see Anděl 1973).

(a) The interaction quantifier is an asymptotical test of the null hypothesis $\delta = 0$ and the alternative hypothesis $\delta > 0$.

(b) Prove

$$Asf_{\sim^3_\alpha}(\langle M, f_1, f_2 \rangle) = 1 \text{ iff } \frac{(\log ad/bc)^2}{\frac{1}{a}+\frac{1}{b}+\frac{1}{c}+\frac{1}{d}} \geq \chi^2_\alpha \text{ and } ad > bc .$$

(Hint: Use the fact that $\frac{(\log ad/bc)^2}{\frac{1}{a}+\frac{1}{b}+\frac{1}{c}+\frac{1}{d}} \geq \chi^2_\alpha$ iff $\frac{|\log ad/bc|}{\sqrt{\frac{1}{a}+\frac{1}{b}+\frac{1}{c}+\frac{1}{c}}} \geq \eta_{\frac{\alpha}{2}}$ and by the symmetry of the normal distribution.)

(c) $\sim^3_\alpha$ and $\sim^2_{\alpha/2}$ are asymptotically equivalent under the null hypothesis $\delta = 0$, i.e., for any increasing sequence $M_1 \subset M_2, \ldots$ of samples

$$\lim_m P(\{\sigma ; f^1_{M_m} \neq f^2_{M_m}\} | \delta = 0) = 0,$$

where $f^1_{M_m} = \dfrac{(\log a_m d_m / b_m c_m)^2}{\frac{1}{a_m}+\frac{1}{b_m}+\frac{1}{c_m}+\frac{1}{d_m}}$ and $f^2_{M_m} = \dfrac{(a_m d_m - b_m c_m)^2\, m}{r_m k_m s_m l_m}$

(For the proof see [Anděl 1974]: use the fact that for each interaction matrix

$$\mathbb{A} = \begin{pmatrix} a_{11}, & a_{12} \\ a_{21}, & a_{22} \end{pmatrix}$$ such that

$$a_{11} + a_{12} = a_{21} + a_{22} = a_{12} + a_{22} = a_{11} + a_{21} = 0$$

(note that then A is determined by a_{11}) Anděl's

$$\frac{d^2(\mathbb{A})}{s^2(\mathbb{A})} = \frac{(\log ad/bc)^2}{\frac{1}{a}+\frac{1}{b}+\frac{1}{c}+\frac{1}{d}} \quad .)$$

(12) Prove that $Asf_{\sim^2_\alpha}(\langle M, f_1, f_2\rangle) = 1$ iff

$$\frac{ad - bc}{\sqrt{rksl}}\sqrt{m} \geq \eta_{\frac{\alpha}{2}}$$

(Hint: $\dfrac{(ad - bc)^2}{rksl}\, m \geq \chi^2_\alpha$ iff $\dfrac{|ad - bc|}{\sqrt{rksl}}\sqrt{m} \geq \eta_{\alpha/2}$.)

Thus, $\varphi_1 \sim^2_\alpha \varphi_2$ is a test on the level $\frac{\alpha}{2}$.

If for $\delta \neq 0$, $\vartheta_m = \dfrac{a_m d_m - b_m c_m}{\sqrt{r_m s_m k_m l_m}}\sqrt{m}$ (if defined, otherwise $\vartheta_m = 0$) has the asymptotically normal distribution with expectation E such that E > 0 iff $\delta > 0$ and E < 0 iff $\delta < 0$, then $\varphi_1 \sim^2_\alpha \varphi_2$ is the asymptotically unbiased test of the null hypothesis $\delta \leq 0$ and the alternative hypothesis $\delta > 0$ on the level $\frac{\alpha}{2}$.

(13) <u>Lemma</u> Let m, r, k be given. Then Fish (a, r, k, m) $\leq \alpha$ implies ad > bc (for each $\alpha \in (0, 0.05)$).

<u>Corollary</u>. If we define the quantifier $\sim^F_\alpha$ of type $\langle 1,1 \rangle$ with the associated function $Asf_{\sim^F_\alpha}(\langle M, f_1, f_2\rangle) = 1$ iff $\mathrm{Fish}(a_{\underline{M}}, r_{\underline{M}}, k_{\underline{M}}, m_{\underline{M}}) \leq \alpha$

then, for any designated open formulae φ_1, φ_2, $\varphi_1 \sim_\alpha \varphi_2$ and $\varphi_1 \sim_\alpha^F \varphi_2$ are logically equivalent. Prove using the above lemma.

Nevertheless, for computation it is better to use $\sim_\alpha$ (computing $\| \varphi_1 \sim_\alpha \varphi_2 \|$), providing ad > bc and thus spare the computation of Fish(a,r,k,m) in many cases .

(14) Consider the random structure $\underline{U} = \langle U, Q_1, \ldots, Q_n \rangle$, $\underline{U} \models \Phi(h_1, \ldots, h_n)$. Denote by $\Phi'(h_1, \ldots, h_n)$ the distributional sentence such that $\underline{U}' \models \Phi'(h_1, \ldots, h_n)$

$$(\underline{U}' = \langle U, P_1^o, \ldots, P_1^{h_1 - 1}, \ldots, P_n^o, \ldots, P_n^{h_n - 1} \rangle)$$

iff $\underline{U}' \models \Phi(1, \ldots, 1)$ and for each

$$o \in U,\ P(\{\sigma ; \sum_{i=0}^{h_1 - 1} P_1^i(o, \sigma) = 1, \ldots, \sum_{i=0}^{h_n - 1} P_n^i(o, \sigma) = 1\}) = 1$$

Then (1) $\underline{U} \models \Phi(h_1, \ldots, h_n)$ iff $\pi(\underline{U}) \models \Phi'(h_1, \ldots, h_n)$

and (2) $P(\{\sigma ; P_i^j(o,\sigma) = 1\}) = P(\{\sigma ; Q_i(o,\sigma) = j\})$ $(j = 0, \ldots, h_i - 1,\ i = 1, \ldots, n)$.

(15) Let φ be a regular formula of a cross-qualitative OMFC. Then $p_\varphi^{\underline{U}}$ has a clear sense. Let $\underline{U}$ be given. Consider pairs of EC and assume $p_{\kappa \& \lambda} > 0$, $p_{\kappa \& \neg\lambda} > 0$, $p_{\neg\kappa \& \lambda} > 0$, $p_{\neg\kappa \& \neg\lambda} > 0$.

We can define

$$\Delta(\kappa, \lambda) = \frac{p_{\kappa \& \lambda} \quad p_{\neg\kappa \& \neg\lambda}}{p_{\kappa \& \neg\lambda} \quad p_{\neg\kappa \& \lambda}}$$

as in 4.4.20. Define a theoretical relation $\langle \kappa, \lambda \rangle \leq_a \langle \kappa', \lambda' \rangle$ and prove that $\underline{U} \models \langle \kappa, \lambda \rangle \leq_a \langle \kappa', \lambda' \rangle$ iff $\underline{U} \models \Delta(\kappa, \lambda) \leq \Delta(\kappa', \lambda')$.

Similarly for $<\kappa,\delta> \leq_{\iota} <\kappa',\delta'>$ and $p_{\delta/\kappa} \leq p_{\delta'/\kappa'}$

(16) Observe the following: (i) If $X \subseteq Y$, then $p_{(X)F} \leq p_{(Y)F}$, (ii) if $\kappa \subseteq \lambda$, then $p_{\kappa} \geq p_{\lambda}$, (iii) if $\kappa \sqsubseteq \lambda$, then $p_{\kappa} \leq p_{\lambda}$, (iv) if $\kappa \lessdot \lambda$, then $p_{\kappa} \geq p_{\lambda}$, (v) if $\delta_1 \lhd \delta_2$, then $p_{\delta_1} \leq p_{\delta_2}$. We can define $\leq_c$ on $[0,1]^2$. Then $<\kappa,\lambda> \lessdot\lessdot <\kappa',\lambda'>$ implies $<p_{\kappa}, p_{\lambda}> \geq_c <p_{\kappa'}, p_{\lambda'}>$.

(17) Prove Theorems 4.5.2 and 4.5.3 for the interaction quantifier.

Chapter V. Rank Calculi

The present chapter is devoted to the description and investigation of a particular class of observational calculi based on statistical procedures called rank tests.

Of what significance is this chapter for the logic of discovery? One finds observational calculi that are intuitively well motivated and are essentially non-two-valued. They might serve as a basis of future methods of mechanized Hypothesis Formation, as we shall outline in Section 4 of Chapter VII. In contradistinction to the calculi with associational quantifiers, the development of methods based on rank calculi is only in an early stage, but it seems to be promising.

As far as statistics is concerned, the present chapter has the following meaning: Statistical procedures investigated here were originally developed using considerations about the behaviour of various statistics on observational data. Later, the investigation of probabilistic properties of these statistics was preferred (cf.J.Hájek and Z.Šidák). But the application of rank tests in the logic of suggestion leads us back to the investigation of observational properties. We feel that this chapter should inspire statisticians to carry out a more detailed investigation of observational properties of these and other procedures.

The above shows that the present chapter should be read by the reader willing to develop actively the logic of discovery, be it from the point of view of logic, statistics or Artificial Intelligence. On the other hand, the chapter can be omitted; the main body of Part B does not presuppose its knowledge.

Recall associational quantifiers: they are characterized by the stability of the associated function w.r.t. a certain quasiordering of models and, consequently, w.r.t. transformations of data preserving the mentioned ordering. Rank tests can also be characterized by their behaviour w.r.t. a certain ordering of models.

They are called "rank tests" because they are invariant w.r.t. transformations of real-valued models consisting of the replacement of any rational-valued function by another rational-valued function which defines the same notion of ranks of objects, i.e., which orders the domain in the same manner. Rank tests will be used as tests for a fixed null hypothesis - the hypothesis of d-homogenity. But this null hypothesis will be joined to various alternative hypotheses, each alternative hypothesis determining some additional requirements concerning appropriate tests (see Section 1). It is easy to see that it suffices to study models with particular rational-valued, in fact natural-number valued functions, namely enumerations of the domain. Here, the null hypothesis reduces to the hypothesis of a uniform distribution of enumerations (see Section 1).

As an example of an alternative hypothesis, we cite the hypothesis of a shift in location. This hypothesis concerns random universes with one real-valued function (quantity) and one two-valued function (property). Roughly speaking, it means that the property divides the universe into two groups such that the first group is characterized by greater values of the quantity. If we restrict ourselves to observational models with enumerations, then the rank statistics for testing this hypothesis define quantifiers having a property of distinctiveness; distinctive quantifiers can be studied from the observational point of view much as associational quantifiers were studied in the preceding chapters. This is done in Section 3.

More generally, we can study stochastical dependence of two random quantities cf. 5.2: we mean positive dependence, i.e., if we consider the alternative hypotheses $Q_1 = Q_1^* + \Delta Z$, $Q_2 = Q_2^* + \Delta Z$, where Q_1^*, Q_2^*, Z are mutually independent, we suppose $\Delta > 0$ see[J.Hájek and Z.Šidák, II.4.11]: we obtain observational correlational quantifiers. Section 3 and the first part of Section 4 are devoted to calculi with the above mentioned quantifiers and models with properties and enumerations: in the second part of Section 4 we generalize our calculi to calculi with rational-valued functions and extend definitions of the quantifiers considered throuthout this chapter, in accordance with statistical rank tests as they are really used.

V.1 Generalized random structures and the hypothesis H_o of d-homogeneity.

In the present and following sections a particular class of statistical tests is introduced. These tests are used for testing a very general null hypothesis against broad alternative hypotheses; from the observational point of view they are thus stable under some transformations of models. We shall describe these tests from the above point of view in Section 3; in Section 4 they will be used as a background for a class of quantifiers in calculi with real-valued models.

First, in the present section we consider $\underline{\Sigma}$ -random structures of type $\langle 1,1 \rangle$ which satisfy some particular assumptions that enable us to introduce the above mentioned tests in a comprehensible way.

In Section 2, these tests are generalized for a more general null hypothesis and alternative hypotheses. This generalization is necessary for the use of such tests as a source of quantifiers for observational calculi useful in automated research.

5.1.1 Discussion and Definition. First, we have to generalize some notions from Section 2 of Chapter IV. Consider an n-tuple $\mathbb{V} = \langle V_1,\ldots,V_n \rangle$, $V_i \subseteq \mathbb{R}$; if M is a non-empty set and, for each $i = 1,\ldots,n$, f_i is a mapping of M into V_i, then the tuple $\underline{M} = \langle M,f_1,\ldots,f_n \rangle$ is called a $\mathbb{V}$ -structure.

In analogy to 4.2.1, we can now define a random $\mathbb{V}$ -structure: Let a probability space $\underline{\Sigma} = \langle \Sigma, \mathcal{R}, P \rangle$ be given, then a $\underline{\Sigma}$ -random $\mathbb{V}$ -structure is any structure $\underline{U} = \langle U,Q_1,\ldots,Q_n \rangle$ where U is a set and Q_i are mappings of $U \times \Sigma$ into V_i. In the following, we shall consider only regular random $\mathbb{V}$ -structures: i.e., random structures for which

(o) U is a recursive set,

(1) each $Q_i(o_i,-)$ as a mapping of Σ to V_i is a random variate (denoted by $\vartheta_{i,o}$ or $\vartheta^{U}_{i,o}$),

(2) for any sequence $o_1,\ldots,o_m$ of elements of U, the sequence of n-dimensional random variates

$$\{< Q_1(o_i,\cdot), \ldots, Q_n(o_i,\cdot)>\}_{i=1,\ldots,m}$$

is stochastically independent,

and (3) V_i are regular sets of values.

Unlike 4.2.3, d-homogeneity will not be automatically assumed.

First, we have to generalize slightly the notion of a distributional sentence. It is good sense to consider for any sequence $\underline{o} = < o_1,\ldots,o_m >$ of elements of U the joint distribution function $D_{\underline{\vartheta},\underline{o}}$ $(\underline{\vartheta} = < \vartheta_{1,o},\ldots, \vartheta_{n,o}>$.

Then, a theoretical sentence Φ is called <u>distributional</u> if for any regular random $\mathbb{V}$ - structures $\underline{U}$ and $\underline{U}'$ of the same type we have the following:

If $\underline{U} \models \Phi$ and if there is a one-to-one mapping ι of U onto U' such that, for each sequence $\underline{o}$ of elements of U, $D^{\underline{U}}_{\underline{\vartheta},\underline{o}} = D^{\underline{U}'}_{\underline{\vartheta},\iota(o)}$, then $\underline{U}' \models \Phi$.

In the present section, we consider Σ -random $\mathbb{V}$ -structures, i.e., different quantities can have different ranges of values. This gives one reason for modifying the notion of a continuous statistic. Further reasons will be clear from considerations following Definition 5.1.2.

5.1.2 <u>Definition</u>. Consider $\mathfrak{M}^{\mathbb{V}} = \{< M, f_1,\ldots,f_n > \;:\; f_i$ is V_i-valued and M is a finite set$\}$

As in 4.2.10, we define $\mathfrak{M}^{\mathbb{V}\cap\mathbb{Q}}$ and, for any given M , $\mathfrak{M}^{\mathbb{V}}_M$ and $\mathfrak{M}^{\mathbb{V}\cap\mathbb{Q}}_M$.

First, we generalize the notion of a continuous computable statistic (cf.4.2.11) :

A mapping $t : \mathfrak{M}^{\mathbb{V}} \rightarrow \mathbb{R}$ is a <u>continuous computable statistic</u> if:

(a) it is invariant under isomorphisms,

(b) for each M , $t \restriction \mathfrak{M}^{\mathbb{V}}_M$ is continuous, and

(c) $t \restriction \mathfrak{M}^{\mathbb{V}\cap\mathbb{Q}}$ is recursive.

Now let Φ be a distributional statement. Consider regular random $\mathbb{V}$ -structures. A mapping $t : \mathfrak{M}^{\mathbb{V}} \rightarrow \mathbb{R}$ is called an <u>almost continuous computable statistic w.r.t. Φ</u> if t is Borel measurable and satisfies conditions (a) and (c) and (b') : For each $\underline{U}$ such that $\underline{U} \models \Phi$ and for each finite

$M \subseteq U$, M finite, $t \upharpoonright \mathfrak{M}^{\mathbb{V}}_{M}$ is continuous on an open set $\mathfrak{M}_{cont} \subseteq \mathfrak{M}^{\mathbb{V}}_{M}$ such that

$$P^{\underline{U}}(\{\sigma : \underline{M}_{\sigma} \in \mathfrak{M}_{cont}\}) = 1 .$$

(Also see problem (9)).

5.1.3 Example. Consider random $\langle\{0,1\}, V\rangle$- structures. Let Φ be true in $\underline{U} = \langle U, Q_1, Q_2\rangle$ iff (a) Q_1 is independent on σ (i.e., for each $o \in M$, $\sigma_1, \sigma_2 \in \Sigma$, $Q_1(o, \sigma_1) = Q_1(o, \sigma_2)$,

(b) $\underline{U}_2 = \langle U, Q_2\rangle$ is d-homogeneous, and (c) $D_{\underline{U}_2}$ is continuous.

Define a mapping t as follows:

$t(M_{\sigma}) = 1$ iff, for each $o_1, o_2 \in M$, $Q_1(o_1) = 1$ and $Q_1(o_2) = 0$ implies $Q_2(o_1,\sigma) \geq Q_2(o_2,\sigma)$. t is an almost continuous computable statistic w.r.t. Φ.

5.1.4 Remark. Note that for each open set A the set $A \cap \mathbb{Q}$ is dense in A. Thus, if t is an almost continuous computable statistic, then, for each $\underline{M} \in \mathfrak{M}_{cont}$, the value $t(\underline{M})$ can be approximated by values of t on rational models - elements of $\mathfrak{M}^{\mathbb{V} \cap \mathbb{Q}} \cap \mathfrak{M}_{cont}$.

One can see that the statistics considered below fulfilla condition of "good" approximation at discontinuity points; but a general description of such conditions is beyond the scope of the present book and will be presented elsewhere.

5.1.5 Discussion and Definition (frame assumptions).

We shall now consider regular random $\mathbb{V}$-structures of type $\langle 1,1\rangle$ i.e., $\underline{U} = \langle U, Q_1, Q_2\rangle$, such that $V_2 = \mathbb{R}$. In all the considerations of the present section we shall suppose the following frame assumptions:

(1) for each $o \in U$, Q_1 does not depend on σ, and

(2) for each $o \in U$, $D_{\vartheta_2, o}$ is continuous.

Conditions (1) and (2) will be called d.c. - conditions (d - Q_1 is deterministic, c - continuity condition concerning Q_2).

Note that (2) is distributional.

5.1.6 Discussion and Definitions (hypotheses).

We can now formulate the hypothesis of d-homogeneity which is usually denoted by H_o in statistical textbooks ; see [J.Hájek],[J.Hájek and D.Vorlíčková]:

$\langle U, Q_2 \rangle$ is d-homogeneous.

This hypothesis will serve as a general null hypothesis. Note that H_o is a distributional theoretical sentence.

We shall test the null hypothesis H_o against different alternative hypotheses. Thus, we shall consider pairs consisting of the null hypothesis H_o of d-homogeneity and an alternative hypothesis.

We now present examples of these alternative hypotheses:

(i) The alternative hypothesis of a shift in location (ASL).

Suppose, moreover, that the frame assumption

(3) $V_1 = \{0,1\}$ holds.

Then ASL can be formulated as follows:

There is a function $F(x)$ such that, for each $o \in U$,

$$D_{\vartheta_2, o} = \begin{cases} F(x) \text{ if } Q_1(o) = 0, \\ F(x - \Delta) \text{ if } Q_1(o) = 1, \end{cases}$$

where $\Delta \neq 0$.

Notice that if we define

$\underline{U}_1 = \langle U \restriction \{o, Q_1(o) = 1\}, Q_2 \rangle$ and

$\underline{U}_o = \langle U \restriction \{o, Q_1(o) = 0\}, Q_2 \rangle$, then $\underline{U}_1$ and $\underline{U}_o$ are d-homogeneous.

If we put $\Delta = 0$, then we obtain the hypothesis H_o.

It is clear that Q_1 divides our observational sample into two subsample groups. ASL states that these groups differ in Q_2 (problem of two samples).

We shall suppose, in constructing tests, that $\Delta > 0$. Then ASL means that the values of Q_2 in $\underline{U}_1$ are expected to be greater than values of Q_2 in $\underline{U}_o$ (see Example 5.1.23). This means that, for each $x \in \mathbb{R}$,

$$P(\{\sigma\ ;\ Q_2(o,\sigma) \geq x\} \mid Q_1(o) = 1) \geq P(\{\sigma\ ;\ Q_2(o,\sigma) \geq x\} \mid Q_1(o) = 0),$$

and for some x the inequality is strict.

(ii) <u>The alternative hypothesis of natural regression in location</u> (ANRL). Suppose, besides (1) and (2), that the frame assumption

(4) $V_1 = \mathbb{N}$ holds.

Then we can formulate ANRL as follows:

There is a function $F(x)$ such that

$$D_{\vartheta_2,o}(x) = F(x - i\Delta) \quad \text{if} \quad Q_1(o) = i,$$

where $\Delta \neq 0$.

(iii) <u>Alternative hypothesis of trend in location</u> (ATL). Suppose, besides (1) and (2), that

(5) $V_1, V_2 \subseteq \mathbb{R}$

ATL then means the following:

for each $o_1, o_2 \in U$,
if $Q_1(o_1) < Q_1(o_2)$, then $D_{\vartheta_2,o_1} < D_{\vartheta_2,o_2}$

(Q_1 can be, e.g., time).

For further alternative hypotheses, see Problem (1).

5.1.7 Discussion. We have seen that the hypothesis H_o of d-homogeneity does not specify the distribution function $D^{U}_{\vartheta_2}$; only the frame assumptions require this function to be continuous. The same holds for the alternative described hypotheses. So we want to have tests for H_o whose basic properties would not depend on particular assumptions concerning the distribution function $D^{U}_{\vartheta_2}$. Such tests are called distribution free tests.

A particular, and the most important, class of such tests are rank tests. To define rank tests and to investigate their basic properties we must mention some facts about enumerations. (These facts are rather trivial; our exposition is based on [J. Hájek and D. Vorlíčková]; see also [J. Hájek].)

5.1.8 Definition. Let M be a finite non-empty set, let card(M) = m. An enumeration of M is a one-to-one mapping of M onto $\{1,\ldots,m\}$. $\mathcal{R}_M$ denotes the set of all enumerations of M; clearly, $\mathcal{R}_M$ has m! elements. Let $\mathcal{P}(\mathcal{R}_M)$ be the power set of $\mathcal{R}_M$.

Let Σ be a random space and let R be a random variate on Σ with values in $\mathcal{R}_M$. Thus, R is called an enumerating random variate for M. R is said to induce a uniform distribution on $\langle \mathcal{R}_M, \mathcal{P}(\mathcal{R}_M)\rangle$ if, for each $\eta \in \mathcal{R}_M$, $P(\{\sigma ; R(\sigma) = \eta\}) = \frac{1}{m!}$

(this means that we have a normalized counting measure on $\langle \mathcal{R}_M, \mathcal{P}(\mathcal{R}_M)\rangle$. We abbreviate $P(\{\sigma ; R(\sigma) = \eta\})$ by $P(R = \eta)$, similarly in other cases. If $o \in M$, then R_o denotes the random variate defined by $R_o(\sigma) = (R(\sigma))(o)$ (the value of $R(\sigma)$ on the object o).

5.1.9 Lemma. Let R be as in 5.1.8 and let R induce a uniform distribution. Then, for any $o_1, o_2 \in M$,

$$P(R_{o_1} = k \text{ and } R_{o_2} = h) = \begin{cases} P(R_{o_1} = k) = \frac{1}{m} & \text{for } o_1 = o_2,\ 1 \leq h = k \leq m, \\ 0 & \text{for } o_1 \neq o_2,\ 1 \leq h = k \leq m, \\ \frac{1}{m(m-1)} & \text{for } o_1 \neq o_2,\ 1 \leq h \neq k \leq m. \end{cases}$$

Proof. Consider the first case. We have a normalized counting measure on $\mathcal{R}_M$ so that

$$P(R_{o_1} = k) = \frac{1}{m!} \text{ card } \{\eta \in \mathcal{R}_M ; \eta(o_1) = k\} = \frac{1}{m!}(m-1)! = \frac{1}{m}.$$

Other cases can be proved in the same way.

5.1.10 Definition. Let $c_o, c_1, \ldots, a_o, a_1, \ldots$ be rational constants and let M be a finite non-empty set of natural numbers.. Then the function l_M defined on $\mathcal{R}_M$ by setting

$$l_M = \sum_{o \in M} c_o a_{\eta(o)}$$

is called a simple linear function on $\mathcal{R}_M$.

5.1.11 Remark.(1) Observe that l is a recursive function on $\{\langle M, \eta\rangle ;$ M a finite set of natural numbers, $\eta \in \mathcal{R}_M\}$.

(2) If R is an enumerating random variate for M (on $\underline{\Sigma}$) and if l_M is as in 5.1.10, then $\mathcal{L} = l_M \circ R$ is a random variate (on $\underline{\Sigma}$). o denotes composition, i.e., $\mathcal{L}(\sigma) = l_M(R(\sigma))$. We shall use simple linear functions for testing H_o.

5.1.12 Lemma (on moments of simple linear functions). Let R be an enumerating random variate (for M), and let R induce the uniform distribution on $\langle \mathcal{R}_M, \mathcal{P}(\mathcal{R}_M)\rangle$. Consider a variate $\mathcal{L} = l_M \circ R$, where l_M is a simple linear function on $\mathcal{R}_M$. Then

(1) $E\mathcal{L} = m\bar{c}\bar{a}$, where $\bar{c} = \frac{1}{m} \sum_{o \in M} c_o$

and $\bar{a} = \frac{1}{m} \sum_{o \in M} a_{\eta(o)}$,

(2) $\mathrm{VAR}\,\mathcal{L} = \frac{1}{m-1} \sum_{o \in M} (c_o - c)^2 \sum_{o \in M} (a_{\eta(o)} - a)^2$;

η is an arbitrary element of $\mathcal{R}_M$.

Proof. Note that $\sum_{o \in M} a_{\eta(o)}$, and hence $\bar{a}$, do not depend on η ; similarly for $\sum_{o \in M} (a_{\eta(o)} - a)^2$.

(1) Remember that $\mathcal{L} = \sum_{o \in M} c_o a_{R_o}$: hence, $E\mathcal{L} = \sum_{o \in M} c_o E a_{R_o}$; for each $o \in M$, we then have

$$E a_{R_o} = \sum_{j=1}^{m} a_j P(R_o = j) = \sum_{j=1}^{m} a_j \frac{1}{m} \quad \text{(counting measure)}.$$

The proof of (2) is then an algebraic exercise using 5.1.9.

5.1.13 Lemma. Let s be any function mapping $\mathcal{R}_M$ into $\mathbb{Q}$. Consider the variate $\mathcal{S} = s \circ R$, where R is an enumerating random variate yielding the uniform distribution on $\langle \mathcal{R}_M, \mathcal{P}(\mathcal{R}_M) \rangle$. Then, for each $A \subseteq \mathbb{Q}$

$$P(\mathcal{S} \in A) = \frac{1}{m!} \operatorname{card} \{ \eta \in \mathcal{R}_M ; s(\eta) \in A \}.$$

Proof. We have

$$P(\{\sigma ; \mathcal{S}(\sigma) \in A\}) = P(\{\sigma ; R(\sigma) \in s^{-1}(A)\}) = \frac{1}{m!} \operatorname{card} \{ \eta \in \mathcal{R}_M ; s(\eta) \in A\}.$$

Use $s^{-1}(A) \in \mathcal{P}(\mathcal{R}_M)$.

5.1.14 Definition and Discussion. Let M be a finite set and let f be a real-valued function on M. For each $o \in M$, define the f - rank of o by

$$\mathrm{rk}_f(o) = \operatorname{card} \{ o_1 \in M ; f(o_1) \leq f(o) \} .$$

We shall also write $f^*(o)$ instead of $\mathrm{rk}_f(o)$. Note that f is an enumeration of M iff f is injective (one-to-one; there are no ties). Denote by $\mathcal{JM}^{\mathbb{R}}$ the set of all real-valued models $< M,f >$ where f is injective and M is finite non-empty (analogously, $\mathcal{JM}^{\mathbb{Q}}$, $\mathcal{JM}^{\mathbb{R}}_M$ etc.)

The objects of our interest are structures of type $< 1, 1 >$. We introduce some particular frame assumptions:

Φ_o is the d.c. -condition as specified in 5.1.5, i.e.,

Φ_o says that the first quantity is deterministic and the second has a continuous distribution function;

Φ_2^- says that $V = \langle V_1, \mathbb{R} \rangle$, where $V_1 \subseteq \mathbb{Q}$, and for each $o \in U$,

that $D_{\underline{U},o}$ is a continuous function of the second variable and, for each $o \in U$ and each $v \in V_1$,

$$P(\{\sigma \; ; Q_1(o,\sigma) = v\}) > 0$$

(the p.c. - condition - positive probability of values of the first quantity and continuous distribution of the second).

Φ_2 says that $V = \langle \mathbb{R}, \mathbb{R} \rangle$ and that $D_{\underline{U},o}$ is a two-dimensional continuous function for each $o \in U$ (the t. c. - condition). In the following, Φ denotes any of the assumptions Φ_o, Φ_2^-, Φ_2.

Two structures $\underline{M}_1 = < M_1, f_1, g_1 >$, $\underline{M}_2 = < M_2, f_2, g_2 >$ are called rank equivalent (w.r.t. the second function) iff the structures $< M_1, f_1, g_1^* >$, $< M_2, f_2, g_2^* >$ are isomorphic. For each $\underline{M} = < M,f,g >$, we put $Rk(\underline{M}) = < M,f,g^* >$.

Let t be a statistic (defined on models of type $< 1,1 >$ and almost continuous w.r.t. some frame assumption Φ).

t is a rank statistic if

$$Rk(\underline{M}_1) = Rk(\underline{M}_2) \quad \text{implies} \quad t(\underline{M}_1) = t(\underline{M}_2) . \qquad (*)$$

5.1.15 Lemma. The conjunction of the following conditions (i), (ii) is sufficient for a Borel measurable function t satisfying $(*)$ to be an almost continuous computable statistic w.r.t. Φ .

(i) For each $\langle M, f_1 \rangle$, t is continuous on the set of all models $\langle M, f_1, f_2 \rangle$ where f_2 is injective.

(ii) t restricted to rational-valued models is recursive.

Proof. Use Lemma 5.1.19.

5.1.16 Remark. Evidently, an almost continuous statistic t is a rank statistic iff there is a function s such that, for each $\underline{M}$, $t(\underline{M}) = s(Rk(\underline{M}))$; s is defined on all models $\langle M, f, \eta \rangle$ where f is V_1-valued and the range of η is included in $\{1, \dots, card(M)\}$. For fixed M and f, we obtain a function $s_{M,f}$ defined on all mappings of M into $\{1, 2, \dots, card(M)\}$.

5.1.17 Discussion. We restrict ourselves to random structures satisfying Φ_o. Suppose, further, that $V_1 \subseteq \mathbb{Q}$. Let $\underline{U} = \langle U, Q_1, Q_2 \rangle$ be such a structure. We shall consider inference rules (for hypothesis testing) based on rank statistics. In all cases, the null hypothesis will be the hypothesis H_o of d-homogeneity. Let A be an alternative hypothesis, and let t be a rank statistic. Our inference will be as follows:

We have an observational structure $\langle M, f_1, f_2 \rangle$ regarded as a sample from a universe $\underline{U}$ satisfying Φ_o. Here, f_1 is deterministic; i.e., whenever $\sigma \in \Sigma$ we have $\underline{M}_\sigma = \langle M, f_1, f_2' \rangle$ for some f_2' (σ influences only the second function). Let $c_\alpha(M, f_1)$ be such that if $\underline{U} \models H_o$ then $P^{\underline{U}}(\{\sigma ; t(\underline{M}_\sigma) \geq c_\alpha(M, f_1)\}) \leq \alpha$. Hence, in accordance with the theory of hypothesis testing, if $\underline{M}_\sigma = \langle M, f_1, f_2 \rangle$ is the observed structure and if $t(\underline{M}_\sigma) \geq c_\alpha(M, f_1)$, then we infer A (the form of the set V_o (critical set, c_α - critical value), i.e., $V_o = [c_\alpha(M, f_1), +\infty)$ is given by the alternative hypothesis). For example, in ASL, we assume $\Delta > 0$ (cf. 5.1.6) and we construct the appropriate test statistic (cf. 5.1.23) for which greater values are expected under the alternative hypothesis than under H_o.

The inference rule then consists of pairs of the form

$$\frac{\Phi_o, \ \varphi[t, c_\alpha]}{A},$$

where $\varphi[t, c_\alpha]$ is an observational sentence true in a model $\underline{M} = \langle M, f_1, f_2 \rangle$ iff $t(\underline{M}) \geq c_\alpha(M, f_1)$.

Note that $t(\underline{M}) = s(Rk(\underline{M}))$ for a function s ; this gives importance to the study of the distribution function of the random variate Rk_M $(Rk_M(\sigma) = Rk(\underline{M}_\sigma))$ for a universe $\underline{U}$ satisfying Φ_o and H_o.

5.1.18 Definition. t is a simple linear rank statistic if there is a function $\underline{a}: \mathbb{N} \rightarrow \mathbb{Q}$ (rational sequence) such that

$$t(\langle M, f_1, f_2 \rangle) = \sum_{o \in M} f_1(o)\, \underline{a}(f_2(o)).$$

If V_1 is $\{0,1\}$, this can be expressed as

$$\sum_{f_2(o) = 1} \underline{a}(f_2(o)).$$

Note that t is an almost continuous statistic under Φ_o.

5.1.19 Lemma. Let $\underline{U}$ be a universe of type $\langle 1 \rangle$ such that $D_{\underline{U},o}$ is continuous for each $o \in U$. Then under the assumption of d-homogeneity (under H_o)

$$P(\{\sigma ; \underline{M}_\sigma \in \mathcal{IM}^R_M\}) = 1$$

for each sample $M \subseteq U$.

Proof. Our aim is to prove that, for each sample $M \subseteq U$ such that card $M = m$,

$$P(\{\sigma ; \text{there is an } i,j \in \{1, \ldots, m\}, i \neq j, \text{ such that } Q(o_i, \sigma) = Q(o_j, \sigma)\}) = 0 .$$

Denote by E the event in question and put $\vartheta_i = Q(o_i, \cdot)$, $\vartheta_j = Q(o_j, \cdot)$.

Then $P(E) \leq \sum_{1 \leq i < j \leq m} P(\vartheta_i = \vartheta_j) = \binom{m}{2} P(\vartheta_1 = \vartheta_2)$

(in the last equality we use H_o). Moreover, $D_{\vartheta_1} = D_{\vartheta_2} = F$ and F is continuous. Now let $x_o = -\infty$, $x_{n+1} = +\infty$ and $x_1, \ldots, x_n$ be a sequence of real numbers such that $F(x_{i+1}) - F(x_i) < \varepsilon$, where $\varepsilon > 0$ is an arbitrary positive number (it is possible to find such a sequence for each $\varepsilon > 0$ because F is continuous). Note that $\{\sigma ; \vartheta_1 = \vartheta_2\} \subseteq \bigcup_{i=0}^{n} \{\sigma ; x_i \leq \vartheta_1, \vartheta_2 < x_{i+1}\}$ so that $P(\vartheta_1 = \vartheta_2) \leq \sum_{i=0}^{n} P(x_i \leq \vartheta_1, \vartheta_2 < x_{i+1})$.

ϑ_1, ϑ_2 are stochastically independent (regularity of $\underline{U}$), hence

$$P(x_i \leq \vartheta_1, \vartheta_2 < x_{i+1}) = P(x_i \leq \vartheta_1 < x_{i+1})\, P(x_i \leq \vartheta_2 < x_{i+1}) =$$

$$= [F(x_{i+1}) - F(x_i)]^2 .$$

Then $P(\vartheta_1 = \vartheta_2) \leq \sum_{i=1}^{n} [F(x_{i+1}) - F(x_i)]^2 \leq$

$$\leq \varepsilon \sum_{i=1}^{n} [F(x_{i+1}) - F(x_i)] = \varepsilon\, F(+\infty) = \varepsilon \quad .$$

5.1.20 Discussion. (1) From the point of view of probability theory, the ties play no role. On the other hand, we observe some $\mathbb{Q}$ -structures in which ties can occur. Ties can occur owing to the following reasons:

(i) Error of rational approximation; had we considered more precise measurements, these ties could have been avoided. Such ties are similar to the case of missing information in 4.5. (ii) The sample was observed in a random state σ for which $\underline{M}_\sigma \in \mathfrak{M}^{\mathbb{R}}_M - \mathfrak{I}\mathfrak{M}^{\mathbb{R}}_M$; we know that such σ have null probability; but this does not imply that there is no such σ . (Both cases will be treated in the same way.) Thus, if we are speaking about observational properties of rank statistics (tests) we are forced to consider ties. (Cf. 5.2.11 and 5.4.14.)

(2) Note that Lemma 5.1.19 holds for each alternative hypothesis under which the distribution function is continuous.

5.1.21 Theorem . Let $\underline{U}$ be a universe of type $\langle 1 \rangle$ such that $D_{\underline{U},o}$ is continuous for each $o \in U$. Suppose that $\underline{U} \models H_o$. For each finite $M \subseteq U$, put $\underline{M}_\sigma = \langle M, f_\sigma \rangle$ and put $R(\sigma) = (f_\sigma)^*$ (i.e., $R(\sigma) = rk_{f_\sigma}$). Then (1) $R \in \mathcal{R}_M$ with probability 1, and (2) R induces on $\langle \mathcal{R}_M, \mathcal{P}(\mathcal{R}_M) \rangle$ the uniform distribution.

Proof. (1) Use Lemma 5.1.19(2) Consider $P(R = \eta_1)$ for arbitrary $\eta_1 \in \mathcal{R}_M$ (more precisely $P(R(\underline{M}_\sigma) = \eta_1))$. Let η_2 be an arbitrary enumeration of M, i.e., $\eta_2 \in \mathcal{R}_M$. Then there is a one-to-one mapping h of M such that $\eta_1 = \eta_2 \circ h$. Denote by $\eta_1^{-1}(j)$ the object $o \in M$ such that $\eta_1(o) = j$ (for $j \in \{1,\dots,m\}$), similarly for η_2^{-1}. Note that

$\eta_1^{-1}(j) = h^{-1}(\eta_2^{-1}(j))$.

Remember the notation $\vartheta_o = Q(o, .)$. Then

$$P(R = \eta_1) = P(\vartheta_{\eta_1^{-1}(1)} < \vartheta_{\eta_1^{-1}(2)} < \cdots) = \qquad (*)$$

$$= P(\vartheta_{h^{-1}(\eta_2^{-1}(1))} < \vartheta_{h^{-1}(\eta_2^{-1}(2))} < \cdots).$$

But from the d-homogeneity (i.e., invariance under one-to-one mappings of M) we know that the right-hand side of (*) equals

$$P(\vartheta_{\eta_2^{-1}(1)} < \vartheta_{\eta_2^{-1}(2)} < \cdots) = P(R = \eta_2).$$

Hence, for each $\eta_2 \in \mathcal{R}_M$, we have $P(R = \eta_1) = P(R = \eta_2)$. From the condition $\sum_{\eta \in \mathcal{R}_M} P(R = \eta) = 1$, we obtain $m!\, P(R = \eta_1) = 1$.

5.1.22 Discussion. (1) Let s be a function as in 5.1.16 (1). Let c be a rational number. Then, for each $\underline{U}$ satisfying our frame assumptions and H_0

$$P(\{\sigma ;\ s(Rk(\underline{M}_\sigma)) \geq c\}) = \frac{1}{m_{\underline{M}}!} \operatorname{card}\{\eta \in \mathcal{R}_M;\ s(\langle M, f_1, \eta \rangle) \geq c\}$$

(where $f_1 = Q_1 \restriction M$). If α is the desired rational significance level, we choose

$$c_\alpha(M,f_1) = \min\{c \in \operatorname{range}(s_{M,f_1});\ \frac{1}{m_M!} \operatorname{card}\{\eta \in \mathcal{R}_M;\ s_{M,f_1}(\eta) \geq c\} \leq \alpha\}. \quad (\#)$$

Note that the range of s_{M,f_1} is finite and, hence, the number $c_\alpha(M,f_1)$

can always be effectively constructed, $c_\alpha(\cdot)$ is a recursive function (see the form of (#) . But, for large m_M, the construction of c can be too complex as a combinatorial problem. This is the reason for using the asymptotical properties of rank statistics.

(2) We have seen that all rank statistics have the test property under H_o (and Φ_o); they can be used as tests of H_o. Thus, the choice of an appropriate test depends on the alternative hypothesis only. In such considerations, particular types of distribution functions play a role. A systematic theory can be found in [Hájek and Šidák] . We give a few examples below, On the other hand, such tests can be characterized from the observational point of view: such an approach is taken in Sections 3 and 4.

5.1.23 Example. For testing H_o against ASL rank statistics of the following form are used:

$s(\langle M, f_1, \eta \rangle) = \sum_{o \in M} f_1(o)\, a(\eta(o))$ where a is a non-decreasing rational mapping of $\mathbb{N}$ into $\mathbb{Q}$ (a non-decreasing rational sequence). Remember that we assume $V_1 = \{0,1\}$ here; hence, we may write - equivalently -

$$s(\langle M, f_1, \eta \rangle) = \sum_{f_1(o)=1} a(\eta(o)).$$

Remember also that we restrict ourselves in ASL to cases with $\Delta > 0$, i.e.,

$$D_{\vartheta_2,o}(x) = \begin{cases} F(x) & \text{if } Q_1(o) = 0, \\ F(x-\Delta) & \text{if } Q_1(o) = 1, \text{ where } \Delta > 0 : \end{cases}$$

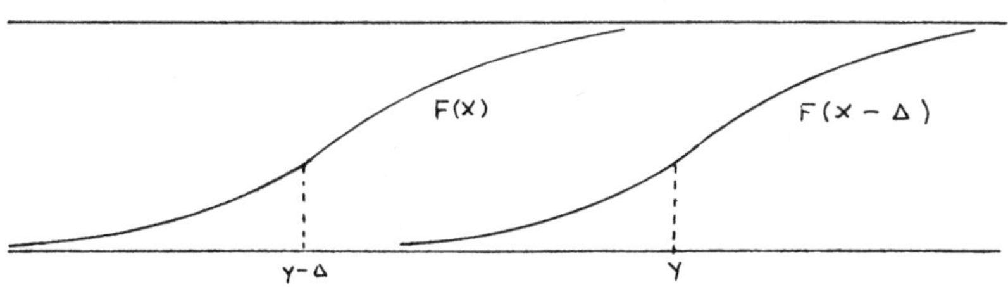

We see that $P(\vartheta_{2,o_1} \geq x \mid Q_1(o_1) = 1) \geq P(\vartheta_{2,o_2} \geq x \mid Q_1(o_2) = 0)$

and that there is an x for which the inequality is strict. Thus, for o_1 such that

$Q_1(o_1) = 1$, the probability of the values of ϑ_{2,o_1} being greater than x is larger than that for o_2 such that $Q_1(o_2) = 1$. Hence, the statistic should attain larger values than under H_o, so that it is reasonable to use the decision rule of the form $s(\langle M, f_1, \eta\rangle) \geq c_\alpha(M, f_1)$. Now let $\sum_{o\in M} f_1(o) = r$, card$(M) = m$. Then, under H_o,

$$E(s_{M,f_1}) = r\,a\ ,\quad VAR(S_{M,f_1}) = \frac{r(m-r)}{m(m-1)} \sum_{i=1}^{m} (a_i - \bar{a})^2$$

This can be proved using Theorem 5.1.21 and Lemma 5.1.12.

We will now introduce particular rank statistics for testing M_o against ASL:

(1) The Wilcoxon statistic: $w(\langle M, f_1, \eta\rangle) = \sum_{f_1(o)=1} \eta(o)$.

Here

$$E(w_{M,f_1}) = \frac{1}{2}\, r(m+1)\ ,\quad VAR(w_{M,f_1}) = \frac{1}{12}\, mr(m-r).$$

Let $\{M_k\}$ be an increasing sequence of samples, denote

$$r_k = \sum_{o\in M} f_1(o) = 1,\ m_k = \text{card}(M_k)\ .\ \text{Then, under } H_o,$$

$$\lim_{k\to+\infty} D_{w'_{M_k,f_1}} = D_{\eta(o,1)} \quad \text{if} \quad \lim_{k\to+\infty} (\min\{r_k, m_k - r_k\}) = +\infty \quad \left(w'_{M_k,f_1} = \frac{w_{M_k,f_1} - E\,w_{M_k,f_1}}{\sqrt{VAR\, w_{M_k,f_1}}} \right).$$

$D_{\eta(0,1)}$ is the normalized normal distribution function. Thus, in inference we use either the exact rule $w(\langle M, f_1, \eta\rangle) \geq c_\alpha(m,r)$ (for $c_\alpha(m,r)$ see 5.1.22) or the asymptotical rule

$$w(\langle M, f_1, \eta\rangle) \geq \left(\eta_\alpha + \frac{1}{2}\, r(m+1)\right)\sqrt{\frac{1}{12}\, rm(m-r)}\ ,$$

where η_α is the $(1-\alpha)$ - quantile of the normalized normal distribution. Such a rule has the asymptotical test property (see 4.3.8).

An analogous way of using the normal approximation is appropriate in other cases, i.e., $s(\langle M, f_1, \eta\rangle) \geq (\eta_\alpha + E\, s_{M,f})\sqrt{VAR\, s_{M,f}}$.

(2) The median statistic:

$$m(\langle M,f,\eta\rangle) = \sum_{f(o)=1} a(\eta(o)), \text{ where } a(\eta(o)) = \begin{cases} 1 & \text{if } \eta(o) > \frac{1}{2}(m+1), \\ \frac{1}{2} & \text{if } \eta(o) = \frac{1}{2}(m+1), \\ 0 & \text{if } \eta(o) < \frac{1}{2}(m+1). \end{cases}$$

Then, under H_o,

$$E(m_{M,f}) = \frac{1}{2} r \quad \text{and} \quad VAR(m_{M,f}) = \begin{cases} \frac{r(m-r)}{4(m-1)} & \text{for m even,} \\ \frac{r(m-r)}{4m} & \text{for m odd.} \end{cases}$$

An analogous asymptotical property holds as in (1).

5.1.24 Example. The most common statistic for testing H_o against ANRL is the following: $s(\langle M,f,\eta\rangle) = \sum_{o \in M} c_o \eta(o)$, where $c_o = f(o)$.
For ATL, we use $s_1(\underline{M}) = \sum_{o \in M} c_o \eta(o)$, where c_o is the rank of o w.r.t. f .

Generally, for testing ANRL every statistic of the form $\sum_{o \in M} c_o \eta(o)$ in which $c_{o_1} > c_{o_2}$ iff $f(o_1) > f(o_2)$ can be used. For the statistics s_1, we have, under H_o, $E(s_{1,M,f}) = \frac{1}{4} m(m+1)^2$,

$$VAR(s_{1,M,f}) = \frac{m^2(m+1)^2(m-1)}{144}.$$

The normal approximation can be used.

5.1.25 Key words: random $\mathbb{V}$-structures, almost continuous statistic; general null hypothesis H_o of d-homogeneity, the frame assumptions $\Phi_o, \bar{\Phi}_2, \Phi_2$ (c.d. - condition), alternative hypotheses ASL, ANRL, ATL; rank statistic, distribution of enumerating random variate under H_o.

V.2 Rank tests of d-homogeneity and independence.

The assumption that one of the random quantities is in fact deterministic (assumption (1)in 5.1.5) is too restrictive and inadequate for the use of rank tests in methods of automated discovery. In this section, we remove the quoted assumption and **correspondingly generalize the** described tests. It is not surprising that we have to strenghten our null hypothesis H_o in an appropriate way and we have to modify our alternative hypotheses similarly. The second aim of this section is to present some tests concerning the dependence of two random quantities - the coefficients of rank correlation.

5.2.1 Definition. Let $\underline{U} = \langle U, Q_1, Q_2 \rangle$ be a regular $\underline{\Sigma}$ - random $\mathbb{V}$ - structure of type $\langle 1,1 \rangle$. Let, moreover, $V_1 \subseteq \mathbb{Q}$ and suppose, for each $o \in U$ and $v \in V_1$, $P(\{\sigma : Q_1(o, \sigma) = v\}) > 0$.

(1) Then, for any $o \in U$, the conditional distribution function of $\vartheta_{2,o} = Q_2(o,.)$ w.r.t. $\vartheta_{1,o}$ is defined as follows:

$$D_{(\vartheta_2/\vartheta_1),o}(x/v) = \frac{P(\{\sigma ; \vartheta_{1,o}(\sigma) = v \ \& \ \vartheta_{2,o}(\sigma) < x\})}{P(\{\sigma ; \vartheta_{1,o}(\sigma) = v\})} .$$

Note that the conditional distribution function is a mapping of $V_1 \times \mathbb{R}$ into $[0,1]$.

(2) We say that $\underline{U}$ is conditionally d-homogeneous (w.r.t. Q_1) if for each $o_1, o_2 \in U$ and $v \in V_1$

$$D_{(\vartheta_2/\vartheta_1), o_1}(x/v) = D_{(\vartheta_2/\vartheta_1), o_2}(x/v)$$

5.2.2 Lemma. Let the assumptions of 5.2.1 hold. (1) If $\underline{U}$ is d-homogeneous, then $\underline{U}$ is conditionally d-homogeneous. (2) The following conditions are equivalent:

(a) For each $o \in U$ and each $v_1, v_2 \in V_1$, $D_{\vartheta_{2,o}}(\cdot,v) = D_{(\vartheta_2/\vartheta_1)}(\cdot,v)$.

(b) For each $o \in U$, the random variates $\vartheta_{1,o}, \vartheta_{2,o}$ are stochastically independent.

(3) Consider, moreover, the condition

(c) The random structure $\underline{U}_1 = \langle U, Q_1 \rangle$ is d-homogeneous.

Let $\underline{U}$ be conditionally d-homgeneous (w.r.t. Q_1) and let $\underline{U}$ satisfy (a) or (b) or (c). Then $\underline{U}_2 = \langle U, Q_2 \rangle$ is d-homogeneous.

Proof.(1) is evident. (2)(b) ⇒(a): From the stochastical independence we have

$$D_{(\vartheta_2/\vartheta_1),o}(x/v) = \frac{P(\{\sigma : \vartheta_{1,o}(\sigma) = v \;\&\; \vartheta_{2,o}(\sigma) < x\})}{P(\{\sigma : \vartheta_{1,o}(\sigma) = v\})} =$$

$$= \frac{P(\{\sigma : \vartheta_{1,o}(\sigma) = v\})\, P(\{\sigma : \vartheta_{2,o}(\sigma) < x\})}{P(\{\sigma : \vartheta_{1,o}(\sigma) = v\})} = D_{\vartheta_{2,o}}(x) .$$

Thus, for arbitrary elements $v_1, v_2 \in V_1$, we have

$$D_{(\vartheta_2/\vartheta_1),o}(x/v_1) = D_{\vartheta_{2,o}}(x) = D_{(\vartheta_2/\vartheta_1),o}(x/v_2) .$$

(a) ⇒(b): By Lemma 4.4.14 we have.

$$D_{\vartheta_{2,o}}(x) = P(\{\sigma : \vartheta_{2,o}(\sigma) < x\}) = \sum_{v \in V_1} D_{(\vartheta_2/\vartheta_1),o}(x/v)\, P(\{\sigma ; \vartheta_{1,o}(\sigma) = v\}) \quad (*)$$

Now, let v_o be an arbitrary element of V_1. Using (a), conclude that the right-hand side of $(*)$ is equal to

$$D_{\vartheta_2/\vartheta_1,o}(x/v_o) \cdot \sum_{v \in V_1} P(\{\sigma : \vartheta_{1,o}(\sigma) = v_o\}) =$$

$$= D_{(\vartheta_2/\vartheta_1),o}(x/v_o).$$

Thus, we have, for each $v \in V_1$, $D_{(\vartheta_2/\vartheta_1),o}(\cdot/v) = D_{\vartheta_2,o}(\cdot)$

which is equivalent to the stochastical independence of ϑ_1 and ϑ_2.

(3) Let $\underline{U}$ be conditionally d-homogeneous and let $\underline{U}$ satisfy (b). Consider two objects $o_1, o_2 \in U$. Then, by (b), for each $v \in V_1$,

$$D_{(\vartheta_2/\vartheta_1),o_1}(\cdot/v) = D_{\vartheta_2,o_1}(\cdot) \text{ and } D_{(\vartheta_2/\vartheta_1),o_2}(\cdot/v) = D_{\vartheta_2,o_2}(\cdot).$$

From the conditional d-homogeneity of $\underline{U}$, we see that the left hand sides of the previous equalities are equal and hence

$$D_{\vartheta_2,o_1}(\cdot) = D_{\vartheta_2,o_2}(\cdot)$$ For (a) we use (2).

It remains to prove under (c) that the assertion holds. By (c), $\underline{U}_1$ is d-homogeneous: hence, $P(\{\sigma ; \vartheta_1(\sigma) = v\})$ is independent of o. By 4.4.14, we have

$$D_{\vartheta_2,o}(x) = \sum_{v \in V_1} D_{(\vartheta_2/\vartheta_1),o}(x/v)\ P(\{\sigma ; \vartheta_{1,o}(\sigma) = v\}).$$

The right-hand side of the previous equality is independent of o ; hence, the same holds for the left-hand side.

5.2.3 Corollary. If $\underline{U}$ is conditionally d-homogeneous (w.r.t. Q_1) and if $\underline{U}$ satisfies (a) and (c) or (b) and (c), then $\underline{U}$ is d-homogeneous.

Proof. By a or (c) we have d-homogenity of $\underline{U}_2$, (c) means the d-homogeneity of $\underline{U}_1$. Use stochastical independence.

5.2.4 Discussion and Definition. Suppose, moreover, that $V_2 = \mathbb{R}$. Now we state the following frame assumptions:

(1) The assumptions of 5.2.1 (positivity),

(2) for each $o \in U$, $D_{(\vartheta_2/\vartheta_1),o}(\cdot/v)$ is continuous for each $v \in V_1$ continuity (i.e., p.c. conditions, Φ_2^-, see 5.1.14).

We are now able to formulate a generalization of the H_o hypothesis for this case. This generalization will be called the hypothesis of independence and d-homogeneity of the second quantity and will be denoted by H_2^-. H_2^- consists of two conditions:

(3) For each $o \in U$, $\vartheta_{1,o}$ and $\vartheta_{2,o}$ are stochastically independent,

(4) $\underline{U}_2 = \langle U, Q_2 \rangle$ is d-homogeneous.

Remember the t.c.-conditions Φ_2^-: For each $o \in U$, $D_{\underline{U}}$ is a continuous function of two variables.

Under such frame assumptions we shall consider a stronger hypothesis H_2 of independence and d-homogeneity; it requires (3) and (5): $\underline{U}$ is d-homogeneous.

5.2.5 Remark. Note that

(1) the conditional d-homogeneity of $\underline{U}$ and (a) from 5.2.2 imply H_2^-,

(2) H_2^- and (c) imply H_2 and

(3): In 5.1.14 we require in Φ_2^- the continuity of $D_{\underline{U}}$ in the second variable. Under our conditions this is equivalent to (2) from 5.2.4.

5.2.6 Discussion. Now, we shall reformulate the alternative hypotheses described in 5.1.6. (i) Assuming $V_1 = \{0,1\}$, the conditional ASL can be formulated as follows:

There is a function $F(x)$ such that, for each $o \in U$,

$$D_{(\vartheta_2/\vartheta_1),o}(x/0) = F(x) \quad \text{and} \quad D_{(\vartheta_2/\vartheta_1),o}(x/1) = F(x - \Delta),$$

where $\Delta \neq 0$ – as in 5.1.6 we restrict ourselves to the case $\Delta > 0$.

(ii) Assuming $V_1 = \mathbb{N}$, we obtain conditional ANRL:

There is a function $F(x)$ such that

$$D_{(\vartheta_2/\vartheta_1),o}(x/i) = F(x - i\Delta), \quad \text{where} \quad \Delta \neq 0 .$$

(iii) Assuming $V_1 \subseteq \mathbb{Q}$, conditional ATL can be stated as follows:

(a) $\underline{U}$ is conditionally d-homogeneous (w.r.t Q_1), and

(b) if $v_1 < v_2$, then $D_{(\mathcal{V}_2/\mathcal{V}_1),o}(\cdot/v_1) < D_{(\mathcal{V}_2/\mathcal{V}_1),o}(\cdot/v_2)$.

5.2.7 Remark. From 5.2.5 we know that if, in the contitional ANRL and conditional ASL, we replace "$\Delta \neq O$" by "$\Delta = O$", we obtain the null hypothesis H_2.

5.2.8 Discussion. Appropriate tests are based on rank statistics, i.e., on functions of the form $s(<M,f_1,f_2>)$ mapping $\check{\mathcal{M}}_M$ into $\mathbb{Q}$. See 5.1.11 and 5.1.12. We infer the alternative hypothesis if

$s(<M,f_1,f_2>) \geq c_\alpha(<M,f_1>)$.

But, now, for given M, both quantities are random, i.e., we have to consider the models $\underline{M}^1_\sigma$ and $\underline{M}^2_\sigma$ obtained from $\underline{U}_1 = <U,Q_1>$ and $\underline{U}_2 = <U,Q_2>$ respectively, by fixing σ and M. Hence, we are going to find an upper bound for the probability $P(\{\sigma;\ s(\mathrm{Rk}\ \underline{M}_\sigma)) \geq c_\alpha(\underline{M}^1_\sigma)\})$. Under our frame assumptions, if, under H_2^-, $P(\{\sigma;\ s(\mathrm{Rk}(\underline{M}_\sigma)) \geq c_\alpha(\underline{M}^1_\sigma)\}/\{\sigma;\ \underline{M}^1_\sigma = <M,f_1>\}) \leq \alpha$ for each $<M,f_1>$, then $P(\{\sigma;\ s(\mathrm{Rk}(\underline{M}_\sigma)) \geq c_\alpha(\underline{M}^1_\sigma)\}) =$

$$\sum_{f_1:M\to V_1} P(\{\sigma;\ s(\mathrm{Rk}(\underline{M}_\sigma)) \geq c_\alpha(\underline{M}^1_\sigma)\}/\{\sigma;\ \underline{M}^1_\sigma = <M,f_1>\})\ P(\{\sigma;\ \underline{M}^1_\sigma = <M,f_1>\}) \leq$$

$$\alpha \sum_{f_1:M\to V_1} P(\{\sigma;\ \underline{M}^1_\sigma = <M,f_1>\}) = \alpha$$

and we obtain the desired upper bound.

On the other hand H_2^- implies H_o (d-homogeneity of $\underline{U}_2$) and, under H_2^-,

$$P(\{\sigma;\ s(\mathrm{Rk}(\underline{M}_\sigma)) \geq c_\alpha(\underline{M}^1_\sigma)\}/\{\sigma;\ \underline{M}^1_\sigma = <M,f_1>\}) =$$

$$\frac{P(\{\sigma;\ s(\mathrm{Rk}(\underline{M}_\sigma)) \geq c_\alpha(<M,f_1>)\ \&\ \underline{M}^1_\sigma = <M,f_1>\})}{P(\{\sigma;\ \underline{M}^1_\sigma = <M,f_1>\})} =$$

$$= P(\{\sigma;\ s(\mathrm{Rk}(\underline{M}_\sigma)) \geq c_\alpha(<M,f_1>)\})$$

and we can apply the results of 5.1.13 - 5.1.22, i.e., find values $c_\alpha(\langle M, f_1\rangle)$ as in 5.1.22. Hence, under our present assumptions Φ_2^-, each quantifier q, of type $\langle 1,1\rangle$ defined by the condition

$\mathrm{Asf}_q(\langle M, f_1, f_2\rangle) = 1$ iff $s(\langle M, f_1, f_2\rangle) \geq c_\alpha(\langle M, f_1\rangle)$ is an observational test of H_2^-.

5.2.9 Discussion and Definition.

Consider the frame assumption Φ_2, i.e., $V_1 = V_2 = \mathbb{R}$ and, for each o, $D_{\underline{U},o}$ is continuous (the t.c. - conditions. Under these frame assumptions we can test the hypothesis H_2. We can say that two models $\underline{M}_1 = \langle M_1, g_1, f_1\rangle$ and $\underline{M}_2 = \langle M_2, g_2, f_2\rangle$ of type $\langle 1,1\rangle$ are weakly rank equivalent if $\langle M_1, g_1^*, f_1^*\rangle$ and $\langle M_2, g_2^*, f_2^*\rangle$ are isomorphic. If $\underline{M} = \langle M, g, f\rangle$, write $\mathrm{Rk}_2(\underline{M})$ for $\langle M, g^*, f^*\rangle$. Then we can define strong rank statistics: they are of the form $t(\underline{M}) = s(\mathrm{Rk}_2(\underline{M}))$ where s is a rational-valued function.

Under H_2, assertions analogous to Lemma 5.1.19 and Theorem 5.1.21 can be formulated.

5.2.10 Example. The most common strong rank statistics for such a case are: (1) Spearmen's rank correlation coefficient:

$$\varrho(\langle M, f_1, f_2\rangle) = \frac{12}{m^2 - m} \sum_{o \in M} \left(f_1^*(o) - \frac{m+1}{2}\right)\left(f_2^*(o) - \frac{m+1}{2}\right)$$

where $m = \mathrm{card}(M)$. (2) Kendall's rank correlation coefficient

$$\tau(\langle M, f_1, f_2\rangle) =$$

$$= \frac{1}{m^2 - m} \sum_{\langle o_1, o_2\rangle,\ o_1 \neq o_2 \in M} \mathrm{sign}(f_1^*(o_1) - f_1^*(o_2))\,\mathrm{sign}(f_2^*(o_1) - f_2^*(o_2)) .$$

(Here

$$\text{sign}(x) = \begin{cases} 1 & \text{if } x > 0 \\ 0 & \text{if } x = 0 \\ -1 & \text{if } x < 0. \end{cases} \quad)$$

Note the following: (i) τ is not a linear function of f_1^*, f_2^* and (ii) $\varrho(\underline{M})$ is, for testing purposes, equivalent to the simpler $s(\langle M, f_1, f_2 \rangle) = \sum_{o \in M} f_1^*(o)\, f_2^*(o)$. In both cases we measure the similarity of rank vectors f_1^*, f_2^*. Higher values of rank correlation coefficients indicate positive stochastical dependence of quantities. Thus, we use tests of the form $\varrho(\langle M, f_1, f_2 \rangle) \geq c_\alpha^1(m_M)$ and $\tau(\langle M, f_1, f_2 \rangle) \geq c_\alpha^2(m_M)$, respectively.

5.2.11 Remark. As mentioned in 5.1.20, tied observations can occur. In such a case, rank statistics are not well defined.

We have to use some supplementary modification of their definition. We now describe briefly two methods. Let us observe, for a $\sigma \in \Sigma$, a structure $\underline{M}_\sigma = \langle M, f_1, f_2 \rangle$. Suppose that for a set $M_t = \{o_1, \ldots, o_t\}$ we have $f_2(o_1) = \ldots = f_2(o_t)$. For the sake of simplicity we assume f_2 be one-to-one on $M - M_t$. Then $f_2^*(o_1) = \ldots = f_2^*(o_t) = \text{card}\{o \in M;\ f_2(o) \leq f_2(o_1)\}$. One has to map M_t (one-to-one) onto the numbers

$\text{card}\{o \in M;\ f_2(o) < f_2(o_1)\} + 1, \ldots, \text{card}\{o \in M;\ f_2(o) < f_2(o_1)\} + t = f_2^*(o_1)$;

in such a manner one obtains an enumeration $f_2^{*)}$. There are $t!$ such mappings. Denote the set of these mappings by $\mathcal{J}_t(f_2)$; $f_2^{*\imath}$ is obtained by $\imath \in \mathcal{J}_t(f_2)$. There are two possible methods of choosing a mapping $\imath \in \mathcal{J}_t(\imath_2)$

(1) Randomization: Enumerate $\mathcal{J}_t(f_2)$ and make an additonal random experiment with the possible outcomes $\{1, \ldots, t!\}$ each of them with probability $\frac{1}{t!}$. If the experiment yields the outcome j, use the mapping $\imath_j \in \mathcal{J}_t$. So we obtain in each case an enumeration $f_2^{*)}$. Under H_o (or H_2^-, H_2), the uniform distribution on $\langle \mathcal{R}_M, \mathcal{P}(\mathcal{R}_M) \rangle$ is preserved. If there are more groups

of tied observations, the process is similar; see [Hájek and Šidák].

(2) Least favourable value. We can treat ties as missing information. We consider a statistic $s(Rk(\underline{M}_\sigma))$ or $s(Rk_2(\underline{M}_\sigma))$; for objects from M_t we have "no" information. By $\iota \in \mathcal{J}_t(f_2)$, we construct a completion $\langle M, f_1, f_2^{*\iota}\rangle$. We use $\iota_0 \in \mathcal{J}_t(f_2)$ such that

$$s(\langle M, f_1, f_2^{*\iota_0}\rangle) = \min_{\iota \in \mathcal{J}_t(f_2)} s(\langle M, f_1, f_2^{*\iota}\rangle) \quad .$$

The test is then $s(\langle M, f_1, f_2^{*\iota_0}\rangle) \geq c_\alpha(\langle M, f_1\rangle)$. The significance level never exceeds the level of the untied procedure, i.e., of the test defined on $\mathcal{TM}$.

See Theorem 4.5.6. For more groups of tied observations the process is similar. We shall go into more detail in Section 4.

5.2.12 Key words: conditional d-homogeneity; the null hypotheses H_2^- and H_2, conditional alternative hypotheses,ties; Spearmen's and Kendall's rank correlation coefficients.

V.3 Function calculi with enumeration models

As we have observed in the preceding section, an important class of tests is related to functions on enumerations. In the present section, we are going to study observational function calculi in which these functions can be dealt with. We generalize slightly the notion of a function calculus by allowing two sorts of functions.

5.3.1 Definition. Let a, b be two abstract symbols. A two-sorted monadic type is a tuple of a's and b's. Let $t = \langle t_1, \ldots, t_n \rangle$ be a type. A (two-sorted monadic) enumeration structure of type t is a tuple $\langle M, f_1, \ldots, f_n \rangle$ where M is a non-empty finite set and, for each i, if $t_i = a$ then f_i is a mapping of M into $\{0,1\}$, and if $t_i = b$ then f_i is an enumeration of M.

The two sorted MOFC with enumeration models of type t are defined as follows:

(1) The set of abstract values is $\mathbb{N}$.

(2) The set $\mathcal{M}$ of models consists of all enumeration structures of the type t which are finite objects.

(3) The language consists of (i) variables, (ii) unary functors $F_1, \ldots, F_n$ (F_i is of sort t_i); (iii) junctors; our junctors will be $\underline{0}$, $\underline{1}$ (nullary), $\neg$ (unary) and &, $\vee$ (binary) with the usual two-valued associated functions, (iv) quantifiers; each quantifier has a type t_q - a k- tuple of a's and b's. If $t_q = t = \langle t_1, \ldots, t_k \rangle$ is the type of q, then Asf_q maps the set of all models of type t_q into $\{0,1\}$. $Asf_q(\underline{M})$ is supposed to be recursive in q and $\underline{M}$.

Formulae are defined as follows: If F_i is function of sort $s \in \{a,b\}$ and if x is a variable, then $F_i(x)$ is a formula of sort s. $\underline{0}, \underline{1}$ are formulae of sort a. If φ is a formula of sort a, then $\neg\varphi$ is as well. If φ, ψ are formulae of sort a, then φ & $\psi, \varphi \vee \psi$ are formulae of sort a as well. If q is a quantifier of type $t = \langle t_1, \ldots, t_k \rangle$, if $\varphi_1, \ldots, \varphi_k$ are formulae, φ_i of sort t_i ($i = 1, \ldots, k$) and if x is a variable, then $(qx)(\varphi_1, \ldots, \varphi_k)$ is a formula of sort a. Free and bound formulae are defined in the usual way. Also, the definition of $\|\varphi\|_M[e]$ (e being a M - sequence for φ) is unchanged; note that if φ is of sort a, then

$\|\varphi\|_{\underline{M}}[e] \in \{0,1\}$, and if φ is of sort b, then $\|\varphi\|_{\underline{M}}[e] \in \{1,2,\ldots,\text{card}(M)\}$.

Note, further, that there are no non-atomic formulae of sort b. In particular, each closed formula is of sort a and, hence, two-valued.

5.3.2 Remark. Let $\mathcal{F}$ be an MOFC with enumeration models as described above. Then: (1) For each open formula φ of sort a distinct from O there is a semantically equivalent formula in conjunctive normal form containing only function and variables which occur in φ. (2) For each closed formula ψ, each finite set $A = \{\varphi_1, \ldots, \varphi_n\}$ of closed formulae, $A \models \psi$ iff $\models \bigwedge A \rightarrow \psi$. (3) $\mathcal{F}$ is axiomatizable (decidable) iff it is strongly axiomatizable (decidable). (Cf. 1.1.12).

5.3.3 Remark. Let t be a two sorted monadic type and let $\hat{t}$ be the sequence resulting from t by the replacement of a by 1 and b by 2. A multiple ordered structure of type $\hat{t}$ is an $\{O, 1\}$ - valued structure $\underline{M} = \langle M, g_1, \ldots, g_n \rangle$ of type $\hat{t}$ such that, for each i such that $\hat{t}_i = 2$, g_i is a characteristic function of a linear ordering of M, i.e., the relation $\prec_i$ defined by ($o_1 \prec_i o_2$ iff $g_i(o_1, o_2) = 1$) is a linear ordering of M. There is a natural one-to-one correspondence of enumeration structures of the two-sorted type t and multiply ordered structures of type $\hat{t}$: If $\underline{M} = \langle M, f_1, \ldots, f_n \rangle$ is an enumeration structure of type

$t = \langle t_1, \ldots, t_n \rangle$, then the corresponding structure $\langle M, g_1, \ldots, g_n \rangle$ is defined as follows:

If $t_i = a$, then $g_i = f_i$

if $t_i = b$, then $g_i(o_2, o_2) = 1$ iff $f_i(o_1) < f_i(o_2)$ and $g_i(o_1, o_2) = 0$ otherwise.

One could formulate some facts concerning the relation between the two-sorted MOFC with enumeration models of type t and the corresponding predicate calculus of type t with multiply ordered models; but we shall not do this. Now, we are going to study some particular kinds of quantifiers.

5.3.4 Definition. Let $\underline{M}_1 = \langle M_1, f_1, f_2 \rangle$, $\underline{M}_2 = \langle M_2, g_1, g_2 \rangle$ be two models of type $\langle a, b \rangle$ such that $\text{card}(M_1) = \text{card}(M_2)$. We say that $\underline{M}_1$ is d-better than $\underline{M}_2$ ($\underline{M}_1 \succeq_d \underline{M}_2$) if there is an isomorphism ι of $\langle M_1, f_1 \rangle$ and $\langle M_2, g_1 \rangle$ such that, for each $o \in M_1$, if $f_1(o) = 1$ then $f_2(o) \geq g_2(\iota o)$ and if $f_1(o) = 0$ then $f_2(o) \leq g_2(\iota o)$.

A quantifier q of type $\langle a, b \rangle$ is called distinctive if, for each $\underline{M}_1$ and $\underline{M}_2$,

$\underline{M}_2 \preceq_d \underline{M}_1$ and $\text{Asf}_q(\underline{M}_2) = 1$ implies $\text{Asf}_q(\underline{M}_1) = 1$.

5.3.5 Definition and Remark. In the sequel, we shall often consider models of type $\langle a, b \rangle$ (i.e., with one unary relation and one enumeration). If $\underline{M} = \langle M, f_1, f_2 \rangle$ is such a model, then we put $m_M = \text{card}(M)$ and $r_M = \text{card}\{o \in M;\ f_1(o) = 1\}$.

5.3.6 Lemma. Let $a: \mathbb{N} \to \mathbb{Q}$ be a non-decreasing recursive sequence of rational numbers; let q be a quantifier of type $\langle a, b \rangle$ such that

$\text{Asf}_q(\langle M, f_1, f_2 \rangle) = 1$ iff

$$\sum_{o \in M} f_1(o)\, a(f_2(o)) = \sum_{f(o)=1} a(f_2(o)) \geq c_\alpha(m_M, r_M).$$

Then q is distinctive (Here, $c: \mathbb{N} \times \mathbb{N} \to \mathbb{Q}$ is a recursive function.)

Proof. Let $\underline{M}_1 \preceq_d \underline{M}_2$ and $Asf_q(\underline{M}_1) = 1$; let

$\underline{M}_1 = \langle M_1, f_1, f_2 \rangle$, $\underline{M}_2 = \langle M_2, g_1, g_2 \rangle$ and let ι be the isomorphism from the definition of $\preceq_d$. Then

$$\sum_{f_1(o)=1,\ o\in M_1} a(f_2(o)) = \sum_{g_1(\iota o)=1,\ o\in M_1} a(f_2(o)) \leq$$

$$\sum_{g_1(\iota o)=1,\ o\in M_1} a(g_2(\iota o)) = \sum_{g_1(o)=1,\ o\in M_2} a(g_2(o)) .$$

5.3.7 Remark. The above mentioned case of distinctive quantifiers corresponds to simple linear tests of ASL (such as the Wilcoxon or median tests). For another case, see Problem (2).

5.3.8 Definition. A quantifier q of type $\langle a, b \rangle$ is called executive if there is an $m_{min} \in \mathbb{N}$ such that

(i) for each $m > m_{min}$, there are models $\underline{M}_1$ and $\underline{M}_2$ such that $m_{M_1} = m_{M_2}$ and $Asf_q(\underline{M}_1) = 1$ and $Asf_q(\underline{M}_2) = 0$, and

(ii) for each $\underline{M}$, $m_M \leq m_{min}$ implies $Asf_q(\underline{M}) = 0$.

5.3.9 Definition. (1) (Auxiliary). If $\underline{M}_o = \langle M, f_1 \rangle$ is a model of type $\langle a \rangle$, let $Exp(\underline{M}_o)$ denote the set of all models $\langle M, f_1, f_2 \rangle$ of type $\langle a, b \rangle$ ($\langle a, b \rangle$ - expansions of $\underline{M}_o$).

Let $\mathcal{G}$ be a class of quantifiers of type $\langle a, b \rangle$; define an equivalence on $Exp(\underline{M}_o)$ by putting $\underline{M}_1 \sim_{\mathcal{G}} \underline{M}_2$ iff $Asf_q(\underline{M}_1) = Asf_q(\underline{M}_2)$ for each $q \in \mathcal{G}$.

$C(\underline{M}_o)$ denotes the minimum of the cardinalities of the equivalence classes in $Exp(M_o)$.

(2) Moreover let $\alpha \in (0,0.5]$. A quantifier $q \in \mathcal{G}$ is said to be <u>of level α</u> (w.r.t. $\mathcal{G}$) if, for each $\underline{M}_o$, the following inequalities hold:

$$\alpha \geq \frac{1}{m!} \operatorname{card}\{\underline{M} \in \operatorname{Exp}(\underline{M}_o);\ \operatorname{Asf}_q(\underline{M}) = 1\} > \alpha - \frac{c(\underline{M}_o)}{m!}, \qquad (*)$$

where $m = m_{\underline{M}_o}$.

(3) By saying that a distinctive quantifier is of level α, we mean that it is of level α w.r.t. all distinctive quantifiers.

5.3.10 <u>Lemma.</u> For distinctive quantifiers

$$C(\underline{M}_o) = C(r_{\underline{M}_o}, m_{\underline{M}_o}) = r_{\underline{M}_o}!\,(m_{\underline{M}_o} - r_{\underline{M}_o})!\,.$$

<u>Proof.</u> Let $\underline{M}_o = \langle M, f_1 \rangle$ be a model of type $\langle a \rangle$, with $m_{\underline{M}_o} = m$ and $r_{\underline{M}_o} = r$. For each $\underline{M} \in \operatorname{Exp}(\underline{M}_o)$, there are exactly $r!(m-r)!$ many models in $\operatorname{Exp} \underline{M}_o$ d-equivalent to $\underline{M}$, since each automorphism of $\underline{M}_o$ (isomorphism of $\underline{M}_o$, $\underline{M}_o$) induces a d-equivalent structure; there are $r!(m-r)!$ many such automorphisms. Hence, $C(\underline{M}_o) = r!(m-r)!$. Cf. [Hájek and Šidák], Theorem IV.1.1.

5.3.11 <u>Corollary.</u> Each distinctive quantifier q with
$\operatorname{Asf}_q(\langle M, f_1, f_2 \rangle) = 1$
iff $s(\langle M, f_1, f_2 \rangle) \geq c_\alpha(\langle M, f_1 \rangle)$,

where s is a function on $\mathcal{M}^{\{0,1\} \times \mathbb{N}}$ and c_α is defined in 5.1.22, is of level α.

5.3.12 <u>Example.</u> In particular, all tests based on simple linear functions with a non-decreasing sequence a (see 5.3.6) and with $c_\alpha(M, f_1)$ defined as in 5.1.22 are of level α (i.e. Wilcoxon, median etc.).

5.3.13 <u>Discussion.</u> In Section 5.1, we have discussed rank tests of H_o: in fact they are based on structures of type $\langle \mathcal{E}, b \rangle$, where $\mathcal{E}$ is an arbitrary sort (corresponding to $\{0,1\}$ or more general). In particular, rank tests of H_o and ASL correspond to our distinctive quantifiers. In 5.1.19 and 5.1.21, we have proved, in fact, that each observational test of H_o against an arbitrary alternative hypothesis A (stable under rank equivalence of models) is related to a quantifier of type $\langle \varepsilon, b \rangle$ satisfying the left-hand side of the inequality (*) (5.3.9). For each rank test not satisfying the right-hand side of (*), there is then a test of H_o against A uniformly more powerful (see 4.3.5). The class is given by the alternative hypothesis A. For ASL, we obtain the class of all distinctive quantifiers.

5.3.14 Denotation and Lemma. Let $\alpha \in (0, 0.5]$ be given; define $m_\alpha = \max\{m \in \mathbb{N};\ \frac{1}{m!} > \alpha\}$. Let q be a quantifier of level α (w.r.t. a class $\mathcal{G}$). Then, for each model $\underline{M}$, $m_{\underline{M}} \leq m_\alpha$ implies $Asf_q(\underline{M}) = 0$.

Proof. Let there be a model $\underline{M}$, $\underline{M} = \langle M, f_1, f_2 \rangle$ and $m_{\underline{M}} \leq m_\alpha$, such that $Asf_q(\underline{M}) = 1$. Then

$$m^+ = \mathrm{card}\{\underline{N} \in \mathrm{Exp}(\langle M, f_1 \rangle);\ Asf_q(\underline{N}) = 1\} \geq 1$$

and from $m_{\underline{M}} \leq m_\alpha$ we have $\frac{1}{m_{\underline{M}}!} > \alpha$. Hence, $\frac{m^+}{m_{\underline{M}}} > \alpha$,

which is a contradiction with the left-hand inequality in (*).

5.3.15 Corollary. Let q be an executive quantifier of level α (w.r.t. a class $\mathcal{G}$). Then $m_{min}(q) \geq m_\alpha$.

5.3.16 Definition and Lemma. Let a class $\mathcal{G}$ be given (for $\underline{M}_o$ and other notations see 5.3.9). Denote

$$C(m) = \min\{C(\underline{M}_o);\ m_{\underline{M}_o} = m\},$$

and $$m_\alpha^{\mathcal{G}} = \sup\{m \in \mathbb{N};\ \frac{C(m)}{m!} > \alpha\}.$$

Let q be a quantifier of level α (w.r.t. $\mathcal{G}$) such that

(i) $\frac{C(m)}{m!}$ is non-increasing and (ii) $\lim\limits_{m \to +\infty} \frac{C(m)}{m!} < \alpha$.

Then q is executive and $m_{min}(q) = m_{\alpha}^{\mathcal{G}}$.

Proof. Condition (i) implies that $m_{\alpha}^{\mathcal{G}} \in N$.
Suppose now that there is an $m > m_{\alpha}^{\mathcal{G}}$ such that, for each $\underline{M}$, $m_{\underline{M}} = m$, $Asf_q(\underline{M}) = 0$. There is an $\underline{M}_o$, $m_{\underline{M}_o} = m$, such that

$$\frac{C(\underline{M}_o)}{m_{\underline{M}_o}!} \leq \alpha \qquad \text{i.e.} \qquad \alpha - \frac{C(\underline{M}_o)}{m_{\underline{M}_o}!} \geq 0 .$$

On the other hand, $\{\underline{M} \in \text{Exp } \underline{M}_o ;\ Asf_q \underline{M} = 1\} = 0$ and we obtain a contradiction with the **right-hand** inequality in $(*)$. Now let $m \leq m_{\alpha}^{\mathcal{G}}$ and $Asf_q(\underline{M}) = 1$ for an $\underline{M}$, $m_{\underline{M}} = m$. There is an $\underline{M}_o$ such that $\underline{M} \in \text{Exp}(\underline{M}_o)$ and $\underline{M}$ is an element of an equivalence class with cardinality greater than or equal to $C(\underline{M}_o) \geq C(m)$. But, by (i),

$$\frac{C(m)}{m!} \geq \frac{C(m_{\alpha}^{\mathcal{G}})}{m_{\alpha}^{\mathcal{G}}!} > \alpha$$

and we have a contradiction with the left-hand side of $(*)$.

5.3.17 Corollary. If $\alpha \in (0, 0.5]$, then each distinctive quantifier q of level α is executive.

Proof. Use Lemma 5.3.10. For distinctive quantifiers we have

$$C(m) = \begin{cases} \left(\frac{1}{2} m\right)!^2 & \text{for } m \text{ even,} \\ \left(\frac{m+1}{2}\right)! \left(\frac{m-1}{2}\right)! & \text{for } m \text{ odd ;} \end{cases}$$

hence, $\lim_{m \to +\infty} \frac{C(m)}{m!} = 0$. Then m_{α}^{d} is the last m for which $\frac{C(m)}{m!} > \alpha$.

5.3.18 Example. For $\alpha = 0.05$ we obtain $m_{\alpha}^{d} = 5$ $\left(\frac{C(5)}{5!} = 0.1,\ \frac{C(6)}{6!} = 0.033\right)$, for $\alpha = 0.025$ we have $m_{\alpha}^{d} = 7$

5.3.19 Definition. An executive distinctive quantifier q is called d-executive if for each $m > m_{min}(q)$ there are $r_{min}(m, q)$ and $r_{max}(m, q) \in \mathbb{N}$ $(r_{min} + 1 < r_{max})$ such that the following holds:

(i) If $\underline{M} = \langle M, f_1, f_2 \rangle$ is such that $m_{\underline{M}} = m$ and $(r_{\underline{M}} \leq r_{min}(m,q)$ or $r_{M} \geq r_{max}(m, q))$, then $Asf_q(\underline{M}) = 0$,

(ii) for each $r \in (r_{min}(m,q), r_{max}(m,q))$ there are models $\underline{M}_1, \underline{M}_2$ such that $Asf_q(\underline{M}_1) \neq Asf_q(\underline{M}_2)$ and $m_{\underline{M}_1} = m_{\underline{M}_2} = m$, $r_{\underline{M}_1} = r_{\underline{M}_2} = r$.

5.3.20 Definition and Lemma. Let $\alpha \in (0, 0.5]$ and let $m > m^d_\alpha$ (m^d_α was defined in 5.3.17). Let $r_\alpha(m)$ be the maximal r such that for $i = 1, \ldots, r$ we have $\frac{1}{\binom{m}{i}} > \alpha$. Let q be a distinctive quantifier of level α. Then q is d-executive with $r_{min}(m,q) = r_\alpha(m)$ and $r_{max}(m,q) = m - r_\alpha(m)$.

Proof. Let $\underline{M}_o = \langle M, f_1 \rangle$ be a model of type $\langle a \rangle$ with $m_{\underline{M}_o} = m$, $r_{\underline{M}_o} = r$. For distinctive quantifiers we have, by 5.3.10, $C(\underline{M}_o) = r!(m-r)!$, so that $\frac{C(\underline{M}_o)}{m!} = \frac{1}{\binom{m}{r}}$. Let a distinctive quantifier q be given. Suppose now that $r \leq r_\alpha(m)$; then $\frac{1}{\binom{m}{r}} > \alpha$. On the other hand, if there is an $\underline{M} \in Exp(\underline{M}_o)$ with $Asf_q(\underline{M}) = 1$, then putting $As(\underline{M}_o) = card\{\underline{M} \in Exp\ \underline{M}_o;\ Asf_q\ \underline{M} = 1\}$ we have, since q is of level α:

$$\alpha \geq \frac{As(\underline{M}_o)}{m!} \geq \frac{C\ \underline{M}_o}{m} = \frac{1}{\binom{m}{r}},$$

which contradicts $\frac{1}{\binom{m}{r}} > \alpha$.

Furthermore, if there is an $r \in (r_\alpha(m), m - r_\alpha(m))$ such that, for each $\underline{M}$ with $r_{\underline{M}} = r$ and $m_{\underline{M}} = m$, $Asf_q(\underline{M}) = 0$, since q is of level α we obtain $0 > \alpha - \frac{1}{\binom{m}{r}}$; which contradicts the definition of $r_\alpha(m)$. On the other hand, if for each $\underline{M}$ with $r_{\underline{M}} = r$, $m_{\underline{M}} = m$, we have $Asf_q(\underline{M}) = 1$, then we obtain $\{\underline{N} \in \text{Exp } \underline{M}_o ; Asf_q(\underline{N}) = 1\} = \text{Exp}(\underline{M}_o)$. But then $As(\underline{M}_o) = m!$, and this contradicts the left-hand inequality in (*) from 5. 3. 9.

5.3.21 Remark. (1) Note that $\frac{1}{\binom{m}{r}} \geq \frac{1}{m!}$ so that the condition of 5.3.20 is stronger that the condition of 5.3.14. If $\frac{1}{\binom{m}{r}} \leq \alpha$, then $m > m_\alpha$.

(2) Note that, e.g., the Wilcoxon two sample test attains exactly the bounds introduced above; see [Pearson and Hartley]. This means that it is of level α (for $\alpha = 0.1$ we have: $m = 10$, $r = 1$, $m = 5$, $r = 2$, ... etc.).

(3) Corollary: Let a calculus $\mathcal{F}$ be given and let q be a distinctive quantifier of the level α. Let $\underline{M}$ be a model, $m_{\underline{M}} = m$. For each of the designated open formulae φ, F of the appropriate sorts,

$\text{card}\{o \in M ; \|\varphi\|_{\underline{M}}[o] = 1\} \notin (r_\alpha(m), m - r_\alpha(m))$

implies $\underline{M} \models \neg q(\varphi, F)$.

Hence, if Extr is a quantifier of type $\langle a \rangle$, with $Asf_{Extr}(\underline{M}) = 1$ iff $r_{\underline{M}} \notin (r_\alpha(m), m - r_\alpha(m))$, then the rule

$$\left\{ \frac{\text{Extr}(\varphi)}{\neg q(\varphi, F)} ; \varphi, F \right\}$$

is sound (Extr(φ) is read " φ is extremely frequented").

5.3.22 Definition. (1) Let q_1, q_2 be two distinctive quantifiers. q_1 is stronger than q_2 if, for each $\underline{M}$, $Asf_{q_2}(\underline{M}) = 1$ implies $Asf_{q_1}(\underline{M}) = 1$. In fact, if for q_1 and q_2 the left-hand inequality from (*) (5.3.9) holds , i.e., q_1 and q_2 are observational tests of H_o , then q_1 is a uniformly more powerful test of H_o and ASL than q_2 in the usual statistical sense.

(2) $\{q_\alpha\}_{\alpha \in A}$ $(A \subseteq (0,0.5])$ is a monotone class of distinctive quantifiers if $\alpha_1 < \alpha_2$ implies that q_{α_2} is stronger than q_{α_1}.

5.3.23 Theorem. Let q_1, q_2 be distinctive quantifiers both of a given level α, and let q_1 be stronger than q_2. Then

$$\{\underline{M};\ Asf_{q_1}(\underline{M}) = 1\} = \{\underline{M};\ Asf_{q_2}(\underline{M}) = 1\}.$$

Proof. Denote $As_{q_i}(\underline{M}_o) = \{\underline{N} \in Exp(\underline{M}_o);\ Asf_{q_i}(\underline{N}) = 1\}$. For each $\underline{M}_o$ of type $\langle a,b \rangle$ with $m_{\underline{M}_o} = m$, $r_{\underline{M}_o} = r$, we have the following:

(i) $As_{q_1}(\underline{M}_o) \supseteq As_{q_2}(\underline{M}_o)$,

(ii) $$\frac{\text{card } As_{q_1}(\underline{M}_o)}{m!} > \alpha - \frac{1}{\binom{m}{r}},$$

(iii) $$\frac{\text{card } As_{q_2}(\underline{M}_o)}{m!} > \alpha - \frac{1}{\begin{matrix} m \\ r \end{matrix}}.$$

Suppose that $m > m_\alpha^d$ and $r \in (r_\alpha(m), m - r_\alpha(m))$. If there is an $\underline{M}_1 \in Exp(\underline{M}_o)$ such that $Asf_{q_1}(\underline{M}_1) = 1$ and $Asf_{q_2}(\underline{M}_1) = 0$, then the same holds for each $\underline{M} \in Exp(\underline{M}_o)$ from the class of d-equivalence determined by $\underline{M}_1$; this class is of cardinality greater than or equal to $r!(m-r)!$ Then

$$\text{card } As_{q_2}(\underline{M}_o) \geq \text{card } As_{q_1}(\underline{M}_o) - r!(m-r)!$$

which contradicts (iii).

5.3.24 Example. Let $\underline{a}: \mathbb{N} \to \mathbb{Q}$ be a non-decreasing sequence of rational numbers (cf. 5.3.6). For a model $\underline{M} = \langle M, f_1, f_2 \rangle$ of type $\langle a,b \rangle$, let $\underline{a}[\underline{M}]$ be $\sum_{f_1(o)=1} \underline{a}(f_2(o))$. Given m and r, take a model

$\underline{M}_1 = \langle M, g_1 \rangle$ of type $\langle a \rangle$ such that $m_{\underline{M}_1} = m$ and $r_{\underline{M}_1} = r$.

For any $\underline{M} \in$ .Exp ($\underline{M}_1$), consider

$\{\underline{N} \in \mathrm{Exp}(\underline{M}_1) ;\ \underline{a}[\underline{N}] \geq \underline{a}[\underline{M}]\} = \mathrm{Greater}(\underline{a}, \underline{M})$.

(Note that if $\underline{U} \models \Phi_o$ and $Q_1 \restriction M = f_1$, then, by 5.2.21 ,

$P^{\underline{U}}(\{\sigma ;\ \underline{a}[\underline{M}] \geq \underline{a}[\underline{M}]\}) = \frac{1}{m!}\ \mathrm{card}(\mathrm{Greater}(a, \underline{M}))$.) Put

$c_\alpha(r,m) = \min\{\underline{a}[\underline{M}] ;\ \underline{M} \in \mathrm{Exp}(\underline{M}_1)$ and $\frac{1}{m!}\ \mathrm{card}(\mathrm{Greater}(\underline{a}, \underline{M})) \leq \alpha\}$

Let q_α be defined by putting $\mathrm{Asf}_{q_\alpha}(\underline{M}) = 1$ iff $\underline{a}[\underline{M}] \geq c_\alpha(r_{\underline{M}}, m_{\underline{M}})$.

Then $\{q_\alpha\}_{(0,0.5] \cap Q}$ is a monotone class.

For example, for $\underline{a}(i) = i$, we have the monotone class corresponding to the Wilcoxon tests for different values of significance level.

5.3.25 <u>Remark.</u> The above results will be used in the logic of suggestion (c.f. 7.4.2).

5.3.26 <u>Theorem.</u> Consider a calculus $\mathcal{F}$ (MOFC with enumeration models), let q be a d-executive quantifier. There are no tautologies of the form $q(\varphi , F)$ (φ, F designated open).

The proof is left to the reader as an easy exercise (use d-executiveness).

5.3.27 <u>Lemma.</u> Let q be a d-executive quantifier and let φ, ψ be two designated open formulae. Suppose that each of the following formulae is satisfiable: $\varphi, \neg\varphi, \psi, \neg\psi$.

Then $q(\alpha , F)$ logically implies $q(\psi, F)$ iff φ is logically equivalent to ψ .

<u>Proof.</u> First, assume $\varphi \not\models \psi$: let u_{10} be a card satisfying $\varphi \ \& \neg \psi$.

(i) If there is a card u_{01} satisfying $\neg\varphi \ \& \ \psi$, then take $m > m_{min}$, r such that $r \in (r_{min}(m), r_{max}(m))$ and let $\underline{M}_1 = \langle M, f_1^+ \rangle$ be such that $m_{\underline{M}} = m$ and $r_{\underline{M}} = m$. If $\underline{M}_o$ is a model with the field $\underline{M}$ in which r objects have the card u_{10} and the remaining objects have the card u_{01}, then $\underline{M}_1 = \langle M, \|\varphi\|_{\underline{M}_o} \rangle$ If we put $f_1^-(o) = 1 - f_1^+(o)$, then $f_1^- = \|\psi\|_{\underline{M}_o}$. Let f_2 be an enumeration of M such that $f_1^+(o_1) = 1$ and $f_i^+(o_2) = 0$ implies $f_2(o_1) > f_2(o_2)$. Then, for each $\underline{M} \in \mathrm{Exp}(\underline{M}_1)$, we have $\underline{M}^+ = \langle M, f_1^+, f_2 \rangle \succeq_d \underline{M}$ so that $\mathrm{Asf}_q(\underline{M}^+) = 1$. Consider $\underline{M}^- = \langle M, f_1^-, f_2 \rangle$. For each $\underline{M} \in \mathrm{Exp}(\langle M, f_1 \rangle)$, we have $\underline{M}^- \preceq_d \underline{M}$; hence, $\mathrm{Asf}_q\, \underline{M}^- = 0$. Summarizing, if $\underline{M}^* = \langle \underline{M}_o, f_2 \rangle$ is the expansion of $\underline{M}_o$ interpreting F as f_2, then $\|q(\varphi, F)\|_{\underline{M}^*} = 1$ and $\|q(\psi, F)\|_{\underline{M}} = 0$: hence, $q(\varphi, F)$ does not imply $q(\psi, F)$.

(ii) If there is no u_{01}, then there is a card u_{00} satisfying $\neg\varphi \,\&\, \neg\psi$ (since $\neg\varphi$ is satisfiable; furthermore, since ψ is satisfiable, $\varphi \,\&\, \psi$ is also satisfiable. Let us form a model $\underline{M}_o$ with the field M of cardinality m in which each object has one of the cards u_{10}, u_{00}, u_{11} and, if their frequencies are denoted m_{10}, m_{00}, m_{11}, respectively, then $m_{11} + m_{10} = r$ but $m_{11} \leq r_{min}(m)$. Put $\|\varphi\|_{\underline{M}_o} = f_1$ and let f_2^+ be such that $\mathrm{Asf}_q(\langle M_1, f_1, f_2^+ \rangle) = 1$. Then $\mathrm{Asf}_q(\langle M, \|\psi\|_{\underline{M}_o}, f_2^+ \rangle) = 0$, since the frequency of $\|\psi\|_{\underline{M}_o}$ is less than **or** equal to $r_{min}(m)$. Put $\underline{M}^* = \langle \underline{M}_o, f_2^+ \rangle$; then $\|q(\varphi, F)\|_{\underline{M}^*} = 1$ and $\|q(\psi, F)\|_{\underline{M}^*} = 0$.

Finally, assume $\varphi \models \psi$ but $\psi \not\models \varphi$; let u_{10} be a card satisfying $\neg\varphi \,\&\, \psi$. Since φ is satisfiable, we have a card u_{11} satisfying $\varphi \,\&\, \psi$, and since $\neg\psi$ is satisfiable we have a card u_{oo} satisfying $\neg\varphi \,\&\, \neg\psi$. Let m_{01}, m_{11}, m_{00} have the obvious meaning; suppose that $\underline{M}_o$ is such that $m_{11} = r$ but $m_{11} + m_{01} \geq r_{max}(m)$ (and, obviously, $m = \mathrm{card}(M) = m_{11} + m_{01} + m_{00}$). Let $f_1 = \|\varphi\|_{\underline{M}_o}$ and let f_2^+ **be** such that $\mathrm{Asf}_q(\langle M, f_1, f_2^+ \rangle) = 1$. Then, putting $\underline{M}^* = \langle M_o, f_1, f_2 \rangle$, we have $\|q(\varphi, F)\|_{\underline{M}^*} = 1$ and $\|q(\psi, F)\|_{\underline{M}^*} = 0$. The proof is thus completed.

5.3.28 Theorem. Let q **be** a d-executive quantifier. If R is a binary relation on factual designated open formulae and if

$I = \left\{ \frac{q(\varphi, F)}{q(\psi, F)} ; \varphi R \psi \right\}$ is sound, then $\varphi R \psi$ implies that φ and ψ are logically equivalent.

Proof. The theorem is a corollary of the previous lemma.

5.3.29 Remark. This means that there is no non-trivial deduction rule of the form just described.

5.3.30 Key words: enumeration models, distinctive quantifiers, executive and d-executive quantifiers, quantifiers of level α ; monotone classes of distinctive quantifiers.

V.4 Observational monadic function calculi with rational valued models.

In the present section, we introduce a class of calculi which enables us to describe some tasks of the logic of suggestion, particularly in situations with real-valued random quantities (cf. Section 4 of Chapter VII). In accordance with Chapter IV and Section V.1, the corresponding observational calculi have rational-valued models.

First, we define a new kind of quantifiers (correlational quantifiers) in observational calculi with enumeration models; the theory of quantifiers of this kind is quite uninteresting except in connection with rational-valued models and the corresponding quantifiers. So we define calculi with such models and the particular class of quantifiers (rank quantifiers).

The relation between distinctive and correlational quantifiers will be considered.

5.4.1 Definition. Let $\underline{M}_1 = \langle M_1, f_1, f_2 \rangle$ and $\underline{M}_2 = \langle M_2, g_1, g_2 \rangle$ be two models of type $\langle b, b \rangle$. $\underline{M}_1$ is c-better than $\underline{M}_2$ ($\underline{M}_1 \succeq_c \underline{M}_2$) if there is a one-to-one mapping τ of $\underline{M}_2$ onto $\underline{M}_1$ such that $|f_1(\tau o) - f_2(\tau o)| \leq |g_1(o) - g_2(o)|$ for each $o \in M_2$.

A quantifier q of type $\langle b, b \rangle$ is called correlational if $Asf_q(\underline{M}_2) = 1$ and $\underline{M}_1 \succeq_c \underline{M}_2$ imply $Asf_q(\underline{M}_1) = 1$.

5.4.2 Discussion. As in the case of quantifiers of type $\langle a, b \rangle$ we can define executive quantifiers of type $\langle b, b \rangle$ and (correlational) quantifiers of level α (w.r.t. all correlational quantifiers).

In the first case, we require that there is a number $m_{min}(q)$ such that for each model $\underline{M}$ of type $\langle b, b \rangle$, $m_{\underline{M}} \leq m_{min}(q)$ implies $Asf_q(\underline{M}) = 0$, and for each $m > m_{min}(q)$ there are two models $\underline{M}_1$, $\underline{M}_2$, $m_{\underline{M}_1} = m_{\underline{M}_2} = m$ such that $Asf_q(\underline{M}_1) = 1$ and $Asf_q(\underline{M}_2) = 0$.

Almost all that is said below (5.4.3 - 5.4.11) is formulated for all quantifiers of type $< b, b >$; but the only class of quantifiers of type $< b, b >$ which is really useful is the class of correlational quantifiers (as far as the authors know).

5.4.3 Definition (auxiliary). Each pair f_1, f_2 of enumerations of a set M determines a permuation of $\{1, \ldots, \mathrm{card}(M)\}$ denoted by π_{f_1, f_2} and defined as follows: $\pi_{f_1,f_2}(i) = f_2(f_1^{-1}(i))$. If $\underline{M} = < M, f_1, f_2 >$ is a model of type $< b, b >$, then we write $\pi_{\underline{M}}$ instead of π_{f_1,f_2}. See Problem (8).

5.4.4 Lemma. Consider two models $\underline{M}_1 = < M_1, f_1, f_2 >$ and $\underline{M}_2 = < M_2, g_1, g_2 >$ of type $< b, b >$. Then $\pi_{\underline{M}_1} = \pi_{\underline{M}_2}$ iff $\underline{M}_1$ and $\underline{M}_2$ are isomorphic.

Proof. Let τ be a one-to-one mapping of M_1 onto M_2 such that $f_1(o) = g_1(\tau o)$ for each o, and let $\pi_{\underline{M}_1} = \pi_{\underline{M}_2}$. Then

$$f_2(o) = \pi_{\underline{M}_1}(f_1(o)) = \pi_{\underline{M}_2}(g_1(\tau o)) = g_2(\tau o)$$

and τ is an isomorphism.

Conversely, let τ be an isomorphiom of M_1 and M_2 . Then

$$\pi_{\underline{M}_1}(i) = f_2(f_1^{-1}(i)) = f_2(\tau^{-1}(g_1^{-1}(i))) = g_2(g_1^{-1}(i)) = \pi_{\underline{M}_2}(i),$$

hence $\pi_{M_1} = \pi_{M_2}$.

5.4.5 Corollary. Let $\mathrm{card}(M) = m$; then structures of type $< b, b >$ with field M decompose into $m!$ isomorphism classes.

5.4.6 Discussion. Let $\underline{U}$ be a random structure (of type $< 1, 1 >$ satisfying the t.c.-condition (see 5.1.14), i.e., $\underline{U} \models \Phi_2$. Given an $M \subseteq U$, $\mathrm{card}(M) = m$, consider $\mathrm{Rk}_2(\underline{M}_\sigma) = < M, f^*_{1\sigma}, f^*_{2\sigma} >$ (cf. 5.2.9). As in 5.1.19 and 5.1.21, one shows that $P(\{\sigma ; \mathrm{Rk}_2(\underline{M}_\sigma)$ is a structure of type $< b, b > \}) = 1$ and under H_2 (independence and d-homogeneity)

$$P(\{\sigma ; \pi_{Rk_2(\underline{M}_\sigma)} = \}) = \frac{1}{m!}$$

for each permutation γ of $\langle 1,\dots, m\rangle$.

5.4.7 Definition. Let a quantifier q of type $\langle b, b\rangle$ be given. Denote $\mathcal{E}_{M,q} = \{\gamma$ permutation of $\langle 1,\dots,m\rangle$; for an $\underline{M}$ with field M, $\pi_{\underline{M}} =$ and $Asf_q(\underline{M}) = 1\}$. Let a number $\alpha \in (0, 0.5]$ be given. We say that q is of level α if for each $m \in \mathbb{N}$ and for each M, card(M) = m,

$$\alpha \geq \frac{\text{card } \mathcal{E}_{M,q}}{m!} > \alpha - \frac{1}{m}.$$

5.4.8 Discussion. (i) Let q be of level α in the sense of 5.4.7. Then q is of level α ' w.r.t. all quantifiers of type $\langle b, b\rangle$.

(ii) The words "for each M, card(M) = m" in the previous definition can be replaced by "for an M such that card(M) = m".

(iii) If we consider a model $\underline{M}_o = \langle M, f_1\rangle$ of type $\langle b\rangle$ with card(M) = m, then Exp $(\underline{M}_o)$ has cardinality m! and contains representatives of all classes of the permutational equivalence on models of type $\langle b,b\rangle$ and of the given cardinality. Moreover, $c(\underline{M}_o) = 1$ (w.r.t. all quantifiers of type $\langle b,b\rangle$) .

Now, we have

$$\text{card } \mathcal{E}_{M,q} = \text{card}\{\underline{M} \in \text{Exp}(\underline{M}_o) ; Asf_q(\underline{M}) = 1\}$$

for each $\underline{M}_o$ of type $\langle b\rangle$ and with field M. Thus, we see the consistency of the present definition with the one in 5.3.9.

(iv) Note that being of level α "w.r.t. all quantifiers of type $\langle b,b\rangle$ " is equivalent to being of level α "w.r.t. all correlational quantifiers". In the following, we shall say simply "a quantifier is of level α ".

5.4.9 Remark. As in 5.3.16, we can prove that if a quantifier q of type $\langle b, b\rangle$ is of level α , then q is executive with $m_{min}(q) = m_\alpha$.

5.4.10 Definition. We can define a monotone class of quantifiers of type $\langle b, b\rangle$ as follows:

A class $\{q_\alpha\}_{\alpha \in A}$, $A \subseteq (0,0.5]$ is a monotone class if

(1) q_α is of level α,

(2) $\alpha_1 < \alpha_2$ implies $\mathcal{E}_{M,q_{\alpha_1}} \subseteq \mathcal{E}_{M,q_{\alpha_2}}$ for each M finite .

5.4.11 Remark. We now summarize some useful but trivial facts concerning executive correlational quantifiers: Let a calculus $\mathcal{F}$ be given:

(1) For each model $\underline{M}$, $m_{\underline{M}} > m_{min}(q)$ implies $\underline{M} \models q(F,F)$.

(2) For each model $\underline{M}$, $m_{\underline{M}} \leq m_{min}(q)$ implies $\underline{M} \models \neg q(F,F)$;

in particular, for quantifiers of level α, if $m_{\underline{M}} \leq m_\alpha$ then $\underline{M} \models \neg q(F,F)$.

(3) Denote by q_{min} a quantifier of the empty type with $Asf_{q_{min}}(\underline{M}) = 1$ iff $m_{\underline{M}} > m_{min}(q)$. Then $q_{min} \rightarrow q(F,F)$ is a tautology.

(4) For $q(F_1,F_2)$ where $F_1 \neq F_2$, there is a model $\underline{M}$ with $m_{\underline{M}} > m_{min}(q)$ such that $\underline{M} \models \neg q(F_1, F_2)$.

5.4.12 Example. (1) Considėr two particular cases of correlational quantifiers:

Spearmen's ϱ:

$$Asf_q(\langle M,f_1,f_2\rangle) = 1 \text{ if } \sum_{o \in M} f_1(o)f_2(o) \geq c_\alpha(m_{\underline{M}}) ,$$

Kendall's τ (w-correlational; see Problem (8, e)):

$$Asf_\tau(\langle M,f_1,f_2\rangle) = 1 \quad \text{if}$$

$$\sum_{o,o' \in M,\, o \neq o'} \operatorname{sign}(f_1(o) - f_1(o'))\operatorname{sign}(f_2(o) - f_2(o')) \geq c'_\alpha(m_{\underline{M}}).$$

For further examples see Problem (9) .

5.4.13 Discussion and definition. Now we shall describe a more general situation. It is usual that in many research situations we investigate real random quantities and $\{0,1\}$ - random quantities together. Corresponding observational calculi must then have models with both rational-valued and $\{0,1\}$ - valued quantities, i.e., such models are $\mathbb{V}$ - structures, where $\mathbb{V} = \langle \{0,1\}^{k_1}, \mathbb{Q}^{k_2} \rangle$ ($k_1\ k_2 \in \mathbb{N}$). Let $\mathcal{M}$ be set of all finite $\mathbb{V}$ - structures. Hence, in such calculi we consider predicates $P_1, \ldots, P_{k_1}$ and rational function symbols $F_1, \ldots, F_{k_2}$, assumed to be monadic. The type of function calculus in question is $\langle \underbrace{a, \ldots, a}_{k_1\text{-times}}, \underbrace{c, \ldots, c}_{k_2\text{-times}} \rangle$. Further, let there be given sets of designated open formulae of two sorts a and c. Suppose that $\|\varphi\|_{\underline{M}}$ is determined by a $\{0,1\}^{k_1}$ - structure $\langle M, \|P_1\|_{\underline{M}}, \ldots, \|P_{k_1}\|_{\underline{M}} \rangle$ if φ is of sort a and by a $\mathbb{Q}^{k_2}$ structure $\langle M, \|F_1\|_{\underline{M}}, \ldots, \|F_{k_2}\|_{\underline{M}} \rangle$ if φ is of sort c. Now, we can consider a reasonable class of quantifiers.

The most important thing now is the definition of a rank quantifier. (Cf. Definition 5.1.14). A quantifier q of type $\langle a,c \rangle$ (or $\langle c,c \rangle$ or $\langle b,c \rangle$) is a rank quantifier if:

$$Asf_q(\underline{M}) = Asf_q(\, Rk(\underline{M}))$$

i.e., if $\underline{M} = \langle M, f_1, f_2 \rangle$, then $Asf_q(\underline{M}) = Asf_q(\langle M, f_1^*, f_2^* \rangle)$.

Analogously, a quantifier q of type $\langle c,c \rangle$ is a strong rank quantifier (cf. 5.1.9) if $Asf_q(\underline{M}) = Asf_q(\, Rk_2(\underline{M}))$, i.e., if $\underline{M} = \langle M, f_1, f_2 \rangle$, then $Asf_q(\underline{M}) = Asf_q(\langle M, f_1, f_2 \rangle)$.

5.4.14 Remark. Consider random $\langle \{0,1\}, \mathbb{R} \rangle$ - structures now; considerations for $\langle \mathbb{R}, \mathbb{R} \rangle$ - structures are similar.

Remember that - by 5.1.19 - if $\underline{U}$ satisfies Φ_0 (d.c.-condition) or Φ_2(p.c. - condition) and if $M \subseteq U$ is a finite sample, then $P(\{\sigma\ ;\ Rk_2(\underline{M}_\sigma)$ is of type $\langle a,b \rangle\}) = 1$, i.e., if

$\underline{M}_\sigma = \langle M, f_{1\sigma}, f_{2\sigma} \rangle$, then $P(f^*_{2\sigma}$ is an enumeration $) = 1$.

Let q_o be a quantifier of type $\langle a,b \rangle$ i.e., Asf_{q_o} is defined on all structures of type $\langle a,b \rangle$. How can q_o be extended to a rank quantifier q?

If $\underline{M} = \langle M, f_1, f_2 \rangle$ is a model of type $\langle a, c \rangle$ and if f_2 is one-to-one, then $Rk(\underline{M})$ is of type $\langle a,b \rangle (f^*_2 \in \mathcal{R}_M)$ and $Asf_q(\underline{M}) = Asf_{q_o}(Rk(\underline{M}))$ is uniquely determined. If this is not the case, then f^*_2 is a mapping of M onto a proper subset of $\{1, \ldots, card(M)\}$ with the following property: If i has k_i pre-images in f^*_2, then $i-1, i-2, \ldots, i-k_i+1$ have no pre-images. Call a function $\eta : M \to \{1, \ldots, card(M)\}$ a <u>pseudoenumeration</u> if η has the property just stated about f^*_2. Say that an enumeration ξ of M <u>linearizes</u> η if, for each i in the range of η, when $\eta^{-1}(i)$ has k_i elements then ξ maps $\eta^{-1}(i)$ onto $\{i-1, \ldots, i-k_i+1\}$. The following table gives an example:

M	a	b	c	d	e	f
	2	2	5	5	5	6
	1	2	3	4	5	6

By 5.1.20, we assume that if in our observed $\underline{M}_\sigma = \langle M, f_{1\sigma}, f_{2\sigma} \rangle$ $f_{2\sigma}$ is not one-to-one, then, in fact, $\underline{M}_\sigma$ is an inexact observation of a structure $\langle M, f_{1\sigma}, \hat{f}_{2\sigma} \rangle$ where $\hat{f}^*_{2\sigma}$ is a linearization of $f^*_{2\sigma}$. Hence, one way **in which to** extend the definition of Asf_q to models $\langle M, f_1, f_2 \rangle$ where f^*_2 is a pseudo-enumeration is to put $Asf_q(\langle M, f_1, f_2 \rangle) = 1$ iff $Asf_{q_o}(\langle M, f_1, \hat{f}_2 \rangle) = 1$ for all linearizations $\hat{f}_2$ of f^*_2. Then Asf_q extends to all models of type $\langle a,c \rangle$ and q is a rank quantifier. This extension can be called the <u>secured extension</u> of q_o, and the obtained rank quantifier q can be called the <u>secured rank quantifier.</u>

Another possibility is to associate with each M a preferred enumeration en_M of M (corresponding, e.g., to the order in which the objects were observed) and associate with each pseudoenumeration η the <u>preferred linearization</u> s defined in accordance with preferred enumeration(for each $o_1\, o_2 \in M$, $\eta(o_1) = \eta(o_2)$ and $en_M(o_1)$ $en_M(o_2)$. Hence, we can define the <u>preferring extension</u> of q_o by putting $Asf_q(\langle M, f_1, f_2 \rangle) = 1$ iff $Asf_{q_o}(\langle M, f_1, \hat{f}_2 \rangle) = 1$ where $\hat{f}_2$ is the preferred linearization of f^*_2.

For strong rank quantifiers (i.e., random $\langle \mathbb{R},\mathbb{R}\rangle$-structures), extensions are completely analogous.

5.4.15 Definition and Lemma. Let q be a rank quantifier of type $\langle a,c\rangle$. Let $\underline{M} = \langle M, f_1, f_2\rangle$ be a model of type $\langle a,c\rangle$. A critical linearization of f_2^* w.r.t. $\underline{M}$ is each $\hat{f}_2$ such that $Asf_q(\langle M,f_1,\hat{f}_2\rangle) = 1$ implies $Asf_q(\langle M,f_1,f_2\rangle) = 1$ for each linearization f_2 of f_2^*.

Thus, for secured rank quantifiers, if f_2^* is a pseudoenumeration, we need to look for a critical linearization.

Lemma. Let q be a secured extension of a distinctive quantifier and let $\underline{M} = \langle M,f_1,f_2\rangle$ be a model for which f_2^* is a pseudoenumeration. Each $\hat{f}_2$ satisfying the following conditions is a critical linearization of f_2^* w.r.t. $\underline{M}$:

(1) If $(f_2^*)^{-1}(i)$ has one element, then $\hat{f}_2^{-1}(i) = (f_2^*)^{-1}(i)$.

(2) If $\mathrm{card}((f_2^*)^{-1}(i)) = k_i > 1$, then $\hat{f}_2$ maps $(f_2^*)^{-1}(i)$ onto $\{i-1, \ldots, i-k_i+1\}$ in such a way that $f_1(o_1) = 1$ and $f_1(o_2) = 0$ imply $\hat{f}_2(o_1) > \hat{f}_2(o_2)$ for arbitrary $o_1, o_2 \in (f^*)^{-1}(i)$.

Proof. $\langle M, f_1, \hat{f}_2\rangle$ is the least element (w.r.t. $\preceq_d$) in the set of all models obtained by the linearization of $Rk(\underline{M})$.

5.4.16 Remark. (1) In a similar way, we obtain the critical linearization of the secured extension q of a correlational quantifier. Then we use the following condition:

If $f_1^*(o_1) > f_1^*(o_2)$, then $\hat{f}_2(o_1) < \hat{f}_2(o_2)$

(i.e., we first linearize f_2^*) and

if $\hat{f}_2(o_1) < \hat{f}_2(o_2)$, then $\hat{f}_1(o_1) > \hat{f}_1(o_2)$.

For example,

f_1^*	f_2^*	$\hat{f}_2$	(or)	$\hat{f}_1$	(or)
2	1	1	1	2	2
2	2	2	2	1	1
5	5	5	5	3	3
5	4	4	3	4	5
5	4	3	4	5	4
6	6	6	6	6	6

Note that both possibilities determine the same permutation.

Obviously, we can first linearize f_1^*; in our example we get

$\hat{f}_1$	$\hat{f}_2$
2	1
1	2
3	5
4	4
5	3
6	6

and we obtain the same permutation.

(2) Remember Theorem 4.5.6 which states the preservation of the test property for secured extensions.

5.4.17 Definition. A strong rank quantifier is a correlational rank quantifier if q restricted to models of type $\langle b,b\rangle$ is a correlational quantifier in the sense of 5.4.1.

5.4.18 Lemma. Let q be a correlational rank quantifier, and let $\mathcal{JM}$ be as in 5.1.14 ($\langle M,f_1,f_2\rangle \in \mathcal{JM}$ iff f_2 are one-to-one). Then, for each $\underline{M} \in \mathcal{JM}$ and all designated open formulas φ_1, φ_2 of sort c,

$$\|q(\varphi_1, \varphi_2)\|_{\underline{M}} = \|q(\varphi_2, \varphi_1)\|_{\underline{M}}.$$

Proof.

$$\|q(\varphi_1,\varphi_2)\|_{\underline{M}} = \mathrm{Asf}_q(\langle M, \|\varphi_1\|_{\underline{M}}, \|\varphi_2\|_{\underline{M}}\rangle) = \mathrm{Asf}_q(\langle M, (\|\varphi_1\|_{\underline{M}})^*, (\|\varphi_2\|_{\underline{M}})^*\rangle).$$

Hence it suffices to prove that $\mathrm{Asf}_q(\langle M,f_1,f_2\rangle) = \mathrm{Asf}_q(\langle M,f_2,f_1\rangle)$ for $\langle M,f_1,f_2\rangle$ of type $\langle b,b\rangle$. But the last fact follows directly from the definition.

5.4.19 Remark. From Kendall's considerations in [Kendall 1955] we can use the definition of general correlation coefficient. General correlation coefficients are functions on $\langle \mathbb{R},\mathbb{R}\rangle$ - structures of the following form:

$$k(\langle M,f_1,f_2\rangle) = \frac{\sum_{i,j\in M} a_{ij} b_{ij}}{\sqrt{\sum_{i,j\in M} a_{ij}^2 \sum_{i,j\in M} b_{ij}^2}},$$

where $a_{ij} = f(f_1(i), f_1(j))$, $b_{ij} = g(f_2(i), f_2(j))$ for some functions f,g satisfying $f(y,x) = -f(x,y)$, $g(y,x) = -g(x,y)$ (in particular, $f(x,x) = g(x,x) = 0$).

5.4.20 <u>Lemma.</u> (Kendall) Let $f(x,y)$, $g(x,y)$ be positive, non-decreasing recursive functions of $|x - y|$. Define a quantifier q as follows: For each $\underline{M} \in \mathfrak{M}^{\langle Q,Q \rangle}$,

$$\mathrm{Asf}_q(\langle M, f_1, f_2 \rangle) = 1 \quad \text{if} \quad k(\mathrm{Rk}_2(\underline{M})) \geq c(m_{\underline{M}}),$$

where c is a recursive function on $\mathbb{N}$. Then the secured extension of q is a <u>w-correlational rank quantifier.</u>

<u>Proof.</u> See Problem (8).

Compare with the form of Spearmen's ϱ and Kendall's τ.

5.4.21 Now we have to interrupt our investigations of rank calculi. For further considerations that are connected with the logic of suggestion we need notions that will be introduced in Chapter VI. We shall continue with some discussions on the practical applicability of rank calculi in Chapter VII, Section 4.

5.4.22 <u>Key words:</u> correlational quantifiers, rank quantifiers, strong rank quantifiers, secured extensions ; correlational, distinctive rank quantifiers.

PROBLEMS AND SUPPLEMENTS TO CHAPTER V.

(1) Consider random $\langle \mathbb{R},\mathbb{R}\rangle$-structures satisfying the d.c.-condition (see 5.1.5). Under this frame assumption we can formulate the alternative of general regression in location (AGRL):

There is a function F (x) such that

$$D_{\vartheta_2,o}(x) = F(x - Q_1(o)\Delta) \quad , \text{ where } \quad \Delta \neq 0.$$

If $\Delta = 0$, we obtain H_o.

Moreover, if one defines general conditional distribution functions (see, e.g., [Burril]), then the conditional AGRL can be stated as follows: There is a function F (x) such that

$$D_{(\vartheta_2/\vartheta_1),o}(x/y) = F(x - \Delta y) \qquad (\Delta \neq 0).$$

If we consider $\Delta > 0$, then we can use tests of the form:

$$\sum_{o \in M} \underline{b}(f_1(o))\underline{a}(f_2(o)) \geq c_\alpha(\langle M,f_1\rangle) \quad , \qquad (*)$$

where $\underline{a}$ and $\underline{b}$ are non-decreasing recursive sequences of rational numbers. Apply to such a case considerations of 5.1.19 - 5.1.22 and 5.2.8 .(Prove that by (*) we can define a regression quantifier in the sense of 7.4.10)

(2) Consider the following statistic (Haga test) for testing ASL (and c.ASL):

$$T(\langle M,f_1,f_2\rangle) = A(\langle M,f_1,f_2\rangle) + B(\langle M,f_1,f_2\rangle) \quad ,$$

where

$$A(\langle M,f_1,f_2\rangle) = \text{card}\,\{o;\ f_1(o) = 1 \text{ and } f_2(o) > \max\{f_2(o');\ f_1(o') = 0\}\}$$

and

$$B(\langle M,f_1,f_2\rangle) = \text{card}\{o;\ f_1(o) = 0 \text{ and } f_2(o) < \min\{f_2(o');\ f_1(o') = 1\}\}.$$

Prove that a quantifier with the associated function

$$\text{Asf}_q(\underline{M}) = 1 \quad \text{if} \quad T(\underline{M}) \geq c(m,r)$$

is a distinctive rank quantifier.

Note that T is not a simple linear function (and T is not asymptotically normal, cf. [Hájek and Šidák]).

(3) Consider $\langle a,b \rangle$ - models. We can define "asymptotical forms" of the Wilcoxon and median quantifiers:

$$\mathrm{Asf}_{w.as.}(\langle M, f_1, f_2 \rangle) = 1 \text{ iff } \sum_{o \in M} f_1(o) f_2(o) \geq c_\alpha^{as}(m_{\underline{M}}, r_{\underline{M}}) ,$$

where

$$c_\alpha^{as}(m,r) = (\quad + \frac{1}{2} r\ (m+1)) \sqrt{1/12 \,.\, rm\,(m-r)} \,.$$

Analogously for m.as. $c_\alpha^{as}(m,r) = \left(+ \frac{1}{2} r \right) \sqrt{\frac{r\,(m-r)}{4\,K}}$,

where $K = \begin{cases} m-1 & \text{for } m \text{ even,} \\ m & \text{for } m \text{ odd.} \end{cases}$

Prove that such quantifiers are distinctive and d-executive.
Are they of level α ?

(4) The general form of simple linear rank statistics for testing H_2 is the following:

$$s(\langle M, f_1, f_2 \rangle) = \sum_{o \in M} \underline{a}(f_1^*(o))\, \underline{a}(f_2^*(o)) ,$$

where $\underline{a}$ is as in Problem (1),
Note that, for models of type $\langle b,b \rangle$,

$$s(\langle M, f_1, f_2 \rangle) = \sum_{o \in M} \underline{a}(i)\ \underline{a}(f_2(f_1^{-1}(o))$$

the so-called dual form, cf [J. Hájek]. Use the considerations of 5.1.21, 5.2.9 and prove that for each M, card M = m , under H_2 we have

$$E\, s(\underline{M}_\sigma) = m\bar{a} ,\ VAR\, s(\underline{M}_\sigma) = \frac{1}{m-1} \Big(\sum_{i=1}^{m} (a(i) - \bar{a})^2 \Big)^2$$

Prove the analogue of Theorem 5. 1. 21.

(5) Prove the correlationality of Spearmen's ρ.
Prove Lemma 5.4.15. (Hint: Consider the behaviour of a k- function under transformations which are improving in the sence of $\leq_c$).

(6) For testing H_2, we can use the following statistic:

$$s(\langle M,f_1,f_2\rangle) = \sum_{o \in M} \text{sign}\left(f_1^*(o) - \frac{1}{2}(m+1)\right)\text{sign}\left(f_2^*(o) - \frac{1}{2}(m+1)\right).$$

(quandrant test). Prove the correlationality of the corresponding quantifier. Is it of level α ?

(7) Prove: Let $\mathcal{F}$ be an MOFC with enumeration models.

(a) Each sentence is logically equivalent to a Boolean combination of pure prenex formulae (Hint: The proof is analogous to that of 5.2.3 .)

(b) Each MOFC with enumerations models is decidable. (Hint: Use the same method as for 5.2.6 via 5.2.5 from 5.2.4 .)

(8) In [Yanagimoto 1969] there are, in fact, considered orderings of models and their relation to correlational rank statistics. These orderings are based on orderings of permutations

$$\pi_{\underline{M}} = \pi_{f_1,f_2}, \quad \text{if } \underline{M} = \langle M,f_1,f_2\rangle .$$

(a) These orderings of permutations are denoted there by $\leq^W$ and $\leq^S$.

$\leq^S$ is stronger than $\leq^W$. Both orderings are based on interchanging members of permutations in accordance with the order of indices (for $\leq^W$ we can use only repeated interchanging of neighbours). Our ordering of models (c-better) generates an ordering of permutations also: denote it by $\leq^C$. For cardinality of samples equal to 3 we obtain for $\leq^C$:

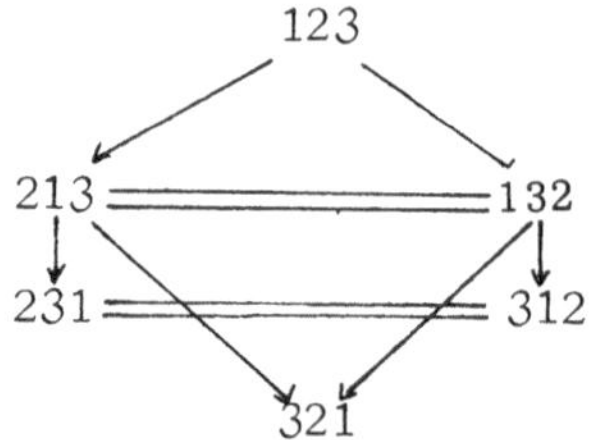

(arrows mean: strictly c-better, double lines: c-equivalent);

For $\leq^W$ we have:

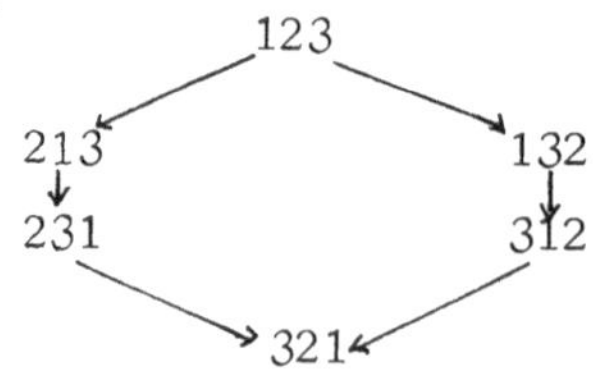

Hence $\leq^c$ bears no relation to $\leq^w$ and $\leq^s$.

(b) If $\pi_1 \leq^w \pi_2$ or $\pi_1 \leq^s \pi_2$ then $\pi_1 >^c \pi_2$ cannot occur.

(c) $\leq^w$ is "nonsymmetrical", i.e. if we have models

$\underline{M}_1 = \langle M, f_1, f_2 \rangle$, $\underline{M}_2 = \langle M_2, g_1 g_2 \rangle$, $\underline{M}_1' = \langle M_1, f_2, f_1 \rangle$, $\underline{M}_2' = \langle M_2, g_2, g_1 \rangle$

and $\pi_{\underline{M}_1} \leq^w \pi_{\underline{M}_2}$ we do not know anything about $\pi_{\underline{M}_1'}$, $\pi_{\underline{M}_2'}$ (note that $\pi^{-1}_{M_i} = \pi_{M_i'}$).

(d) $\leq^c$ is "symmetrical": clearly, if $\underline{M}_1 \leq^c \underline{M}_2$ then $\underline{M}_1' \leq^c \underline{M}_2'$. This justifies our usage of $\pi_{\underline{M}}$.

(e) Nevertheless, we could base the notion of correlational quantifiers on $\leq^w$ or $\leq^s$, but we claim that $\leq^c$ is more reasonable. With respect to $\leq^c$ the models

$\begin{pmatrix} 1 & 2 \\ 2 & 3 \\ 3 & 1 \end{pmatrix}$ and $\begin{pmatrix} 1 & 3 \\ 2 & 1 \\ 3 & 2 \end{pmatrix}$ are equivalent but they are incomparable in $\leq^w$

and $\leq^s$. The question whether $\begin{pmatrix} 1 & 2 \\ 2 & 3 \\ 3 & 1 \end{pmatrix}$ is better than $\begin{pmatrix} 1 & 3 \\ 2 & 2 \\ 3 & 1 \end{pmatrix}$ seems to be less important.

(f) Does Theorem 6.2 from [Yanagimoto] remain true under our ordering?

(9) We conclude with some remarks concerning acc-statistics.

(a) Note that the condition of Borel measurability in the definition of acc-statistics is superflous. Let $\underline{U}$ be a $\langle \Sigma, \mathcal{R}, P \rangle$ - random structure. Then we have the induced probability measure, say μ_P, on Borel sets of $\mathfrak{M}^V_M$ (for each M). Remember the notion of μ_P-measurable sets.

(i) If now $\mathfrak{M}_o \subseteq \mathfrak{M}^V_M$ is an open set, such that $\mu_P(\mathfrak{M}_o) = 1$, then we have the following:

for each $B \in \mathcal{M}_M^V$, B is μ_P - measurable iff

$B \cap \mathcal{M}_o$ is μ_P-measurable.

(Hint: Note that, if μ_P^* is the outer measure generated by μ_P, then $\mu_P^*(B) = \mu_P^*(\mathcal{M}_o \cap B)$.

(ii) If f is now an arbitrary function satisfying conditions (a), (c) and (b′) from 5.1.2 , then (for each sample M):

for each X Borel, $f_M^{-1}(X)$ is μ_P-measurable (under Φ). It is exactly this property that we need (not Borel measurability). Such a "generalized" statistic f need not be a statistic in the classical sense.

(Hint: f is continuous on $\mathcal{M}_{cont} \subseteq \mathcal{M}_M^V$, $\mu_P(\mathcal{M}_{cont}) = 1$. Hence, for each X Borel, $f_M^{-1}(X) \cap \mathcal{M}_{cont} = f_M^{-1}(X \cap f(\mathcal{M}_{cont}))$ is Borel and, consequently, μ_P-measurable. Then $f^{-1}(X)$ is μ_P-measurable.

(b) Let k be a natural number and $\mathbb{R}^k$ the metric space of k-tuples of reals with the metric ϱ. Let λ be the Lebesgue measure.

(i) Theorem (Luzin; cf [Oxtoby 1971]). A real function f on $\mathbb{R}^k$ is Lebesgue measurable iff for each $\varepsilon > 0$ there is an $A \subseteq \mathbb{R}^k$ such that $\lambda(A) < \varepsilon$ and $f \upharpoonright \mathbb{R}^k - A$ is continuous.

(ii) Let Φ be a distributional sentence. Suppose that $\underline{U} \models \Phi$ implies that, for each $M \subseteq U$, μ_P is absolutely continuous. Then each statistic satisfying conditions (a) and (c) from 5.1.2 is an acc-statistic w.r.t. Φ. All computable statistics are acc-statistics w.r.t. any Φ such that μ_P is absolutely continuous under Φ. (Hint: If μ_P is absolutely continuous then the Borel measurability of a statistic f implies by (i) that for each n there is an $A_n \subseteq \mathcal{M}_M^V$ such that $\mu_P(A_n) \geq 1 - 1/n$ and $f \upharpoonright A_n$ is continuous. Define $\mathcal{M}_{cont} = \bigcup_{n=1}^{\infty} A_n$. Note that $\mu_P(A_1 \cup \ldots \cup A_n) > 1 - 1/n$,

hence

$P(\{\sigma ; \underline{M}_\sigma \in \mathcal{M}_{cont}\}) = \mu_P(\mathcal{M}_{cont}) = \lim_{n \to +\infty} \mu_P(A_1 \cup \ldots \cup A_n) = 1$.)

Part B
A Logic of Suggestion

Chapter VI. Listing of Important Observational Statements and Related Logical Problems

Let us begin with a quotation from Novalis, which stands as a motto in [Popper]: Hypothesen sind Netze; nur der fängt, wer auswirft. The reader found in Part A an analysis of observational and theoretical languages of science that resulted in a study of classes of some observational and theoretical calculi and their relationships. But he may object that the study of Part A was too static in character and thus ignored hypothesis formation, i.e. "the process of discovery" [Buchanan]. This is indeed the case and corresponds to our notion of a logic of induction as an answer to the questions (L0) -(L2) in Chapter I. Bear in mind questions (L3)-(L4) (cf. 1.1.5), we are now going to develop a logic of suggestion as a possible answer to the latter questions. Since our investigation belongs to AI rather than to the psychology of scientific thinking we shall not be forced to simulate the process of the scientist's guessing hypotheses but will feel free to respect and utilize the differences between human and computer skills. Furthermore, we shall not attempt to mechanize the whole process of arriving at hypotheses but only one of its substantial parts, namely the process of intelligent observation of data. Our aims are explained in detail in Section 1 of this Chapter; the main notions are of a problem and its solution. This is in accordance with the concept of scientific discovery as the solution of problems sui generis. "We speak of a problem, or a problem-solving situation, if there is something undecided, something which is an obstacle to activity and is to be overcome, etc. One important thing is that a problem is not just anything unknown, but something unknown, undiscovered, undecided ... etc. Accordingly, a problem is the question which for one reason or other we want, need or have to answer". [Tondl].

Problem solving has become a well-developed part of AI particularly in connection with the robot's plan formation (cf. [Nilsson], [Kowalski]);

in comparison with the usual terminology of problem solving our notions will be rather specific. To avoid confusion, we shall speak of observational research problems, briefly, r-problems.

Section 1 is devoted to informal derivation of our main notions, namely r-problems, their solutions and GUHA-methods (as methods for constructions of solutions of r-problems). Furthermore, Section 1 contains some discussion concerning realizability of GUHA-methods and some particular results, mainly concerning computational complexity. Note that Section 1 is a continuation of the investigation of Chapter I and does not suppose any knowledge of Part A. Section 2 is devoted to some quantifiers (called helpful) which occur naturally when one wishes to look for indirect solutions (in a sense to be defined). Section 3 studies helpful quantifiers in connection with associational and implicational quantifiers. The final short Section 4 is devoted to some specific problems concerning helpful quantifiers in connection with associational quantifiers in cross-nominal calculi. Results of Sections 3,4 will be used in the next chapter where we shall describe in detail a rather complex GUHA-method. The final chapter is devoted to some statistical questions arising in immediate connection with GUHA-methods but having general importance for logics of discovery similar to our own.

VI.1 Observational research problems and their solutions

6.1.1 As we have already mentioned in Chapter I, many philosophers of science deny the possibility of formulating a logic of suggestion. "These authors assume that there are no rational methods for the formulation of hypotheses, that hypotheses are merely happy guesses or leaps out of the reach of methods as Whewell says" [Buchanan] . On the other hand, Meltzer [1970] **emphasizes** the possibility of Hypothesis Formation in the spirit of AI. There is elaborate work on this subject; [Plotkin],[Meltzer 1970b] and[Morgan]can serve as selected important examples. All the papers mentioned use the first order predicate calculus and understand induction as a sort of inverse deduction. In a slightly different context, Kowalski claims that "predicate logic is a useful language for representing knowledge". Reeken [1971] is an interesting paper considering possibilities of mechanized statistical inference. Possible criticism of Hypothesis Formation based on logic should also be mentioned. (a) The discussion of [Rabin 1974]concerning AI is relevant also for Hypothesis Formation. The main observation concerning AI, Rabin points out, is that its projects often contain components whose complexity grows too rapidly. (b) Minsky argues that "traditional logic cannot deal very well with realistic, complicated problems because it is poorly suited to represent approximations to solutions - and these are absolutely vital". (c) Critism can come also from statisticians; Van Reeken says: ".. those toys are dangerous in the hands of nonstatisticians ... I sincerely hope this possibility of misuse will not be offered to anybody who merely asks for it, at least not before **it is** "foolproof". Otherwise it would become true that: there are lies, dammed lies and statistics". We shall formulate some comments on this criticism below (in 6.1.11).

6.1.2 The logic of suggestion developed in the rest of the present book is motivated by the principal idea of the GUHA-method, which can be found already in [Hájek, Havel and Chytil 1966a] . This idea can be formulated as the task generaling automatically <u>all interesting hypotheses</u> based on given data. Take note of the contrasting character of "all" (exhaustiveness) and "interesting" (minimization). A very similar idea was formulated idependently by Leinfelner [1965] : he imagines a machine producing all hypotheses "wahllos einfache Hypothesen bilden" but retaining only all interesting hypotheses "auf keinen Fall ohne nachherige Selektion" . Note that Leinfellner thought an "Induktionsmachine" to be "heute noch fiktiv" whereas [Hájek and al. 1966 a] contains already a realized even if simple method.

6.1.3 Let us stress the fact that by saying "hypotheses" we mean scientific hypotheses. We have accepted the distinction between observational and theoretical languages; in Part A we gave possible formalizations of both sorts of languages as well as of the statistical inference rules bridging the gap between them.

We claim that the scientist has to choose his observational and theoretical language and inference rules; they are not determined by his evidence. Certainly this choice is (or at least, can be) creative in character; but even if the conceptual frame has been chosen, the task remains to formulate and justify the hypotheses. We have already stressed in Chapter I that it is an intelligent observation of the data (important observational statements) that leads to justified theoretical hypotheses, not the data themselves. Thus the task of formulating interesting justified hypotheses has a subtask of formulating important (interesting) observational statements. It can be seen from the investigations of Chapter IV that there are many inference rules such that observational sentences of a certain form are in one-one correspondence with the respective theoretical sentences; if such a rule has been chosen then the task of finding interesting justified hypotheses is reduced to the subtask just formulated. In the present context, the idea of the GUHA method can be specified as the task of the automatic listing of all important observational statements. Cf. [Hájek 1973]. Naturally, by "important" one means "important at a certain stage of scientific research in a certain branch and relative to certain data (evidence)".

6.1.4 We need some more detailed informal discussion so as to arrive at appropriate mathematical notions. Suppose the scientist has chosen (elaborated) his conceptual frame, i.e. observational and theoretical language and inductive inference rule and has collected some data - an observational model (in the sense of observational semantic systems). **Remember that this problem** is not just anything unknown but something unknown. Consequently, observational sentences can be classified (in principle) as relevant or irrelevant w.r.t. the general scientist's problem. Let us distinguish between relevant observational questions and relevant observational truths. An observational sentence φ is a relevant observational question if the decision whether φ is true in given data **or not,** is valuable since (i) we do not know whether φ is true and (ii) if φ is true then it leads, via the inference rule, to an interesting theoretical hypothesis which is justified since φ is true. We call φ a relevant observational truth if it is a relevant observational question and is true in the data.

The computer should help us to convert all relevant questions true in $\underline{M}$ into relevant statements. **We want** to have the whole relevant truth at our disposal. But, evidently, should the computer only list all the relevant questions true in a given model, the resulting output would be a formidably long list of unorganized truths and therefore of little value. So we shall suppose that the scientist and the computer have a (sound) observational deduction rule and can draw immediate conclusions. We prefer the notion of immediate conclusion rather then the notion of deductive consequence (provability) since the former notion can be considered as a formalization of the scientist's ability to see some consequences at a glance. This ability can be used to construct a handy representation of the set of all relevant observational truths, namely by an appropriate set X of true observational statements such that each relevant truth is an immediate conclusion from X. The set X can be optimalized in various directions; if optimalized, then its elements can be called important statements (or better, X is an important set of statements).

We now give an exact definition of an r-problem and of its solution. Remember that, given a semantic system, $Tr_{V_o}(\underline{M})$ denotes the set of all sentences V_o-true in $\underline{M}$ i.e.

$$Tr_{V_o}(\underline{M}) = \{\varphi \in \text{Sent};\ \underline{M} \models_{V_o} \varphi\}.$$

6.1.5 Definition. Let $\mathcal{S} = \langle \text{Sent}, \mathfrak{M}, V, Val \rangle$ be an observational semantic system. An observational research problem (briefly r-problem) in $\mathcal{S}$ is a triple $\mathcal{P} = \langle RQ, V_o, I \rangle$, where RQ is a non-empty recursive subset of Sent, V_o is a non-empty recursive subset of V and I is a recursive inference rule on Sent V_o-sound w.r.t. $\mathcal{S}$. RQ is called the set of relevant questions,

V_o is called the set of designated values, I is the deduction rule.

In addition, let $\underline{M} \in \mathcal{M}$; a solution of $\mathcal{P}$ in $\underline{M}$ is an arbitrary $X \subseteq Tr_{V_o}(\underline{M})$ such that $RQ \cap Tr_{V_o}(\underline{M}) \subseteq I(X)$.

6.1.6 Example. Recall the semantic system $\mathcal{S}_n$ of 1.2.6 (1): models of $\mathcal{S}_n$ are matrices of zeros and ones with n columns and for each $e \subseteq \{1,\ldots,n\}$ we have a sentence φ_e saying "the properties $\{P_i ; i \in e\}$ are incompatible". Put

$RQ_n = \{\varphi_e ; e \subseteq \{1,\ldots,n\}\}$ and put $V_o = \{1\}$ (the designated value is 1 - truth). Let I_n be the deduction rule defined in 1.2.9., i.e.

$$I_n = \left\{ \frac{\varphi_e}{\varphi_{e'}} ; e \subseteq e' \right\}.$$

Then $\mathcal{P} = \langle RQ, V_o, I_n \rangle$ is an r-problem. Let $\underline{M}$ be a model and let $\varphi_e \in RQ$; call φ_e $\mathcal{P}$-prime in $\underline{M}$ if (i) $\|\varphi_e\|_{\underline{M}} = 1$ but (ii) for each proper subset $e' \subset e$ we have $\|\varphi_{e'}\|_{\underline{M}} = 0$. It is an easy exercise to show that the set of all sentences $\mathcal{P}$-prime in $\underline{M}$ is a solution of $\mathcal{P}$ in $\underline{M}$.

6.1.7 Discussion. Remember our question (L3): What are the conditions for a theoretical statement or set of theoretical statements to be interesting (important) with respect to the task of scientific cognition? We must admit that so far we have not given any explicit answer to this question. But we have achieved at least two things: First, in Part A we have analysed theoretical and observational languages together with inductive inference rules so that we have a relatively broad variety of possibilities for answering the question in a more specific context of one science or of one general research area . Second, we have stressed the role of interesting observational statements as a necessary step towards interesting hypotheses (cf. 6.1.3, 1.1.6, 4.4.17 4.4.27 and have formulated the notion of an r-problem and its solution as a possible formalization of the notion "an interesting set of observational truths". This leads to a modification of the question L4, to the following question:

(L'4) Are there methods for constructing good solutions to r-problems?

The desired methods facilitate good answers to (L3) in each particular case. In other words, they should offer a satisfactorily broad frame for answering (L3). Each particular answer will be a result of the collaboration of a mathematician and a scientist. The following aspects should be respected:

(1) One must be able to choose an appropriate type of questions ; a method must allow satisfactorily variable syntactical descriptions of sets of relevant questions.

(2) The notion of interest should depend on the length and complexity of sentences: if no other criteria apply a shorter (simpler) sentence is more interesting than a long one .

(3) Some properties of objects may be declared as more important than some other properties; sentences referring to more important properties are more important than others (when no other criteria apply).

(4) If an observational sentence names a test for a statistical hypothesis then the significance of the test is relevant for the interest of the sentence as an observational statement (cf. critical strengthening, 8.2.5).

These criteria may be combined in particular cases in various ways to obtain a definition of an "interest" - quasiordering. On the other hand, there are other factors, hardly formalizable, such as surprise, beauty, etc.; thus we cannot rely too much on any one interest - quasiordering. Let us repeat what we said in 1.1.6: we want to construct methods aiding the choice of the best hypothesis.

Such methods will be called GUHA-methods. The formal notion of a GUHA-method together with a supply of particular realizable examples will form the first part of our logic of suggestion (Chapters VI, VII). Since we have paid our main attention to inductive rules of a statistical nature, we are faced with several statistical questions concerning the statistical properties of solutions of r-problems. Chapter VIII is devoted to this topic; that chapter provides additional information on the concept of interesting theoretical statements (hypotheses) and completes our answer to (L3) and (L4').

Intuitively, a method for constructing solutions of r-problems is something that, having obtained a particular observational model and information specifying an r-problem (i.e. determining relevant questions, designated values and deduction rule - everything with respect to an observational semantic system), produces a solution. Now, r-problems acceptable by such a method vary over a system $\{ \mathcal{P}(p);\ p \text{ parameter}\}$ where the parameter p may be identified with the information specifying $\mathcal{P}(p)$ and the corresponding observational semantic system. The method defines a mapping X associating with each p and each model $\underline{M}$ a solution of $\mathcal{P}(p)$ in $\underline{M}$. Hence we have the following formal definition:

6.1.8 Definition. (1) Let Par be a recursive set (of parameters). A GUHA-method is a parametric system $\Xi = \{ \mathcal{S}(p), \mathcal{P}(p), X_p \}_{p \in Par}$ where each $\mathcal{S}(p) = \langle Sent(p), \mathfrak{M}(p), V(p), Val(p) \rangle$ is an observational semantic system and $\mathcal{P}(p) = \langle RQ(p), V_0(p), I(p) \rangle$ is an r-problem in $\mathcal{S}(p)$; X_p is a function associating with each $\underline{M} \in \mathfrak{M}(p)$ a solution $X_p(\underline{M})$ of $\mathcal{P}(p)$.

(2) A GUHA-method $\Xi = \{ \mathcal{S}(p), \mathcal{P}(p), X_p \}_{p \in Par}$ is realizable (in principle) if for each parameter p and each model $\underline{M} \in \mathfrak{M}(p)$ the set $X_p(\underline{M})$ is a finite set of sentences and the function X associating with each $p \in Par$ and each $\underline{M} \in \mathfrak{M}(p)$ their solution $X_p(\underline{M})$ is a partial recursive function of p and $\underline{M}$.

(3) Ξ is realizable in polynomial time if there is a Turing machine operating in polynomial time and computing the function X. (Note that one assumes an appropriate encoding of all necessary objects.)

6.1.9 Example. We illustrate the definition by a very simple example. Further examples will be presented in Section 2; and the whole of Chapter VII will be an extensive example of a complex GUHA method.

It is convenient to describe a GUHA method by describing **successively the** set Par of parameters and a certain structure on it corresponding to the variety of things one has to decide to determine this semantic system, r-problem and, for each model, the solution.

In our example, each parameter p decomposes into two parts: TYPE and SYNTR. TYPE is a positive natural number; if TYPE is n then our observational system will be $\mathcal{S}_n$ of 1.2.6 (1) mentioned in example 6.1.6. (Thus input models are matrices of zeros and ones with n columns.) Our relevant questions will be some sentences φ_e determined by syntactical restrictions SYNTR. SYNTR consists of a subset $\hat{e}$ of $\{1,\dots,n\}$ and a positive natural number $b \leq n$. A sentence φ_e is a relevant question iff $e \subseteq \hat{e}$ and if e has at most b elements. (Say, φ_e contains only interesting predicates and is simple enough.) Thus RQ is specified. Since we work with two values 0, 1, our designated value is 1 (truth). Hence we have only to specify our deduction rule and the problem $\mathcal{P}(p)$ will be defined. We take the rule I_n mentioned in 6.1.6.

For each model $\underline{M}$, we put $X_p(\underline{M}) = \{\varphi_e \in RQ;\ \varphi_e \text{ prime in } \underline{M}\}$.

(Prime sentences were defined in 6.1.6.) Obviously, $X_p(\underline{M})$ is a solution of $\mathcal{P}(p)$ in $\underline{M}$; hence we have described a GUHA method Ξ_o. We shall call it the Baby-GUHA. This method is certainly realizable in principle (in fact, our example is a simplification of the first GUHA-method as described in [Hájek and al, 1966]); experience shows that if b=n=15 and the model has about 1000 rows then the computer finds the whole solution in a reasonable time. If the number is larger (say, about 30) then it is reasonable to change the definition of relevant questions, e.g. put b=5 < n=30.

We shall show later on in this section that after similar natural restrictions we can obtain methods realizable in polynomial time.

6.1.10 Remark. We neglect here the choice of a theoretical language and of an inductive inference rule. Note that, on the one hand, **the notion of** a GUHA-method does not depend on such a choice. On the other hand, the question whether a particular GUHA-method is useful (adequate) at a particular stage of particular scientific research does depend on the whole conceptual frame, including a theoretical language and inductive inference rules that must be specified at least implicitly. Furthermore, after one has decided to use a particular GUHA-method, the general theoretical problem one wants to solve is responsible for the choice of an appropriate value of the parameter. In our example, the parameter defines the set RQ of relevant observational questions by syntactical means - this is typical. Warning: The present example is certainly oversimplified and would hardly be used in practice. See the next chapter.

6.1.11 Now that we have outlined our logic of suggestion we shall try to formulate some remarks on possible criticism as given in 6.1.1.

(a) As far as the question of complexity is concerned, notice that there are two notions of complexity important in the present context: the computational complexity - i.e. the time and space necessary for the construction of the solution and the structural complexity of the solution as a list of sentences (plus various additional information). It is also true that some problems closely related to the notion of GUHA-methods are as difficult as some famous combinatorial problems considered in the theory of computational complexity. For the reader familiar with Cook's paper we shall present below some universal NP-problems concerning solutions of r-problems. This is the negative part of our answer. On the other hand, it is very important that there are natural restrictions or modifications of those well-motivated GUHA-methods that make them realizable in polynomial time. We shall show this in the present section for our simple example and in the next chapter for the GUHA-method described there. This corresponds to some of Rabin's suggestions; other suggestions, e.g. the possibility of some "randomization", have not yet been analysed. Concerning the structural complexity of solutions, we can add to the fact stated above that we shall study efforts to minimalize solutions by allowing indirect solutions (in a sense) in Sections 2 - 4 of the present Chapter.

(b) Let us formulate some comments on the role of logic in Hypothesis Formation in the style of GUHA-methods. First, scientific cognition as a solution of problems differs from thinking in general; the specific features of scientific cognition make the role of logical means in the formation of scientific hypotheses different from their role e.g. in robot plan formation.

Second, we use logical means in a broad meaning of the word; on the one hand, we have considered various non-traditional calculi and, on the other hand, we have not based our considerations on iterated deduction. We stressed explicit semantics; but, on the observational level, we have dealt with non-iterated conclusions (seeing at a glance) and on the theoretical level we have used quite general inference rules.

Third, a word about consistency. Minsky says that "the preoccupation with consistency, so valuable for Mathematical Logic, has been incredibly destructive to those working on models of the mind". In our context, on the observational level we have trivial consistency: the set of sentences true in an observational model is consistent in any possible meaning of the word. On the theoretical level, we do not have any strict requirement of consistency; our rationality conditions do not guarantee the consistency of the set of all inferences made from elements of a solution of an r-problem. See Chapter VIII for more information.

(c) The statistical criticism is fully justified; some suggestions on how to carefully treat results obtained by our (and similar) "toys" are contained in Chapter VIII. On the other hand, this criticism does not concern the idea of Hypothesis Formation, but merely the misuse of the constructed methods.

(d) Let us add some short comments on the relevance of GUHA methods of hypothesis formation for Artificial Intelligence. There seems to be **no doubt** about the AI-relevance of GUHA methods from the point of view of what they do: they suggest hypotheses, and hypothesis formation is a branch of Artificial Intelligence. Moreover, we show that our GUHA procedures are realizable in practice.

6.1.12 Definition. Let I be an inference rule on a set Sent and let $X, Y, Z \subseteq$ Sent. (1) X is Y-sufficient (w.r.t. I) if $Y \subseteq I(X)$, i.e. if for each $\varphi \in X$ either $\varphi \in Y$ or there is a $e \subseteq Y$ such that $\frac{e}{\varphi} \in I$.

(2) X is Z-independent (w.r.t. I) if for each $\varphi \in X$ we have $Z \cap I(X - \{\varphi\}) \neq Z \cap I(X)$.

6.1.13 Remark. Let $\mathcal{S} = \langle \text{Sent}, \mathfrak{M}, V, Val \rangle$ be a semantic system, let $\mathcal{P} = \langle RQ, V_o, I \rangle$ be an r-problem in $\mathcal{S}$ and let $\underline{M} \in \mathfrak{M}$ be a model. For each $X \subseteq Tr_{V_o}(\underline{M})$ we have the following: X is a solution of $\mathcal{P}$ in $\underline{M}$ iff X is $RQ \cap Tr_{V_o}(\underline{M})$- sufficient (w.r.t. I). Elements of $RQ \cap Tr_{V_o}(\underline{M})$ were called "relevant truths" ; hence X is a solution iff it consists of some sentences V_o-true in $\underline{M}$ and is sufficient for all relevant truths.

6.1.14 Remark. Let X be a solution of $\mathcal{P}$ in $\underline{M}$; then $I(X) \cap RQ = RQ \cap Tr_{V_o}(\underline{M})$. Consequently, X is RQ-independent iff no proper subset of X is a solution (say, X is a $\subseteq$ -minimal solution).

6.1.15 Definition and Remark. Let I be an inference rule on Sent and let $X \subseteq$ Sent. X is weakly independent if, for each $\varphi \in X$,

$$I(X - \{\varphi\}) \neq I(X);$$

X is strongly independent if $\varphi \notin I(X - \{\varphi\})$ for each $\varphi \in X$.

One can immediately see that (1) X is weakly independent iff X is I(X)-independent and that (2) X is strongly independent iff X is X-independent. For transitive rules we have the following simple fact:

6.1.16 Theorem. Let I be a transitive inference rule on Sent. Then for each $X \subseteq$ Sent and each $Z \subseteq$ Sent such that $X \subseteq Z$ the following holds: X is strongly independent iff X is Z-independent. Hence X is strongly independent iff X is weakly independent.

The proof is obvious.

On the other hand, one can ask how the GUHA procedures work. Some authors claim that heuristic elements, the possibility of learning and formulating subtasks are indispensable features of AI-procedures. Now, if the reader observes our examples of GUHA methods, especially the method of Chapter VII, Sections 1 - 3 , he will probably agree that these methods are quite complex, practically realizable, and include some modest heuristic elements (cf. Problem (4) of Chapter VII) but in principle are realizable by routine programming work, even if the programs are quite extensive. Then whether the reader qualifies GUHA procedures as AI-procedures or not, we claim the following:

The logical analysis and formalization of statistical hypothesis testing is an indispensable step towards mechanized formation of statistical hypotheses. Suitable formal calculi for this task are developed here. The notion of an r-problem and its solution is a useful formal model of the scientist's suggestion of hypotheses; and GUHA methods as methods of the construction of solutions of r-problems are practicaly realizable - even by routine programming. The heuristic approach should be applied to GUHA methods. This remains a task of further investigation; it will probably be useful to make the notion of relevant questions variable during the computation and dependent on previous results.

We now turn to some mathematical considerations. Our aim is to introduce some useful notions concerning inference rules and solutions of r-problems and to present some considerations relating those notions to the theory of computational complexity. The reader not familiar with this theory may omit 6.1.20 - 6.1.28.

6.1.17 Remark. Let $\mathcal{P}$ be an r-problem in a semantic system $\mathcal{S}$. Let Sent be the set of sentences of $\mathcal{S}$ and let I be the deduction rule of $\mathcal{P}$. Let $\underline{M}$ be a model. Notice that an $\subseteq$-minimal solution (of $\mathcal{P}$ in $\underline{M}$) is weakly independent w.r.t. I, but a weakly independent solution need not be $\subseteq$-minimal since it is possible that X is a solution, X is weakly independent, but, for some φ, $X - \{\varphi\}$ is also a solution since the difference $I(X) - I(X - \{\varphi\})$ consists only of sentences not in RQ.

Similarly, a strongly independent solution can be diminished by omitting a $\varphi \in$ Sent - RQ provided $X - \{\varphi\}$ is $(RQ \cap Tr_{V_o}(\underline{M}))$-sufficient.

6.1.18 Definition. Let $\mathcal{P}$ be an r-problem in $\mathcal{S}$, let RQ be the set of relevant questions of $\mathcal{P}$ and let $\underline{M}$ be a model. A solution X of $\mathcal{P}$ in $\underline{M}$ is direct if $X \subseteq RQ$.

6.1.19 Remark. (1) A strongly independent direct solution is $\subseteq$-minimal. (2) It is often reasonable to deal with indirect solutions. Section 2 will be devoted to this matter.

6.1.20 Remark. We are now going to consider the simple example of the Baby-GUHA Ξ_o; its computational properties are typical. (Cf. the conciderations of Chapter VII, Section 3.) The following simple result is due to Pudlák:

6.1.21 Lemma. Pudlák [1975c]. For each $n > 0$ and each $m \geq n$, there is a model $\underline{M}$ with m objects and 2n properties which has 2^n prime sentences in the sense of the Baby-GUHA method.

Proof. One can easily construct a model $\underline{M}_o$ with m objects and n properties with exactly one prime sentence (cf. [Chytil 1975]): $\underline{M}_o$ as a matrix consists only of rows containing exactly one zero (and n-1 ones) while each n-tuple containing exactly one zero occurs in $\underline{M}_o$ at least once. Let $\underline{M}$ be a model with m rows and 2n columns such that, for each $i = 1, \ldots, n$, the (n+i)-th column coincides with the i-th column. Then for each n-tuple $\langle \varepsilon_1, \ldots, \varepsilon_n \rangle$ of zeros and ones, the sentence

$$(\forall x)(\neg P_{1+\varepsilon_1 n} \vee \neg P_{2+\varepsilon_2 n} \vee \ldots \vee \neg P_{n+\varepsilon_n n})$$

is prime. ($(\forall x)(\neg P_1 \vee \ldots \vee \neg P_n)$ is prime; for each $i = 1,\ldots,n$, one can replace P_i by P_{i+n}.)

6.1.22 Discussion. Hence if we assume that the algorithm has to "print" each prime sentence and that "printing" each sentence takes at least one step then the Baby-GUHA in full generality is exponentially complex. We show a natural restriction that makes the Baby-GUHA polynomially complex. Let us measure the complexity of the input by m and n (the number of rows and columns of the input model $\underline{M}$). Let $\leq_p$ be the natural linear ordering of RQ(p) ($\mathrm{card}(e_1) \leq \mathrm{card}(e_2)$ implies $\varphi_{e_1} \leq_p \varphi_{e_2}$; for $\mathrm{card}(e_1) = \mathrm{card}(e_2)$ we use the lexicographic ordering). The algorithm realizing the Baby-GUHA can be described by the following simple flow-diagram:

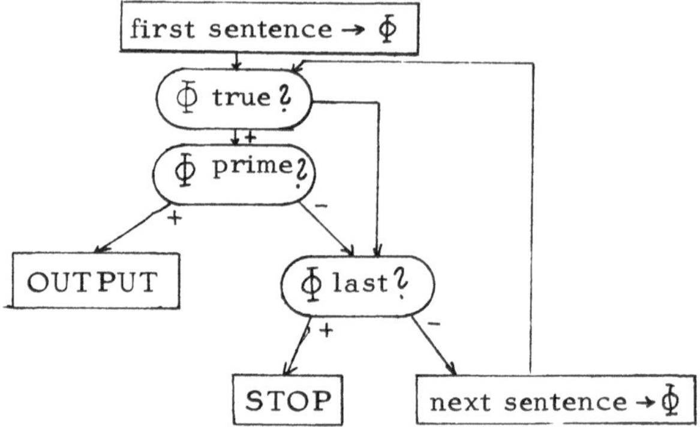

(We ignore the possibility of further optimalizations here; they do not affect our considerations.) It can easily be shown that - independently of a particular formalization of the notion of an algorithm - the realization of each single item of the above flow diagram **requires a time which is bounded from above by a** polynomial in m and n. (Note that a true sentence φ_e is prime iff there is no element i of e such that $\varphi_{e-\{i\}}$ is true.)
Hence we come to the following conclusion:

6.1.23 (Discussion continued). If we change the definition of the Baby-GUHA in such a way that the number of relevant questions depends polynomially on n then the changed method is realizable in polynomial time.

We show that there is a certain natural supplementary assumption on the set Par of parameters which implies that the cardinality of RQ p depends polynomially on the TYPE part of p. **Even if we allow m and n to be** arbitrarily large in theory we postulate that there is an upper bound b such that if the complexity of φ_e is larger than b then φ_e is not intelligible (**comprehensible**). The complexity of φ_e is identified with the cardinality of e. (Naturally, you can have a model with 500 properties; but if you obtain the sentence: "the properties 1, 7, 11, 13, 29, 31, 57, 121, 124, 200, 201, 294, 430, 444 and 491 are incompatible"
then you will **probably** have difficulty in understanding it as a single observational sentence. If not, make a more complicated example.) Thus we make the following:

6.1.24 Definition. The Baby-GUHA Ξ_o^* with an intelligibility bound $\hat{b}$ is the restriction of Ξ_o to the set $Par_{\hat{b}} \subseteq$ Par of parameters $p = \langle TYPE, SYNTR \rangle$ satisfying the following: $SYNTR = \langle e, b \rangle$ and $b \leq \min(n, \hat{b})$.
(Thus, the maximal length of a relevant sentence in each problem is bounded by $\hat{b}$.)

6.1.25 Theorem. The Baby-GUHA with an intelligibility bound is realizable in polynomial time.

Proof. This follows from 6.1.23. Let the intelligibility bound be $\hat{b}$. Given n, the cardinality of RQ is bounded by
$\sum_{i=1}^{\min(n,\hat{b})} \binom{n}{i}$, which gives a polynomial bound.

Our next aim is to exibit some universal NP-problems related to r-problems and their solutions. The following results are due to Pudlák [1975c].

6.1.26 Lemma. Let CHOICE SET be the following problem: Given an undirected graph $\underline{G} = \langle G,R \rangle$ and a natural number k, to determine whether there is a $Y \subseteq G$, Y of cardinality $\leq$ k and such that for each $u \in G$, $u \in Y$ or $\{u,v\} \in R$ for some $v \in Y$. (Such a Y is called a choice set for $\underline{G}$.) CHOICE SET is a universal NP-problem.

Proof. Evidently CHOICE SET is NP. By [Karp 1972], the following problem called NODE COVER is a universal NP-problem: Given an **(undirected)** graph $\underline{G} = \langle G,R \rangle$ and a number k, to determine whether there is a node cover $Y \subseteq G$ of cardinality $\leq k$, i.e. a set Y such that for each edge $\{x,y\}$ either $x \in Y$ or $y \in Y$. We show that this problem is reducible to the problem CHOICE SET. Let $\underline{G} = \langle G,R \rangle$ be an undirected graph and let $k \in \mathbb{N}$, $G \cap R = \emptyset$. Define G', R' and k_1 as follows:

$G' = G \cup R$; $R' = R \cup \{\{u,\{u,v\}\}$; $\{u,v\} \in R$; $k_1 = k + \text{card}(G - \text{dom}(R))$.

We show that $\underline{G}$ has a node cover of cardinality k iff $\underline{G}'$ has a choice set of cardinality k_1.

($\Rightarrow$) : If $Y \subseteq G$ is a node cover for $\underline{G}$, card$(Y) \leq k$, then $Y \cup (G - \text{dom}(R)) = Y_1$ is a choice set for G', card$(Y_1) \leq k_1$.

($\Leftarrow$) : If $Y_1 \subseteq G'$ is a choice set for G', card$(Y_1) \leq k_1$,

then Y_1 can be written as a disjoint union $(G - \text{dom}(R)) \cup Y_1'$:

Y_1' consists of some elements of G and some elements of R.

Construct Y as follows: For each $u,v \in G$,

(a) if $u \in Y_1'$, put u into Y,

(b) if $\{u,v\} \in Y_1'$, then put one of the elements u,v, into Y. Then Y has $\leq$ k elements and is a node cover for $\underline{G}$.

6.1.27 Theorem [Pudlák 1975c]. The following two problems are universal NP-problems:

(1) TRUE SENTENCE . Given a matrix $\underline{M}$ of zeros and ones with m rows and n columns and a number k, to determine whether there is a sentence φ_e true in $\underline{M}$ (in the sense of 1.2.6 (1)) such that e has at most k elements.

(2) SUFFICIENT SET OF SENTENCES. Given a finite set Sent of sentences, an inference rule I on Sent and a number k, to determine whether there is a $X \subseteq$ Sent, X of cardinality $\leq k$, such that X is Sent-sufficient (w.r.t. I).

Proof. Both problems are NP. We show that CHOICE SET is reducible both to TRUE SENTENCE and to SUFFICIENT SET OF SENTENCES. Let $\underline{G}$ and k be given.

(1) Consider a square matrix $\underline{M} = (m_{i,j})_{i,j \in G}$ indexed by elements of G and such that $m_{i,j} = 0$ iff $i=j$ or $\{i,j\} \in R$. Evidently, $\underline{M}$ has a true sentence of length $\leq k$ iff $\underline{G}$ has a choice set with $\leq k$ elements.

(2) Put Sent = G, $I = \left\{\frac{u}{v} \; ; \{u,v\} \in R\right\}$. Then $X \subseteq$ Sent is Sent-sufficient iff X is a choice set for $\underline{G}$.

6.1.28 Remark. (1) The problem TRUE SENTENCE can be interpreted as follows: Before starting to process the model, one would like to know quickly whether there will be some results. The preceding theorem shows that should one find a quick (deterministic polynomial) test, one would solve positively the P-NP-problem.

(2) Concerning the meaning of the problem SUFFICIENT SET OF SENTENCES, remember that a solution of an r-problem $\mathcal{P} = \langle RQ, V_o, I \rangle$ in $\underline{M}$ is a set $X \subseteq Tr_{V_o}(\underline{M})$ which is $(RQ \cap Tr_{V_o}(\underline{M}))$-sufficient (w.r.t. I).

6.1.29 Key words: r-problem (relevant questions, designated values, deduction rule), solution of an r-problem, GUHA-methods, realizability, realizability in polynomial time; sufficient set, independent set (of sentences), direct solution, Baby-GUHA with an intelligibility bound, universal NP-problems concerning inference rules and r-problems.

VI.2 Indirect solutions

Let $\mathcal{P}$ be an r-problem and let RQ be its set of relevant questions; let $\underline{M}$ be a model. We know that a solution X of $\mathcal{P}$ in $\underline{M}$ is indirect if X is not a subset of RQ, hence it contains some auxiliary truths, sentences true in $\underline{M}$ but not elements of RQ. In the present section we first present two simple examples of problems having reasonable indirect solutions. We shall see that in the examples, auxiliary truths are sentences with some auxiliary quantifiers, quantifiers not occurring in relevant questions. We shall arrive at a general notion of helpful quantifiers - quantifiers helpful in constructing indirect solutions. In this section we first deal with monadic predicate calculi, later with monadic $\times$ -predicate calculi.

6.2.1 We describe a GUHA method Ξ_1 with auxiliary equivalences. Each parameter p decomposes into four parts: TYPE, QUANT, SYNTR and AUX. TYPE is a positive natural number determining the type of the OPC $\mathcal{F}(p)$ to be used. If it is n then the OPC determined by p will have n unary predicates $P_1, \ldots, P_n$. Furthermore , $\mathcal{F}(p)$ has two quantifiers of type $\langle 1,1 \rangle$, $\sim$ and $\Leftrightarrow$. The quantifier $\Leftrightarrow$ is the equivalence quantifier, i.e. $Asf_{\Leftrightarrow}(\langle M,f,g \rangle) = 1$ iff $f = g$. The semantic of $\sim$ is determined by QUANT. It depends on the implementation which quantifiers can really be used and how QUANT determines the corresponding associated function.

There are no restrictions concerning the properties of the quantifiers admitted except the following weak satisfiability condition: For each $\underline{M}$ and each pair φ, ψ of designated formulae, $\underline{M} \models \varphi \sim \psi$ implies that both φ and ψ are satisfiable in $\underline{M}$, i.e. $\underline{M} \models (\exists x)\ \varphi \ \&\ (\exists x)\ \psi$.

To give a particular example, imagine that $\sim$ may be either the Fisher quantifier $\sim_\alpha$ for a rational $\alpha \in (0, 0.5]$ or the following presence quantifier pr_α :

$$Asf_{pr_\alpha}(\underline{M}) = 1 \text{ iff } a_{\underline{M}} \geq \alpha m_{\underline{M}},\ b_{\underline{M}} \geq \alpha m_{\underline{M}},\ c_{\underline{M}} \geq \alpha m_{\underline{M}},\ d_{\underline{M}} \geq \alpha m_{\underline{M}}$$

for a rational $\alpha \in (0,0.25]$,

or some other possibilities.

(The presence quantifier says that all four cards are frequented enough. It may seem unnatural but it serves as an example of a non-associational quantifier obeying the satisfiability condition.) Denote the quantifier determined by a particular choice of QUANT by q; relevant questions will be some formulae $(qx)(\varphi_1(x), \varphi_2(x))$ or, briefly, $\varphi_1 \sim \varphi_2$ where φ_1, φ_2 are two designated elementary conjunctions formed by some of the predicates $P_1, \ldots, P_n$ and the variable x such that φ_1, φ_2 have no predicates in common.

As in Baby-GUHA, SYNTR determines syntactical restrictions concerning the length of φ_1, φ_2, occurrence of particular predicates in φ_1 and/or φ_2 (e.g. P_1 never in φ_2, P_2 only without negation etc.). The deduction rule is specified by AUX. AUX asks whether we allow auxiliary questions: it can be YES or NO. If it is NO then we require a direct solution, i.e. consisting only of some (true) elements of RQ. Assume that there is no (reasonably simple) rule formed only by relevant questions and sound for all the quantifiers admitted by possible choices of QUANT. Then our rule is empty: the unique solution is the whole of $\{\chi \in \text{RQ}; \underline{M} \vdash \chi\}$. If AUX is YES then we consider auxiliary questions of the form $\kappa_0 \Leftrightarrow \kappa$ where κ is an elementary conjunction and κ_0 is its subconjunction (notation: $\kappa_0 \subseteq \kappa$).
The rule I is then

$$\frac{\varphi \sim \psi, (\varphi \Leftrightarrow \varphi'')\&(\psi \Leftrightarrow \psi'')}{\varphi' \sim \psi'}$$

where $\varphi \subseteq \varphi' \subseteq \varphi''$, $\psi \subseteq \psi' \subseteq \psi''$ are elementary conjunctions. This rule is obviously sound for any quantifier $\sim$ of type $\langle 1,1\rangle$. At this moment, we have specified the problem $\mathcal{P}(p)$ determined by p. (Naturally, $V_o = \{1\}$.) It remains to describe, for each model $\underline{M}$, a solution $X_p(\underline{M})$ (for the case of AUX being YES); this will conclude the description of the GUHA method Ξ_1.
Let $\underline{M}$ be a model. Call $\varphi' \sim \psi'$ <u>M-obtainable</u> from $\varphi \sim \psi$ if the equivalences $\varphi \Leftrightarrow \varphi'$, $\psi \Leftrightarrow \psi'$ are true in $\underline{M}$. Call a relevant question

$\varphi \sim \psi$ p-prime in $\underline{M}$ if it is true in $\underline{M}$ and is not $\underline{M}$-obtainable from any relevant question $\varphi_0 \sim \psi_0$ different from $\varphi \sim \psi$ and simpler than $\varphi \sim \psi$ (i.e. such that $\varphi_0 \subseteq \varphi$ and $\psi_0 \subseteq \psi$).

6.2.2 Lemma. For each designated EC κ and each $\underline{M}$ such that κ is satisfiable in $\underline{M}$, there is a uniquely determined maximal designated EC $\bar{\kappa} \supseteq \kappa$ such that $\underline{M} \models \kappa \Leftrightarrow \bar{\kappa}$; $\bar{\kappa}$ is the conjunction of all literals L such that $\underline{M} \models \kappa \Rightarrow L$.

Proof. Denote the conjunction of all such literals by $\bar{\kappa}$; evidently, $\underline{M} \models \kappa \Leftrightarrow \bar{\kappa}$ and $\bar{\kappa}$ is the largest conjunction of designated literals equivalent to κ in $\underline{M}$. Now, $\bar{\kappa}$ is an elementary conjunction **since it** is satisfiable and hence there is no F such that $\underline{M} \models \kappa \Rightarrow F(x)$ and $\underline{M} \models \kappa \Rightarrow \neg F(x)$.

6.2.3 Notation. If κ is a designated EC satisfiable in $\underline{M}$ then we denote the EC $\bar{\kappa}$ from 6.2.2 by $\mathrm{Reg}_{\underline{M}}(\kappa)$.

6.2.4 We continue the description of the GUHA method Ξ_1 ; we describe the case in which AUX isYES. Let $\underline{M}$ be a model and let $X_p(\underline{M})$ contain, for each relevant question $\varphi \sim \psi$ p-prime in $\underline{M}$, both $\varphi \sim \psi$ itself and the formula $(\varphi \Leftrightarrow \varphi'') \& (\psi \Leftrightarrow \psi'')$ where $\varphi'' = \mathrm{Reg}_{\underline{M}}(\varphi)$ and $\psi'' = \mathrm{Reg}_{\underline{M}}(\psi)$.

Obviously, $X_p(\underline{M})$ is a solution of $\mathcal{P}(p)$ in $\underline{M}$; hence the description of the GUHA method Ξ_1 is completed.

6.2.5 We shall modify the GUHA method just described and obtain another example - the GUHA method Ξ_2 with associational quantifiers and auxiliary equivalences. Each parameter p decomposes into TYPE, QUANT, SYNTR and AUX. TYPE and SYNTRare as above; QUANT determines the semantics of $\sim$, but here we require $\sim$ to be an associational quantifier obeying the following satisfiability condition: If $\underline{M} \models \varphi \sim \psi$ then $\underline{M} \models (\exists x)(\varphi \& \psi)$ (i.e., φ, ψ are simultaneously satisfiable: think of $\sim_\alpha$ and $\Rightarrow^!_{p,\alpha}$). Then everything is as in Ξ_1 when AUX is NO .

When AUX is YES we use the rule

$$I = \left\{ \frac{\varphi\sim\psi, (\varphi \,\&\, \psi) \Leftrightarrow (\varphi'' \,\&\, \psi'')}{\varphi' \sim \psi'} ; \varphi \subseteq \varphi' \subseteq \varphi'', \psi \subseteq \psi' \subseteq \psi'' \right\} .$$

(φ, ψ etc. denote designated EC's .)

This completes the description of $\mathcal{P}_{(p)}$; we must show that I is sound.

6.2.6 Lemma. I is sound for each associational quantifier.

Proof. First we notice that if $\underline{M} \models (\varphi \,\&\, \psi) \Leftrightarrow (\varphi'' \,\&\, \psi'')$ then $\underline{M} \models (\varphi \,\&\, \psi) \Leftrightarrow (\varphi' \,\&\, \psi')$ thanks to the above inclusions. Let a,b,c,d be the frequencies of cards in $\underline{M}_1 = \langle M, \|\varphi\|_{\underline{M}}, \|\psi\|_{\underline{M}} \rangle$ and a', b', c', d' the corresponding frequencies in $\underline{M}_2 = \langle M, \|\varphi'\|_{\underline{M}}, \|\psi'\|_{\underline{M}} \rangle$. Assume $\underline{M} \models (\varphi \sim \psi)$ and $\underline{M} \models (\varphi \,\&\, \psi) \Leftrightarrow (\varphi' \,\&\, \psi')$. Then $a = a'$. Since $\underline{M} \models \varphi' \Rightarrow \varphi$ we have $a + b \geq a' + b'$, hence $a + b \geq a + b'$ and $b \geq b'$. Similarly, $c \geq c'$; consequently, $d \leq d'$. Hence $\underline{M}_2$ is a-better than $\underline{M}_1$ and we obtain $\underline{M} \models \varphi' \sim \psi'$.

6.2.7 We complete the definition of Ξ_2. Define the notions "$\underline{M}$-obtainable" and "p-prime" as above, but with respect to the new rule. Define $Reg^+_{\underline{M}}(\varphi, \psi)$ as the maximal pair $\langle \varphi'', \psi'' \rangle$ of EC's such that $\underline{M} \models (\varphi \,\&\, \psi) \Leftrightarrow (\varphi'' \,\&\, \psi'')$ (for each $\underline{M}$ in which $\varphi \,\&\, \psi$ is satisfiable). Note that in general φ'' and ψ'' have common literals. Define $X_p(\underline{M})$ as the set containing, for each sentence $\varphi \sim \psi$ p-prime in $\underline{M}$, both $\varphi \sim \psi$ itself and the formula $(\varphi \,\&\, \psi) \Leftrightarrow (\varphi'' \,\&\, \psi'')$ where $\langle \varphi'', \psi'' \rangle = Reg^+_{\underline{M}}(\varphi, \psi)$. Then $X_p(\underline{M})$ is a solution of $\mathcal{P}(p)$ in M. This completes the definition of Ξ_2.

6.2.8 Lemma. Suppose that p determines the same associational quantifier and the set of relevant questions both in Ξ_1 and in Ξ_2. Let M be a model. Denote by X^i the solution of the problem $\mathcal{P}^i p$ determined by p in Ξ_i, $i = 1,2$. Then $card(X_2) \leq card(X_1)$.

Proof. It suffices to show that each sentence p-prime in M in the sense of Ξ_2 is p-prime in M in the sense of Ξ_1. This follows from the fact $(\varphi \Leftrightarrow \varphi'') \& (\psi \Leftrightarrow \psi'')$ logically implies $(\varphi \& \psi) \Leftrightarrow (\phi \& \psi'')$.

6.2.9 Remark. The lesson of the preceding lemma is that the complexity of the solution produced by Ξ_2 is less than the complexity of the solution produced by Ξ_1. This is natural since Ξ_2 makes use of the fact that one works with associational quantifiers.

(2) One could discuss the question of computational complexity; but as the situation is similar to that in the Baby-GUHA and since the present question is included in the discussion concerning the complexity of the method in Chapter VII, we therefore refer the reader to Chapter VII.

(3) A further reasonable question is, what is the relation between the direct and indirect solutions (for a given set RQ of relevant questions)? This question makes sense both for Ξ_1 and Ξ_2 but this, too is deferred to Chapter VII 7.2.11(c) where we give a simple answer in a wider context. Roughly speaking, the indirect solution cannot be much worse than the direct one but the direct solution can be arbitrarily worse than the indirect one.

(4) It can be seen that the sentence $(\varphi \Leftrightarrow \varphi') \& (\psi \Leftrightarrow \psi')$ can be expressed as $Eq(\varphi, \varphi', \psi, \psi')$ where Eq is a quantifier of type $\langle 1,1,1,1 \rangle$. Eq can be said to be helpful for pairs of designated EC's and for the class of all quantifiers obeying the weak satisfiability condition since it yields the indirect solution. Similarly for the quantifier Eq^+ such that $Eq^+(\varphi, \varphi', \psi, \psi')$ is equivalent to $(\varphi \& \varphi') \Leftrightarrow (\psi \& \psi')$ and for associational quantifiers obeying the satisfiability condition.

We shall give a general definition of helpful quantifiers. From now on, our formulations make sense also for $\times$-predicate calculi; this will be utilized later.

6.2.10 Definition. Let a $\times$-predicate calculus $\mathcal{F}$ be given, let PF be a set of pairs of designated open formulae and let $\subseteq$ be an ordering of PF such that any two elements of PF have the supremum. Let $\ll$ be a quantifier of type $\langle 1^4 \rangle$.

(1) $\ll$ <u>satisfies modus ponens</u> w.r.t. a quantifier $\sim$ of type $\langle 1,1\rangle$ on $(PF,\subseteq)$ if the following rule is $\{1\}$-sound:

$$\left\{ \frac{\varphi\sim\psi,\ (\varphi,\psi)\ll(\varphi',\psi')}{\varphi'\sim\psi'}\ ;\ \langle\varphi,\psi\rangle,\langle\varphi',\psi'\rangle\in PF\ ,\ \langle\varphi,\psi\rangle\subseteq\langle\varphi',\psi'\rangle \right\}.$$

(2) Let C be a class of quantifiers of type $\langle 1,1\rangle$. $\ll$ is C-<u>improving</u> on $(PF,\subseteq)$ if $\ll$ satisfies modus ponens w.r.t. each quantifier from C on $(PF,\subseteq)$.

(3) $\ll$ is a <u>closure quantifier</u> on $(PF,\subseteq)$ if $\Phi\ll\Phi$ is a $\{1\}$-tautology for each $\Phi\in PF$ and if the following rules are 1-sound:

(i) $\dfrac{\Phi\ll\Psi,\ \Psi\ll\Omega}{\Phi\ll\Omega}$

(ii) $\dfrac{\Phi\ll\Omega}{\Phi\ll\Psi}\ ,\ \dfrac{\Phi\ll\Omega}{\Psi\ll\Omega}$

for (i), (ii): $\Phi\subseteq\Psi\subseteq\Omega$

(iii) $\dfrac{\Phi\ll\Psi_1\ ,\ \Phi\ll\Omega}{\Phi\ll\Omega}$ $\quad \Phi\subseteq\Psi_1\ ,\ \Phi\subseteq\Psi_2\ ,\ \Omega=\sup(\Psi_1,\Psi_2)$.

(4) is C-helpful on $(PF,\subseteq)$ if it is a closure quantifier on $(PF,\subseteq)$ and if it is C-improving on $(PF,\subseteq)$.

6.2.11 <u>Remark</u>. Concerning our examples of GUHA methods Ξ_1, Ξ_2, let EC^* be the set consisting of all designated EC's and of the formula $\underline{0}$; extend the subformula ordering $\subseteq$ to EC^* making $\underline{0}$ be the largest element. Then all suprema exist. Extend $\subseteq$ to $PF = EC^*\times EC^*$ putting $\langle\varphi,\psi\rangle\subseteq\langle\varphi',\psi'\rangle$ iff $(\varphi\subseteq\varphi'$ and $\psi\subseteq\psi')$. We have the following:

(1) The quantifier Eq such that Eq $(\varphi,\psi,\varphi',\psi')$ is logically equivalent to $(\varphi\Leftrightarrow\varphi')\ \&\ (\psi\Leftrightarrow\psi')$ is C_1-helpful on $(PF,\subseteq)$, where C_1 is the class of all quantifiers.

(2) The quantifier Eq^+ such that $Eq^+(\varphi,\psi,\varphi',\psi')$ is logically equivalent to $(\varphi\ \&\ \psi)\Leftrightarrow(\varphi'\ \&\ \psi')$ is C_2-helpful on $(PF,\subseteq)$, where C_2 is the class of all associational quantifiers.

The verification is straightforward and is left to the reader.

6.2.12 Convention. We shall say "a-helpful" for "C_2-helpful" where C_2 is the class of all associational quantifiers and we shall say "i-helpful" for "C_3-helpful" where C_3 is the class of all implicational quantifiers.

6.2.13 Definition. Let $\leq$ be an ordering on a set X. A subset $Y \subseteq X$ is a lower tuft if

(i) $a \leq b \leq c$ and $a, c \in Y$ implies $b \in Y$,

(ii) Y has a least element.

Similarly, one defines an upper tuft replacing "least" by "largest".

6.2.14 Lemma. If $\ll$ is a closure quantifier on $(PF, \subseteq)$ then, for each $\underline{M}$, PF can be expressed as a union of a system of pairwise disjoint upper tufts (w.r.t. $\subseteq$) such that, for arbitrary $\Phi, \Psi \in PF$, Φ, Ψ are in the same tuft iff $Reg_{\underline{M}}(\Phi) = Reg_{\underline{M}}(\Psi)$. (Obvious.)

6.2.15 Theorem. Let $\ll$ be a quantifier of type $\langle 1,1 \rangle$. The quantifier $\ll$ is a-improving on $(PF, \subseteq)$ iff the following holds:

Whenever $\langle \varphi, \psi \rangle, \langle \varphi', \psi' \rangle \in PF$, $\langle \varphi, \psi \rangle \subseteq \langle \varphi', \psi' \rangle$ and $\|(\varphi, \psi) \ll (\varphi', \psi')\|_{\underline{M}} = 1$, then $\langle M, \|\varphi'\|_{\underline{M}}, \|\psi'\|_{\underline{M}} \rangle$ is a-better than $\langle M, \|\varphi\|_{\underline{M}}, \|\psi\|_{\underline{M}} \rangle$.

Proof. $\Leftarrow$. If the condition holds then evidently $\ll$ satisfies modus ponens w.r.t. each associational quantifier on $(PF, \subseteq)$ and hence is a-improving.

$\Rightarrow$. If the condition does not hold then let $\underline{M}_1 = \langle M, \|\varphi\|_{\underline{M}}, \|\psi\|_{\underline{M}} \rangle$, $\underline{M}_2 = \langle M, \|\varphi'\|_{\underline{M}}, \|\psi'\|_{\underline{M}} \rangle$ and suppose that $\|(\varphi, \psi) \ll (\varphi', \psi')\|_{\underline{M}} = 1$ and $\underline{M}_2$ is not a-better than $\underline{M}_1$. According to 3.2.4, there is an associational quantifier $\sim$ such that $Asf_{\sim}(\underline{M}_1) = 1$ and $Asf_{\sim}(\underline{M}_2) \neq 1$;

$\ll$ does not satisfy modus ponens w.r.t. $\sim$ and hence $\ll$ is not a-improving.

6.2.16 Remark. (1) Obviously, if we replace "a-improving" by "i-improving" and "a-better" by "i-better" then the theorem remains valid ; cf. 3.2.1.

(2) One could investigate properties of indirect solutions of r-problems in which auxiliary sentences contain a helpful quantifier. We defer this subject to Chapter VII. Our next aim will be to investigate a-helpful and i-helpful quantifiers.

6.2.17 Key words: The GUHA method with auxiliary equivalences, the GUHA method with associational quantifiers and auxiliary equivalences; a quantifier satisfies modus ponens, a closure quantifier, a helpful quantifier ; tufts.

VI.3 Helpful quantifiers in $\times$-predicate calculi

In this section we shall study a-helpful and i-helpful quantifiers, i.e. quantifiers helpful w.r.t. all associational (implicational) quantifiers.

We have already said that each associational (implicational) quantifier $\sim$ in a $\times$-predicate calculus is secured, i.e. $Asf_{\sim}(\underline{M}) = 1$ iff for each two-valued completion $\underline{M}'$ of $\underline{M}$ $Asf_{\sim}(\underline{M}') = 1$ and similarly for 0. This corresponds to the concept of incomplete information: the truth of a formula $Px \sim Qx$ in $\underline{M}$ must guarantee that it is true in all completions (among them is the "right" one). On the other hand, helpful quantifiers form an auxiliary means for the construction of solutions of r-problems and therefore may express some facts about our present knowledge of the "right" completion, i.e. about the three-valued model, rather than about the "right" completion itself. Hence when studying helpful quantifiers we do not require securedness.

6.3.1 Definition. A quantifier q of type $\langle 1^k \rangle$ is universally definable if there is a set $U \subseteq \{0, \times, 1\}^k$ such that

(i) $Asf_q(\underline{M}) = 1$ iff cards of all objects in $\underline{M}$ belong to U,

(ii) $Asf_q(\underline{M}) = 0$ otherwise.

We shall restrict ourselves to universally definable helpful quantifiers since they are rather simple but yield a sufficient number of possibilities.

6.3.2 Discussion. We shall be specific on particular sets PF we want to study. Our main interest will be devoted (i) to pairs of elementary conjunctions and (ii) to pairs $\langle \kappa, \delta \rangle$ where κ is an EC or the formula $\underline{1}$ and δ is an ED. A formula $\kappa \sim \lambda$, where κ, λ are EC's and $\sim$ is an associational quantifier, is thought of as expressing some association (connection) between κ and λ; the meaning of the word "association" is made precise by the associated function of $\sim$. Elementary conjunctions expressing the simultaneous presence of some properties can be well understood by non-mathematicians as well; they form a reasonable compromise between single literals and disjunctions of (pseudo) elementary conjunctions (cf. 3.4.18). In particular, if $\sim$ is implicational, then $\kappa \sim \lambda$ can be thought of as expressing some "causal" connection between κ and λ. In this connection ($\sim$ implicational), a formula $\kappa \sim \delta$, where δ is an ED, is also commonly understood: κ "causes" the occurrence of at least one of the properties joined in δ. If, in addition, κ is $\underline{1}$ (i.e., identically true), then $\kappa \sim \lambda$ expresses the fact that λ is a "rather frequented" property ("caused" by the empty condition).

For technical reasons, we allow pseudoelementary conjunctions and disjunctions (to have enough suprema); but we take care of ED's and EC's. As far as orderings of pairs of such formulae are concerned, we shall use the orderings $\subseteq$, $\sqsubseteq$, $\leftarrow$, $\lhd$ on psEC's (psED's) (contained in, poorer than, hoops, is hidden in). Cf. 3.4.21. These orderings can be trivially extended to the set consisting of all psEC's and of $\underline{1}$ (empty conjunction): $\underline{1} \subseteq \kappa$, $\underline{1} \lhd \kappa$, $\underline{1} \leftarrow \kappa$ for each κ, and $\underline{1} \sqsubseteq \kappa$ iff κ is $\underline{1}$. We have the following easy generalization of 3.4.22:

6.3.3 Lemma. (1) Let κ, λ be psEC's or $\underline{1}$. If $\kappa \leftarrow \lambda$ (in particular, if $\kappa \subseteq \lambda$), then $\|\kappa\|_{\underline{M}}[o] \geq \|\lambda\|_{\underline{M}}[o]$ for each $\underline{M}$ and each $o \in M$.

(2) Let γ, δ be psED's. If $\gamma \lhd \delta$ (in particular, if $\gamma \subseteq \delta$), then $\|\gamma\|_{\underline{M}}[o] \leq \|\delta\|_{\underline{M}}[o]$ for each $\underline{M}$ and each $o \in M$. (Remember that $0 < \times < 1$; remember also Cleave's notion of logical implication for three-valued logic, cf. 3.3.16.)

6.3.4 Definition. (1) A (pseudo-)conjunctive pair of formulas (psCPF or CPF) is a pair $\langle \kappa, \lambda \rangle$ where κ, λ are (ps)EC's; a (**pseudo-**) **elementary** pair of formulae (psEPF or EPF) is a pair $\langle \kappa, \delta \rangle$ where κ is either a (pseudo) elementary conjunction or the formula $\underline{1}$ and δ is a (ps)ED. Occasionally, we use psCPF and psEPF to denote the set of all pseudo-conjunctive (pseudoelementary) pairs of formulae, similarly for CPF and EPF.

(2) We introduce "product" orderings of psCPF and psEPF

$\langle \kappa, \lambda \rangle \subseteq\subseteq \langle \kappa', \lambda' \rangle$ iff ($\kappa \subseteq \kappa'$ and $\lambda \subseteq \lambda'$),

the same for psEPF's,

$\langle \kappa, \lambda \rangle \ll \langle \kappa', \lambda' \rangle$ iff $(\kappa \lessdot \kappa'$ and $\lambda \lessdot \lambda')$

$\langle \kappa, \delta \rangle \lessdot\!\lhd \langle \kappa', \lambda' \rangle$ iff $(\kappa \lessdot \kappa'$ and $\delta \lhd \delta')$.

(3) Introduce "product orderings" $\leq_c$ and $\leq_e$ on $\{0, \times, 1\}^2$:

$\langle u,v \rangle \leq_c \langle u',v' \rangle$ iff $(u \leq u'$ and $v \leq v')$,

$\langle u,v \rangle \leq_e \langle u'v' \rangle$ iff $(u \leq u'$ and $v \geq v')$.

(4) Put $\langle u,v \rangle \ \& \ \langle u',v' \rangle = \langle u \& u', v \& v' \rangle$

(& on the right-hand side denotes the associated function of the conjunction),

$$\langle u,v \rangle (\& v) \langle u', v' \rangle = \langle u \& u', v \& v' \rangle$$

6.3.5 Lemma. (1) If $\langle \kappa, \lambda \rangle \ll \langle \kappa', \lambda' \rangle$, then

$$\langle \|\kappa\|_{\underline{M}}[o], \ \|\lambda\|_{\underline{M}}[o] \rangle \ \geq_c \ \langle \|\kappa'\|_{\underline{M}}[o], \ \|\lambda'\|_{\underline{M}}[o] \rangle .$$

(2) If $\langle \kappa, \delta \rangle \lessdot\!\lhd \langle \kappa', \delta' \rangle$, then

$$\langle \|\kappa\|_{\underline{M}}[o], \ \|\delta\|_{\underline{M}}[o] \rangle \ \geq_e \ \langle \|\kappa'\|_{\underline{M}}[o], \ \|\delta'\|_{\underline{M}}[o] \rangle$$

6.3.6 Remark and Definition. (1) Our choice of psEPF and psCPF as possible sets of relevant pairs of formulae does not mean that other sets (such as, for instance, pairs of psED's, etc.) would not be interesting. But we find the pseudoconjunctive and pseudoelementary pairs to be very typical examples and, moreover, most useful from a practical point of view (cf. Chapter VII).

(2) Our next aim is to analyze and classify universally definable closure quantifiers for our sets of relevant pairs of formulae. Such a quantifier is defined by a set $U \subseteq \{0, \times, 1\}^4$ which can be viewed as a binary relation on $\{0, \times, 1\}^2$. Call a $U \subseteq \{0, \times, 1\}^4$ a closure set for $(PF, \subseteq)$ if the quantifier universally defined by U is a closure quantifier for $(PF, \subseteq)$

(3) We shall first consider closure sets for pseudoconjunctive pairs of formulae and $\ll$. By Lemma 6.3.5, we have the following:

6.3.7 Lemma. If $\mathcal{U} \subseteq \{0, \times, 1\}^4$ is a closure set for (psCPF, $\ll$), then $\hat{\mathcal{U}} = \{\langle \underline{u}, \underline{v}\rangle ; \langle \underline{u}, \underline{v}\rangle \in \{0, \times, 1\}^2, \underline{u}\,\mathcal{U}\,\underline{v}, \underline{u} \geq_c \underline{v}\}$ is also a closure set for (psCPF, $\ll$) ; if $\ll$ and $\hat{\ll}$ are the corresponding quantifiers and if $\langle \kappa, \lambda\rangle \ll \langle \kappa', \lambda'\rangle$, then $(\kappa, \lambda) \ll (\kappa', \lambda')$ is logically equivalent to $(\kappa, \lambda) \hat{\ll} (\kappa', \lambda')$.

6.3.8 Remark and Definition. Hence, we may restrict ourselves to closure sets $\mathcal{U}$ such that $\underline{u}\,\mathcal{U}\,\underline{v}$ implies $\underline{u} \geq_c \underline{v}$ for each $\underline{u}, \underline{v} \in \{0, \times, 1\}^2$. Such a $\mathcal{U}$ will be called an economical closure set (for psCPF, $\ll$). The following lemma is an easy consequence of the definition of an (economical) closure set.

6.3.9 Lemma. $\mathcal{U} \subseteq \{0, \times, 1\}^4$ is an economical closure set for (psCPF, $\ll$) iff the following holds for each $\underline{u}, \underline{v}, \underline{w}$:

(a) $\underline{u}\,\mathcal{U}\,\underline{v}$ implies $\underline{u} \geq_c \underline{v}$;

(b) $\underline{u}\,\mathcal{U}\,\underline{u}$;

(c) $\underline{u}\,\mathcal{U}\,\underline{v}$ and $\underline{v}\,\mathcal{U}\,\underline{w}$ implies $\underline{u}\,\mathcal{U}\,\underline{w}$;

(d) $\underline{u}\,\mathcal{U}\,\underline{w}$ and $\underline{u} \geq_c \underline{v} \geq_c \underline{w}$ implies $\underline{u}\,\mathcal{U}\,\underline{v}$ and $\underline{v}\,\mathcal{U}\,\underline{w}$:

(e) $\underline{u}\,\mathcal{U}\,\underline{v}$ and $\underline{u}\,\mathcal{U}\,\underline{w}$ implies $\underline{u}\,\mathcal{U}(\underline{v} \,\&\, \underline{w})$.

Proof. If (a) - (e) are satisfied, then one easily shows that $\mathcal{U}$ is an economical closure set. Conversely, let $\mathcal{U}$ be an economical closure set ; then (a) is obvious. Let $\underline{M}$ be a model with a unique object o having the card $\langle 1, \times, \ldots\rangle$. Put $\kappa_1 = (\{1\})F_1$, $\kappa_2 = (\{1\})F_1 \,\&\, (\{1\})F_2$, $\kappa_3 = (\emptyset)F_1 \,\&\, (\{1\})F_2$; then $\kappa_1 \lessdot \kappa_2 \lessdot \kappa_3$, $\|\kappa_1\|_{\underline{M}}[o] = 1$, $\|\kappa_2\|_{\underline{M}}[o] = \times$ and $\|\kappa_3\|_{\underline{M}}[o] = 0$. Assume, e.g., not $(\langle \times, 0\rangle \mathcal{U} \langle \times, 0\rangle)$, then $\|(\kappa_2, \kappa_3) \ll (\kappa_2, \kappa_3)\|_{\underline{M}} = 0$. Similarly, assume e.g., $\langle 1, \times\rangle \mathcal{U} \langle \times, 0\rangle$ but not $(\langle 1, \times\rangle\, \mathcal{U} \langle 1, 0\rangle)$; then $\|(\kappa_1, \kappa_2) \ll (\kappa_2, \kappa_3)\|_{\underline{M}} = 1$ but $\|(\kappa_1, \kappa_2) \ll (\kappa_1, \kappa_3)\|_{\underline{M}} = 0$ (whereas $\langle \kappa_1, \kappa_2\rangle \ll \langle \kappa_1, \kappa_3\rangle \ll \langle \kappa_2, \kappa_3\rangle$). Similarly for the other cases one shows that if one of the conditions (b) - (e) is not satisfied, then $\mathcal{U}$ does not define a closure quantifier.

6.3.10 Definition. If a set X is decomposed into a system $Y = Y_1, \dots, Y_k$ of disjoint subsets, then we call $u, v \in X$ Y-equivalent if they belong to the same set Y_i.

6.3.11 Theorem. $U \subseteq \{0, \times, 1\}^4$ is an economical closure set for (psCPF, $\ll$) iff $\{0, \times, 1\}^2$ can be expressed as a union of a system $Y_1, \dots, Y_k$ of pairwise disjoint lower tufts w.r.t. $\leq_c$ such that, for each $\underline{u}, \underline{v} \in \{0, \times, 1\}^2$,

(*) $\underline{u}\ U\ \underline{v}$ iff ($\underline{u} \geq_c \underline{v}$ and $\underline{u}, \underline{v}$ are Y-equivalent).

Proof. If the condition of the theorem holds, **one verifies easily (a)-(e)** of 6.3.9. Conversely, suppose that (a) - (e) are satisfied. Consider U as a graph on $\{0, \times, 1\}^2$ and let $Y_1, \dots, Y_k$ be all the components of this graph. (I.e., $\underline{u}, \underline{v}$ are in the same set Y_i iff there exist $\underline{u}_1, \dots, \underline{u}_n$ such that $\underline{u}_o = \underline{u}$, $\underline{u}_n = \underline{v}$ and, for each i, either $\underline{u}_i\ U\ \underline{u}_{i+1}$ or $\underline{u}_{i+1}\ U\ \underline{u}_i$.) If $\underline{u}\ U\ \underline{v}$, then $\underline{u} \geq_c \underline{v}$ and $\underline{u}, \underline{v}$ are obviously Y-equivalent. Conversely, if $\underline{u}, \underline{v}$ are Y-equivalent and $\underline{u}_o, \dots, \underline{u}_n$ are as above, then using 6.3.9 one easily shows $\underline{u}\ U(\underline{u} \& \underline{u}_i)$ by induction; hence, if further we have $\underline{u} \geq_c \underline{v}$ then we obtain $\underline{u}\ U\ \underline{v}$. This proves (*) . It remains to verify that each Y_i is a tuft. First, we have just proved that if $\underline{u}, \underline{v} \in Y_i$ then $\underline{u}\ U(\underline{u} \& \underline{v})$ and $\underline{v}\ U(\underline{u} \& \underline{v})$; hence $(\underline{u} \& \underline{v}) \in Y_i$ and $\underline{u} \& \underline{v}$ is the infimum of $\underline{u}, \underline{v}$. Hence, Y_i is closed under the infimum and has a least element . **Finally, if** $\underline{u} \geq_c \underline{v} \geq_c \underline{w}$ and $\underline{u}, \underline{w} \in Y_i$, then $\underline{u}\ U\ \underline{w}$ and, consequently, $\underline{u}\ U\ \underline{v}$ (by 6.3.9 (d)); hence $\underline{v} \in Y_i$.

6.3.12 Examples. The preceding theorem enables us to represent a closure set as an appropriate decomposition on $\{0, \times, 1\}^2$. The set $\{0, \times, 1\}^2$ is represented as a square matrix where the first row/column corresponds to the value 1, the second to $\times$ and the third to 0. Heavy lines define subsets of $\{0, \times, 1\}^2$ which constitute the **decomposition. The decomposition** (a) - (d) define closure sets while (e) and (f) do not; in (e), $\langle 1, \times \rangle$ and $\langle \times, 0 \rangle$ are equivalent, $\langle 1, \times \rangle \geq_c \langle 1, 0 \rangle \geq_c \langle \times, 0 \rangle$ but $\langle 1, \times \rangle$ and $\langle 1, 0 \rangle$ are not equivalent. In (f), the decomposition is $\{\langle 0, 0 \rangle\} \cup Y_1$

where Y_1 contains all pairs except $\langle 0,0\rangle$ but Y_1 has no least element.

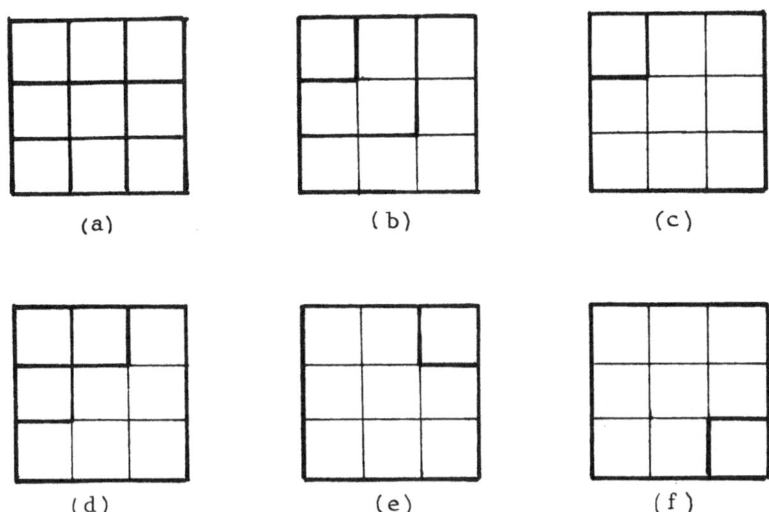

(a) (b) (c)

(d) (e) (f)

6.3.13 Definition. Let $\ll$ be a closure quantifier for (psCPF, $\leqslant\!\leqslant$). Define, for each $\underline{M}$,

$$\| \mathrm{Ant}(\varphi_1, \varphi_2, \psi)\|_{\underline{M}} = \|(\varphi_1, \varphi_2) \ll (\varphi_1 \& \psi, \varphi_2)\|_{\underline{M}},$$

$$\| \mathrm{Suc}(\varphi_1, \varphi_2, \psi\|_{\underline{M}} = \|(\varphi_1, \varphi_2) \ll (\varphi_1, \varphi_2 \& \psi)\|_{\underline{M}}.$$

(Hence Ant and Suc are quantifiers of type $\langle 1^3\rangle$ called the antecedent quantifier and the succedent quantifier corresponding to $\ll$.)

6.3.14 Theorem. If $\ll$ is a closure quantifier for (psCPF, $\leqslant\!\leqslant$) and if κ, λ are psEC's, then $\mathrm{Reg} \ll_{\underline{M}}(\kappa, \lambda) = \langle\bar{\kappa}, \bar{\lambda}\rangle$, where $\bar{\kappa}$ is the conjunction of all pseudoliterals $(X)F_i$ such that (X) is the smallest coefficient (Z) such that $\|\mathrm{Ant}(\kappa, \lambda, (Z)F_i)\|_{\underline{M}} = 1$ and $Z \neq V_i$. Similarly, $\bar{\lambda}$ is the conjunction of all pseudoliterals $(X)F_i$ such that (X) is the smallest coefficient (Z) such that $\| \mathrm{Suc}(\kappa, \lambda, (Z)F_i\|_{\underline{M}} = 1$ and $Z \neq V_i$.

Proof. Put $\mathrm{Reg} \ll_{\underline{M}}(\kappa, \lambda) = \langle\bar{\bar{\kappa}}, \bar{\bar{\lambda}}\rangle$ and let $\bar{\kappa}, \bar{\lambda}$ be defined as in the theorem. Evidently, $\|(\kappa,\lambda) \ll (\bar{\kappa}, \bar{\lambda})\|_{\underline{M}} = 1$, hence, $\bar{\kappa} \leqslant \bar{\bar{\kappa}}$ and $\bar{\lambda} \leqslant \bar{\bar{\lambda}}$. On the other hand, let $(X)F_i$ be a literal from $\bar{\bar{\kappa}}$; then

$\| \mathrm{Ant}(\kappa, \lambda, (X)F_i \|_{\underline{M}} = 1$ by the definition of Ant. Suppose that there is an $X_o \subset X$ such that $\| \mathrm{Ant}(\kappa, \lambda, (X_o) F_i \|_{\underline{M}} = 1$. Then, obviously,

$\| (\kappa, \lambda) \ll (\bar{\bar{\kappa}}[(X_o) F_i /(X)F_i], \bar{\bar{\lambda}}) \|_{\underline{M}} = 1$ ($\bar{\bar{\kappa}}[\cdots]$ results from $\bar{\bar{\kappa}}$ by replacing $(X)F_i$ by $(X_o) F_i$). But $\bar{\bar{\kappa}}[(X_o) F_i /(X)F_i]$ logically implies $\bar{\bar{\kappa}}$ which contradicts $\langle \bar{\bar{\kappa}}, \bar{\bar{\lambda}} \rangle = \mathrm{Reg} \ll_{\underline{M}} (\kappa, \lambda)$. Hence, (X) is the smallest coefficient (Z) such that $\| \mathrm{Ant}(\kappa, \lambda, (Z)F_i) \|_{\underline{M}} = 1$ and $(X) F_i$ is in $\bar{\kappa}$. This proves $\bar{\kappa} = \bar{\bar{\kappa}}$. The proof of $\bar{\lambda} = \bar{\bar{\lambda}}$ is similar.

6.3.15 Remark. The last theorem shows that one can find $\mathrm{Reg} \ll_{\underline{M}} (\kappa, \lambda)$ quickly; one has to consider separate pseudoliterals and not pairs $\langle \bar{\kappa}, \bar{\lambda} \rangle$ such that $\langle \kappa, \lambda \rangle \leftarrow\leftarrow \langle \bar{\kappa}, \bar{\lambda} \rangle$.
(Cf. Chapter VII Section 3.)

6.3.16 Remark. We are interested in EC's; we are dealing with psEC's since they are closed under the supremum w.r.t. $\leftarrow\leftarrow$. Having a closure quantifier, the natural question is: If $\langle \kappa, \lambda \rangle$ is a CPF (i.e., a pair of EC's), is $\mathrm{Reg} \ll_{\underline{M}} (\kappa, \lambda)$ a pair of EC's ? The following theorem gives some information.

6.3.17 Theorem. Let $\mho$ be an economical closure set for (psCPF, $\leftarrow\leftarrow$). The following are equivalent:

(i) $\langle 1, 1, u, v \rangle \in \mho$ implies $u = v = 1$.

(ii) For each $\underline{M}$ and for each CPF $\langle \kappa, \lambda \rangle$ such that $\kappa \& \lambda$ is satisfiable in $\underline{M}$ (i.e., for some $o \in M$ we have $\| \kappa \& \lambda \|_{\underline{M}}[o] = 1$) $\mathrm{Reg} \ll_{\underline{M}} (\kappa, \lambda)$ is a CPF.

Proof. Suppose that (i) is valid. Let $\| (\kappa, \lambda) \ll (\kappa \& X F, \lambda \|_{\underline{M}} = 1$ and $\| \kappa \& \lambda \|_{\underline{M}}[o] = 1$. We want to show that $X = \emptyset$ is impossible. Let $X = \emptyset$. Then $\| \kappa \& (0)F \|_{\underline{M}}[o] = 1$, hence, $\langle \| \kappa \|_{\underline{M}}[o], \| \lambda \|_{\underline{M}}[o], \| \kappa \& (\emptyset)F \|_{\underline{M}}[o], \| \lambda \|_{\underline{M}}[o] \rangle \in \mho$, which is a contradiction.

Suppose that not (i) is valid . Then either $\langle 1,1,1,\times\rangle \in \mathcal{U}$ or $\langle 1,1,\times,1\rangle \in \mathcal{U}$; Suppose $\langle 1,1,1,\times\rangle \in \mathcal{U}$. Let F be a functor not in κ, λ and let $\underline{M}$ be a model in which, for each object o, $\|\kappa\|_{\underline{M}}[o] = \|\lambda\|_{\underline{M}}[o] = 1$ and $\|F\|_{\underline{M}}[o] = \times$. Then $\|(\kappa,\lambda) \ll (\kappa \&(\emptyset)F;\lambda)\|_{\underline{M}} = 1$, hence, $\mathrm{Reg} \ll_{\underline{M}} (\kappa, \lambda)$ is not a pair of EC's.

6.3.18 Remark and Theorem. Consider pseudoelementary pairs (psEPF), i.e., pairs $\langle \kappa, \delta \rangle$ where κ is either an EC or the formula $\underline{1}$ and δ is an ED. We have the ordering $\lessdot\!\lhd$ which is related to $\geq_e$ for psEPF's exactly as $\lessdot\!\lessdot$ is related to $\geq_c$ for psCPF's (cf. 6.3.4). Hence, we can prove an analogue of 6.3.9, namely the following:

(Theorem.) $\mathcal{U} \subseteq \{0,\times,1\}^4$ is an economical closure set for (psEPF, $\lessdot\!\lhd$) iff $\{0,\times,1\}^2$ can be decomposed into a system $Y_1,\ldots,Y_k$ of pairwise disjoint lower tufts w.r.t. $\leq_e$ such that, for each $\underline{u},\underline{v} \in \{0,\times,1\}^2$, $\underline{u}\,\mathcal{U}\,\underline{v}$ iff $\underline{u} \geq_e \underline{v}$ and $\underline{u},\underline{v}$ are Y-equivalent.

Here "economical" means that $\underline{u}\,\mathcal{U}\,\underline{v}$ implies $\underline{u} \geq_e \underline{v}$.

6.3.19 Remark and Definition. We return to pseudoconjunctive pairs. We want to describe a-helpful (i-helpful) universally defined quantifiers. Helpful means closure + improving. Obviously, a closure quantifier for (psCPF, $\lessdot\!\lessdot$) defined universally by $\mathcal{U} \subseteq \{0,\times,1\}^4$ is a-improving iff $\underline{u}\,\mathcal{U}\,\underline{v}$ implies that $\underline{u}$ is a-improved by $\underline{v}$ (cf. 3.3.22) and similarly for "i-" instead of "a-". Furthermore, it is obvious that one can restrict **oneself** to economical closure sets. Hence, let us present the following definition:

(Definition.) A set $\mathcal{U} \subseteq \{0,\times,1\}^4$ is an economical a-helpful set for (psCPF, $\lessdot\!\lessdot$) if $\mathcal{U}$ is an economical closure set for (psCPF, $\lessdot\!\lessdot$) such that $\underline{u}\,\mathcal{U}\,\underline{v}$ implies that $\underline{v}$ a-improves $\underline{u}$ (for $\underline{u},\underline{v} \in \{0,\times,1\}^2$). Similarly for "i-".

6.3.20 Theorem. (1) An economical closure set $\mathcal{U}$ for (psCPF, $\lessdot\!\lessdot$) is a-improving iff the following quadruples are not in $\mathcal{U}$:

$\langle 1,1,1,\times\rangle$, $\langle \times,1,\times,\times\rangle$, $\langle 1,1,\times,1\rangle$, $\langle 1,\times,\times,\times\rangle$.

(2) $\mathcal{U}$ is i-improving iff the following quadruples are not in $\mathcal{U}$:

$$\langle 1,1,1,\times\rangle, \quad \langle \times,1,\times,\times\rangle , \quad \langle 1,1,\times , 1\rangle .$$

Proof. Let $\mathcal{U}$ be an economical closure set. $\mathcal{U}$ is a-improving iff the following holds: If $\underline{u} \geq_c \underline{v}$ and if $\underline{v}$ does not improve $\underline{u}$, then not $\underline{u}\,\mathcal{U}\,\underline{v}$. Similarly for "i-" . Hence, the theorem follows by inspection of the a-improving (i-improving) ordering, cf. 3.3.24 . Use the tables of 3.3.25. The a-improvement relation is expressed by the following table: the long arrows indicate the $\geq_c$ - ordering. Hence, one must take care of forbidden transitions from the left to the right and from above to below. These transitions are just $(1,1) \rightarrow (1,\times)$, $(\times,1) \rightarrow (\times,\times)$ and $(1,1) \rightarrow (\times,1)$, $(1,\times) \rightarrow (\times,\times)$ (indicated by short arrows).

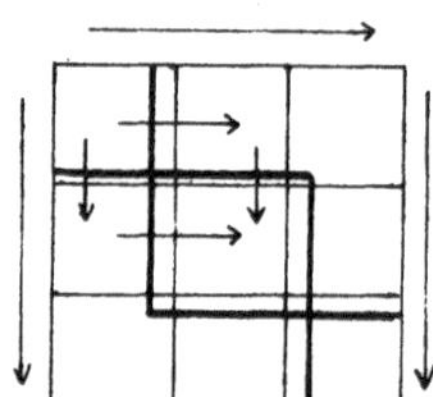

6.3.21 Corollary and Remark. There are four maximal economical a-helpful sets for (psCPF, $\ll$); they are given by the following tables:

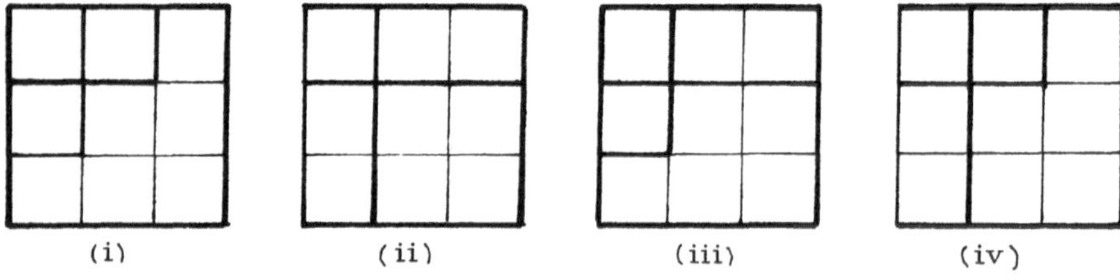

(i) (ii) (iii) (iv)

"Maximal" concerns inclusion. Note that all these sets satisfy 6.2.14 (i).

We want U as large as possible since, obviously, if $U_1 \subseteq U_2$ are economical a-helpful sets and if $<<^1$, $<<^2$ are the corresponding quantifiers, then $(\kappa,\lambda) <<^1 (\bar{\kappa},\bar{\lambda})$ logically implies $(\kappa,\lambda) <<^2 (\bar{\kappa},\bar{\lambda})$ and, hence $\mathrm{Reg} <<^1_{\underline{M}}(\kappa,\lambda) \Leftarrow\Leftarrow \mathrm{Reg} <<^2_{\underline{M}}(\kappa,\lambda)$. Hence, if $\| \kappa \sim \lambda \|_{\underline{M}} = 1$ where $\sim$ is an associational quantifier, then having found $\mathrm{Reg} <<^2_{\underline{M}}(\kappa,\lambda)$ we have more consequences then when using $<<^1$.

6.3.22 Corollary. There are two maximal economical i-helpful sets for $(\mathrm{psCPF}, \Leftarrow\Leftarrow)$, they are given in the following tables:

(i) 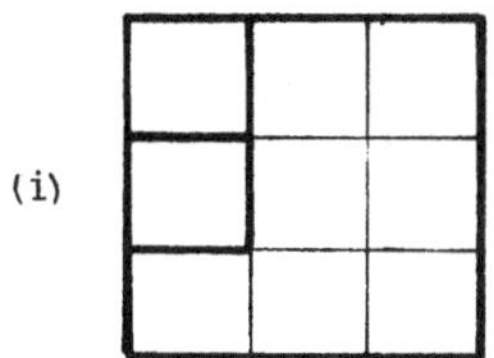

(ii) 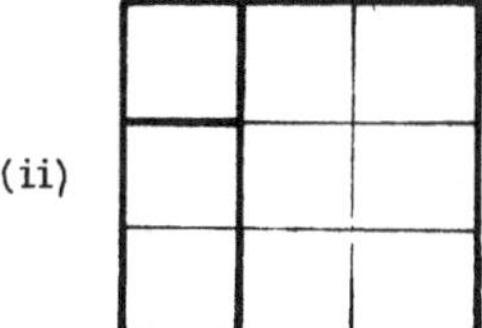

Both of them satisfy 6.2.14 (i).

6.3.23 Remark. For psEPF and $\Leftarrow\lhd$ (instead of psCPF and $\Leftarrow\Leftarrow$) the situation is completely analogous. One has only to replace $\geq_c$ by $\geq_e$ and change accordingly the meaning of "economical". Thus, we have the following theorem:

6.3.24 Theorem. An economical closure set U for $(\mathrm{psEPF}, \Leftarrow\lhd)$

(1) is a-improving iff the following quadruples are not in U :

$\langle 1,1,\times,1\rangle$, $\langle 1,\times,\times,\times\rangle$, $\langle \times,0,\times,\times\rangle$, $\langle 0,0,0,\times\rangle$,

(2) is i-improving iff $\langle 1,1,\times,1\rangle$ is not in U .

Proof.

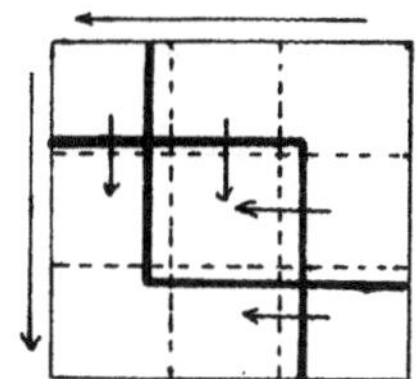

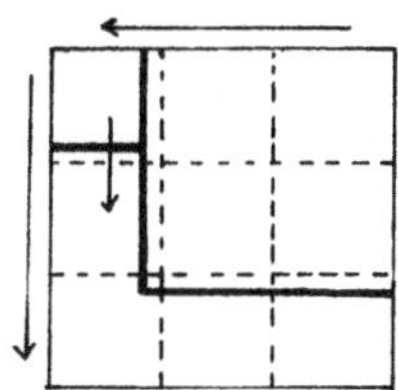

6.3.25 Corollary. 1 There are two maximal economical a-helpful sets for (psEPF, $\leftarrow\lhd$) :

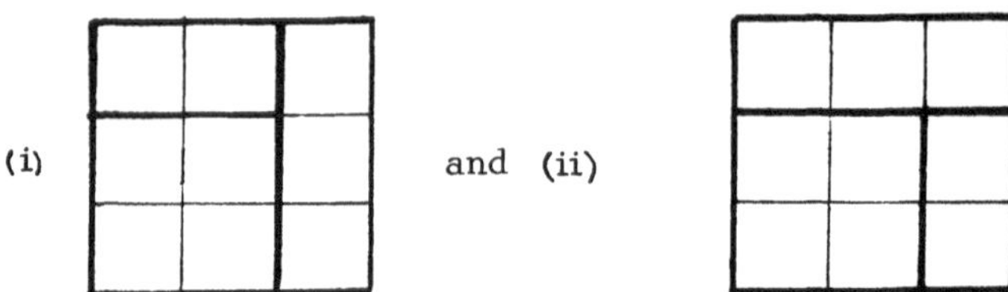

2 There are three maximal economical i-helpful sets for (psEPF, $\leftarrow\lhd$) :

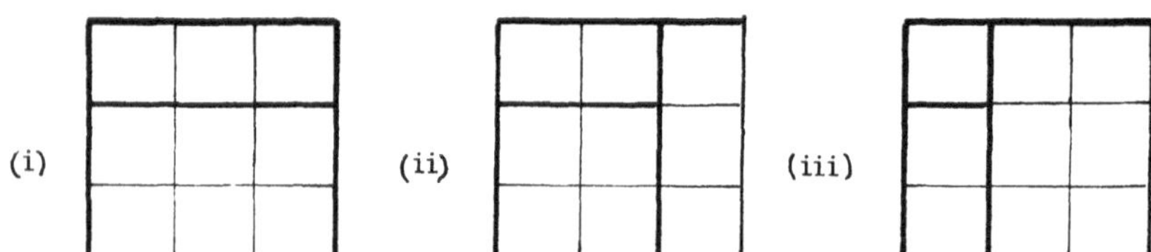

6.3.26 Remark. In accordance with 6.2.12, we shall study psEPF's (i.e., pairs $\langle\kappa,\delta\rangle$ where κ is a psEC or $\underline{1}$ and δ is a psED) in connection with implicational quantifiers. Then we obtain a non-trivial rule exactly as in 3.2.19 - 3.2.22. First, note that the following rules are sound for each cross-qualitative calculus with an implicational quantifier $\sim$:

$$\left.\begin{array}{c}\dfrac{\varphi\sim\psi}{\varphi\sim(\psi\vee\chi)}\\[2em]\dfrac{(\varphi\,\&\,\neg\chi)\sim\psi}{\varphi\sim(\psi\vee\chi)}\end{array}\right\}\quad \varphi,\psi,\chi \quad \text{non-atomic designated open.}$$

Thus, we may obtain the despecifying-dereducing rule as in 3.2.20 : moreover, we can generalize the definition of reduction for psED's as follows:

6.3.27 Definition. (1) A psEPF $<\kappa,\delta>$ is disjointed if κ, δ are disjoint formulae, i. e. if they have no functors in common. Similarly for psCPF.

(2) Let $<\kappa_1, \delta_1>, <\kappa_2, \delta_2>$ be (disjointed) psEPF's. $<\kappa_1, \delta_1>$ despecifies to $<\kappa_2, \delta_2>$ if either $<\kappa_1, \delta_1>$ coincides with $<\kappa_2, \delta_2>$ or if there is a psED δ_o having no functors in common with either κ_2 or δ_1 such that κ_1 is con $(\kappa_2$, neg $(\delta_o))$ and δ_2 is dis (δ_1, δ_o)

(3) $<\kappa_1, \delta_1>$ dereduces to $<\kappa_2, \delta_2>$ if κ_1 is κ_2 and $\delta_1 \triangleleft \delta_2$.

(4) The pair $<\kappa_1, \delta_1>$ is acuter than $<\kappa_2, \delta_2>$ if $<\kappa_2, \delta_2>$ results from $<\kappa_1, \delta_1>$ by successive despecification and dereduction, i.e., if there is a $<\kappa_3, \delta_3>$ such that $<\kappa_1, \delta_1>$ despecifies to $<\kappa_3, \delta_3>$ and $<\kappa_3, \delta_3>$ dereduces to $<\kappa_2, \delta_2>$. (We then write $<\kappa_1, \delta_1> \propto <\kappa_2, \delta_2>$.

6.3.28 Theorem. Let SpRd be the rule

$$\left\{ \frac{\kappa_1 \sim \delta_1}{\kappa_2 \sim \delta_2} ; \quad <\kappa_1, \delta_1> \text{ is acuter than } <\kappa_2, \delta_2> \right\}.$$

Then SpRd is sound in any $\times$-qualitative calculus in which $\sim$ is an implicational quantifier. Furthermore, this rule is transitive.

(The first part is obvious from 6.3.27, cf. 3.2.21 , transitivity is proved as in 3.2.22 .)

6.3.29 Remark. How can the rule SpRd be combined with rules using helpful quantifiers? To answer this question we first analyse the composition of the relations $\leftarrow \triangleleft$ and $\propto$; then we show that it always suffices first to use a helpful quantifier (in a particular manner) and then to use SpRd.

6.3.30 Lemma. (1) The composition of the relations $\lessdot\!\lhd$ and $\propto$ is an ordering of the set of all disjointed psEPF's.

(2) In more detail, whenever

$$\langle \kappa_1, \delta_1 \rangle \propto \langle \kappa_2, \delta_2 \rangle \lessdot\!\lhd \langle \kappa_3, \delta_3 \rangle ,$$

then there is a κ^+ such that

$$\langle \kappa_1, \delta_1 \rangle \lessdot\!\lhd \langle \kappa^+, \delta_1 \rangle \propto \langle \kappa_3, \delta_3 \rangle .$$

Proof. Clearly, (2) implies (1). We prove (2). We have the following relations: $\kappa_2 \subseteq \kappa_1$, $\delta_1 \lhd \delta_2$, $\kappa_2 \lessdot \kappa_3$, $\delta_2 \lhd \delta_3$; hence $\delta_1 \lhd \delta_3$. Put $\kappa^+ = \underline{\text{con}}(\kappa_1, \kappa_3)$. Then $\kappa_1 \lessdot \kappa^+$; we prove $\langle \kappa^+, \delta_1 \rangle \propto \langle \kappa_3, \delta_3 \rangle$. Let κ_1 be $\underline{\text{con}}(\kappa_2, \kappa_2')$ (κ_2, κ_2' disjoint, κ_2' may be $\underline{1}$) and let $\delta_2 \rhd \underline{\text{dis}}(\delta_1, \underline{\text{neg}}(\kappa_2'))$. Let κ_3 be $\underline{\text{con}}(\kappa_{31}, \kappa_{32})$ where κ_{31} is poorer than κ_2 and κ_{32} is disjoint from κ_2. Then κ_{32} is disjoint from κ_2' since κ_3, δ_3 are disjoint and $\delta_3 \rhd \underline{\text{neg}}(\kappa_2')$. Hence, κ^+ is $\underline{\text{con}}(\kappa_3, \kappa_2')$ and κ_3, κ_2' are disjoint. This together with $\delta_3 \rhd \underline{\text{dis}}(\delta_1, \text{neg}(\kappa_2))$ proves the lemma.

6.3.31 Corollary. Under the above notation, if $\langle \kappa_i, \delta_i \rangle$ are disjointed EPF's then κ^+ is an EC (i.e., no coefficient is empty) and, hence, $\langle \kappa^+, \delta_1 \rangle$ is an EPF.

Proof. We showed that κ^+ is $\underline{\text{con}}(\kappa_3, \kappa_2')$ where κ_3, κ_2' are disjoint ; κ_3 is an EC and κ_2' as a subconjunction of κ^+ is also an EC.

6.3.32 Lemma. Denote the i-helpful quantifiers defined by 6.3.36 (i), (ii), (iii) by $\ll^1, \ll^2, \ll^3$ respectively.

(1) If $\langle \kappa_1, \delta_1 \rangle \propto \langle \kappa_2, \delta_2 \rangle \lessdot\!\lhd \langle \kappa_3, \delta_3 \rangle$ ($\langle \kappa_j, \delta_j \rangle$ disjointed psEPF) and if κ^+ is as in the preceding lemma (i.e., $\kappa^+ = \underline{\text{con}}(\kappa_1, \kappa_3)$),

then $\|(K_2,\delta_2) \ll^{\gamma} (K_3,\delta_3)\|_{\underline{M}} = 1$ implies $\|(K_1,\delta_1) \ll^{\gamma} (K_1^+,\delta_1)\|_{\underline{M}} = 1$ for each $\underline{M}$ ($\gamma = 1, 2, 3$).

(2) $\|(K_1,\delta_1) \ll^1 (K^+,\delta_1)\|_{\underline{M}} = 1$ implies $\|(K_1,\delta_1) \ll^2 (K^+,\delta_1)\|_{\underline{M}} = 1$, which implies $\|(K_1,\delta_1) \ll^3 (K^+,\delta_1)\|_{\underline{M}} = 1$.

Proof. (1) (i) We have to verify: If $\|K_1\|_{\underline{M}}[o] = 1$, then $\|K^+\|_{\underline{M}}[o] = 1$. Let $\|K_1\|_{\underline{M}}[o] = 1$. Then $\|K_2\|_{\underline{M}}[o] = 1$ since $K_2 \subseteq K_1$; hence, by $\|(K_2,\delta_2) \ll^1 (K_3,\delta_3)\|_{\underline{M}} = 1$, we have $\|K_3\|_{\underline{M}}[o] = 1$. Thus, $\|K^+\|_{\underline{M}}[o] = 1$. (ii) We have to verify: If $\|K_1\|_{\underline{M}}[o] = 1$ and $\|\delta_1\|_{\underline{M}} > 0$ then $\|K^+\|_{\underline{M}}[o] = 1$. Let $\|K_1\|_{\underline{M}}[o] = 1$. Then $\|K_2\|_{\underline{M}}[o] = 1$ and hence, by $\ll^2$ we have $\|K_3\|_{\underline{M}}[o] = 1$. Analogously for (iii): Verify that if $\|K_1 \& \delta_1\|_{\underline{M}} = 1$ then $\|K^+\|_{\underline{M}} = 1$.

(2) follows from the reformulations of $\|(K_1,\delta_1) \ll^{\gamma} (K^+,\delta_1)\|_{\underline{M}} = 1$ just given.

6.3.33 Conclusion. Suppose one has a model $\underline{M}$ such that $\|K \sim \delta\|_{\underline{M}} = 1$ (where $\sim$ is an implicational quantifier and $\langle K,\delta\rangle$ is a disjointed EPF). Let $con_{\underline{M}}(K,\delta)$ be the set of all sentences $\bar{K} \sim \bar{\delta}$ obtainable from $K \sim \delta$ by the iterated application of SpRd and the "helpful" rule

$$\left\{ \frac{K_1 \sim \delta_1, (K_1,\delta_1) \ll (K_2,\delta_2)}{K_2 \sim \delta_2} ; \langle K_1,\delta_1\rangle \leftarrow\!\triangleleft \langle K_2,\delta_2\rangle \right\},$$

where $\ll$ is an i-helpful quantifier (universally definable) and where all sentences $(K_1,\delta_1) \ll (K_2,\delta_2)$ true in $\underline{M}$ are at one's disposal. Then $con_{\underline{M}}(K,\delta)$ is the set of all $\bar{K} \sim \bar{\delta}$ such that there is a $\langle K_1,\delta_1\rangle$ satisfying $\langle K,\delta\rangle \leftarrow\!\triangleleft (K_1,\delta_1) \leftarrow\!\triangleleft \langle \bar{K},\bar{\delta}\rangle$ and $\langle K_1,\delta_1\rangle \propto \langle \bar{K},\bar{\delta}\rangle$.

We shall use this fact in the next chapter.

6.3.34 <u>Key words:</u> Economical closure sets, antecedent and succedent quantifiers, economical a-helpful sets.

VI.4 Incompressibility

In the present short section we shall study notions of incompressibility of pseudo elementary conjunctions in cross-qualitative calculi. An EC is viewed as a certain description of a set of objects, namely, the set of all objects having simultaneously all the properties expressed by the literals involved. An incompressible conjunction is an "economical" description. Our main aim is to consider the relation of incompressibility to various helpful quantifiers.

6.4.1 Definition. A psEC κ is incompressible in a model $\underline{M}$ (or: $\underline{M}$ -incompressible) if there is no $\kappa_o \neq \kappa$ poorer than κ and equivalent to κ in $\underline{M}$ (i.e., such that $\|\kappa \Leftrightarrow \kappa_o\|_{\underline{M}} = 1$).

6.4.2 Remark. The notion of incompressibility is obviously redundant in predicate calculi since each EC is $\underline{M}$ -incompressible in each $\underline{M}$.

6.4.3 Lemma. (1) Each subconjunction of an $\underline{M}$ -incompressible psEC is $\underline{M}$ -incompressible. (2) For each psEC κ and each $\underline{M}$, there is a uniquely determined $\kappa_o \sqsubseteq \kappa$ which is $\underline{M}$ -incompressible and equivalent to κ in $\underline{M}$.

Proof. (1) Let $\kappa_o \subsetneqq \kappa_1, \kappa_1 \subseteq \kappa$, $\|\kappa_o \Leftrightarrow \kappa_1\|_{\underline{M}} = 1$; form $\bar{\kappa}$ by adding to κ_o literals from κ with function symbols not in κ_o. Then $\bar{\kappa} \subsetneqq \kappa$ and $\|\bar{\kappa} \Leftrightarrow \kappa\|_{\underline{M}} = 1$; hence, κ is not $\underline{M}$- incompressible.

(2) Let $\kappa = \bigwedge_I (X_i) F_i$ and let $\kappa_j = \bigwedge_I (X_i^j) F_i$ $(j = 1,\ldots,k)$ be all the conjunctions poorer than κ and $\underline{M}$ -equivalent to κ. Put $\kappa_o = \bigwedge_I (\bigcap_{j=1}^{k} (X_i^j) F_i)$; then $\kappa_o \sqsubseteq \kappa$, $\|\kappa_o \Leftrightarrow \kappa\|_{\underline{M}} = 1$ and κ_o is $\underline{M}$ -incompressible.

6.4.4 Remark. Consider a quantifier $\ll$ which is a closure quantifier w.r.t. (psEC, $\ll$) . We have the following natural questions: Let κ, λ be psEC's and let $\mathrm{Reg} \ll_{\underline{M}}(\kappa, \lambda) = \langle \bar{\kappa}, \bar{\lambda} \rangle$. Are $\bar{\kappa}, \bar{\lambda}$ $\underline{M}$-incompressible? Is $\underline{\mathrm{con}}(\bar{\kappa}, \bar{\lambda})$ $\underline{M}$-incompressible?

6.4.5 Theorem. (1) If $\ll$ is a closure quantifier for (psCPF , $\ll$) and if $\langle \kappa, \lambda \rangle$ is a psCPF (pair of pseudoelementary conjunctions), then for each $\underline{M}$ $\mathrm{Reg} \ll_{\underline{M}}(\kappa, \lambda)$ is a pair of $\underline{M}$-incompressible psEC's.

(2) Let $\ll$ be universally defined by an economical closure set $\mathcal{U}$. The following are equivalent:

(i) $\langle 1,0,0,0 \rangle$, $\langle 0,1,0,0 \rangle \in \mathcal{U}$ (and thus all pairs containing at least one 0 are Y-equivalent in the sense of 6.3.11).

(ii) For each $\underline{M}$ and each psCPF $\langle \kappa, \lambda \rangle$, if we put $\mathrm{Reg} \ll_{\underline{M}}$ then $\bar{\kappa} \,\&\, \bar{\lambda}$ is $\underline{M}$ - incompressible.

Proof. (1) follows from 6.4.3 (2) .

(2) Suppose that (i) is valid . It suffices to verify the following: If

$\|(\kappa, \lambda) \ll (\kappa \,\&\, (X) F, \lambda\|_{\underline{M}} = 1$ and $\|\kappa \,\&\, \lambda \,\&\, (X) F \Leftrightarrow \kappa \,\&\, \lambda \,\&\, (X_o) F\|_{\underline{M}} = 1$

for an $X_o \subseteq X$, then $\|(\kappa, \lambda) \ll (\kappa \,\&\, (X_o) F, \lambda)\|_{\underline{M}} = 1$ (and similarly for κ, $\lambda \,\&\, (X) F$). Indeed, if for an object o the value of $\kappa \,\&\, \lambda \,\&\, (X) F$ is 1, then the value of $\kappa \,\&\, \lambda \,\&\, (X) F$ is also 1 and thus $\|(X_o) F\|_{\underline{M}}[o] = \|(X) F\|_{\underline{M}}[o] = 1$, hence the quadruple $\langle u,v,\bar{u},\bar{v} \rangle$ of the values of κ, λ, $\kappa \,\&\, (X) F, \lambda$ is equal to the quadruple $\langle u,v,\bar{\bar{u}},\bar{\bar{v}} \rangle$ of the values of $\kappa, \lambda, \kappa \,\&\, (X_o) F, \lambda$ and thus the latter is in $\mathcal{U}$. If the value of $\kappa \,\&\, \lambda \,\&\, (X) F$ is $\times$, then <u>either</u> $\|F\|_{\underline{M}}[o] = \times$ and hence $\|(X) F\|_{\underline{M}}[o] = \|(X_o) F\|_{\underline{M}}[o] = \times$ <u>or</u> $\|(X) F\|_{\underline{M}}[o] = 1$ and then $\|(X_o) F\|_{\underline{M}} = 1$, which follows from

$$\|(\kappa \,\&\, \lambda \,\&\, (X) F \Leftrightarrow \kappa \,\&\, \lambda \,\&\, (X_o) F\|_{\underline{M}} = 1 .$$

If $\|\kappa \,\&\, \lambda \,\&\, (X) F\|_{\underline{M}} = 0$, then $\|(X_o) F\|_{\underline{M}}[o]$ can be different from

$\|(X)F\|_{\underline{M}}[o]$ and we still have $\|(\kappa \,\&\, \lambda \,\&(X)F) \Leftrightarrow \kappa \,\&\, \lambda \,\&\, (X_o)F\|_{\underline{M}} = 1$

But in the present case we have ($\bar{u} = 0$ or $\bar{v} = 0$) and ($\bar{\bar{u}} = 0$ or $\bar{\bar{v}} = 0$), hence $\langle\bar{\bar{u}},\bar{v}\rangle$ is Y-equivalent to $\langle\bar{\bar{u}},\bar{\bar{v}}\rangle$, which together with $\langle u,v\rangle \; \mathcal{U} \; \langle\bar{u},\bar{v}\rangle$ yields $\langle u,v,\bar{\bar{u}},\bar{\bar{v}}\rangle \in \mathcal{U}$.

Suppose that not (i) is valid and let, e.g., $\langle 1,0\rangle$ be not Y-equivalent to $\langle 0,0\rangle$. Let $\underline{M}$, κ, λ, $X_o \subset X$ be such that for each o

$\|\kappa\|_{\underline{M}}[o] = \|(X)F\|_{\underline{M}}[o] = 1$, $\|\lambda\|_{\underline{M}}[o] = \|(X_o)F\|_{\underline{M}}[o] = 0$. Then

$$\|(\kappa,\lambda) \ll (\kappa \,\&(X)F, \lambda)\|_{\underline{M}} = 1 \,, \|(\kappa,\lambda) \ll (\kappa \,\&(X_o)F, \lambda)\|_{\underline{M}} = 0 \,,$$

$\|\kappa \,\&\, \lambda \,\&(X)F \Leftrightarrow \kappa \,\&\, \lambda \,\&(X_o)F\|_{\underline{M}} = 1$.

Hence, if $\langle\bar{\kappa},\bar{\lambda}\rangle = \mathrm{Reg} \ll_{\underline{M}} (\kappa,\lambda)$, then $\bar{\kappa} \,\&\, \bar{\lambda}$ is $\underline{M}$-compressible.

6.4.6 Remark. Recall 6. 3. 17 where we have shown that the regularization of a pair of EC's is a pair of EC's iff

(i) $\langle 1,1\rangle$ forms a one-element tuft.

The condition in 6.4.5 requires that

(ii) $\langle 0,0\rangle$ lies in a tuft containing at least all pairs with at least one zero.

Consider now the quantifier of 6.3.21: we know that all of them satisfy (i). But only the first one satisfies (ii). Hence, we prefer the first quantifier. Similarly for the quantifiers of 6.3.22 - we prefer the first quantifier.

6.4.7 Theorem. If both 6.4.6 (i) and (ii) hold and if $\langle\kappa,\lambda\rangle$ is a disjointed pair of EC's, then for $\mathrm{Reg} \ll_{\underline{M}} (\kappa,\lambda) = \langle\bar{\kappa},\bar{\lambda}\rangle$ we have: Whenever $(X)F$ occurs in $\bar{\kappa}$ and $(Y)F$ occurs in $\bar{\lambda}$, then $X = Y \neq \emptyset$.

Proof. The fact that all coefficients are non-empty follows from (i). If $\|(\kappa,\lambda) \ll (\kappa \,\&(X)F, \lambda \,\&(Y)F\|_{\underline{M}} = 1$, observe that

$\|(\kappa \,\&\, \lambda \,\&(X)F \,\&(Y)F) \Leftrightarrow (\kappa \,\&\, \lambda \,\&(X)F \,\&(X \cap Y)F)\|_{\underline{M}} = 1$

so that, by the proof of 6.4.5, we have $\|(\kappa,\lambda) \ll (\kappa \,\&(X \cap Y)F, \lambda)\|_{\underline{M}} = 1$

Similarly, we obtain $\|(\kappa,\lambda) \ll (\kappa \,\&(X \cap Y)F, \lambda \,\&(X \cap Y)F)\|_{\underline{M}} = 1$.

6.4.8 Theorem. Let $\ll$ be a closure quantifier for (psEC, $\Leftarrow\!\Leftarrow$) defined by an economical closure set $\mathcal{U}$.

(1) Suppose, first, that $\mathcal{U}$ is the identity relation on $\{0, \times, 1\}^2$, thus $(\kappa,\lambda) \ll (\bar{\kappa},\bar{\lambda})$ is equivalent to $(\kappa \Leftrightarrow \bar{\kappa})$ & $(\lambda \Leftrightarrow \bar{\lambda})$. If κ,λ is a pair of M-incompressible psEC's and if $\langle\bar{\kappa},\bar{\lambda}\rangle =$ $=$ Reg $\ll_{\underline{M}}(\kappa,\lambda)$, then $\kappa \subseteq \bar{\kappa}$ and $\lambda \subseteq \bar{\lambda}$.

(2) Suppose that $\langle u,v,\bar{u},\bar{v}\rangle \in \mathcal{U}$ implies $u \,\&\, v = \bar{u} \,\&\, \bar{v}$ (thus $(\kappa,\lambda) \ll (\bar{\kappa},\bar{\lambda})$ logically implies $(\kappa \,\&\, \lambda) \Leftrightarrow (\bar{\kappa} \,\&\, \bar{\lambda})$. If con (κ,λ) is M-incompressible and if $\langle\bar{\kappa},\bar{\lambda}\rangle$ = Reg $\ll_{\underline{M}}(\kappa,\lambda)$, then con $(\kappa,\lambda) \subseteq$ con $(\bar{\kappa},\bar{\lambda})$.

Proof. Let $\langle\bar{\kappa},\bar{\lambda}\rangle$ = Reg $\ll_{M}(\kappa,\lambda\rangle$; then $\langle\kappa,\lambda\rangle \Leftarrow\!\Leftarrow \langle\bar{\kappa},\bar{\lambda}\rangle$.

Let $\kappa \sqsupseteq \kappa_o \subseteq \bar{\kappa}$ and $\lambda \sqsupseteq \lambda_o \subseteq \bar{\lambda}$ (i.e., κ_o is the subconjunction of $\bar{\kappa}$ with the same function symbols as κ; similarly for λ_o).

(1) In the first case we have $\| \kappa \Leftrightarrow \bar{\kappa} \|_{\underline{M}} = \| \lambda \Leftrightarrow \bar{\lambda} \|_{\underline{M}} = 1$, whence $\| \kappa \Leftrightarrow \kappa_o \|_{\underline{M}} = \| \lambda \Leftrightarrow \lambda_o \|_{\underline{M}} = 1$, which implies $\kappa_o = \kappa$, $\lambda_o = \lambda$ by incompressibility.

(2) In the second case we have $\|$ con $(\kappa,\lambda) \Leftrightarrow$ con $(\bar{\kappa},\bar{\lambda}) \|_{\underline{M}} = 1$, so that $\|$ con $(\kappa,\lambda) \Leftrightarrow$ con $(\kappa_o,\lambda_o) \|_{\underline{M}} = 1$, which implies con (κ,λ) = con (κ_o,λ_o) by incompressibility. Hence, con $(\kappa,\lambda) \subseteq$ con $(\bar{\kappa},\bar{\lambda})$.

6.4.9 Remark. Obviously, the quantifier of 6.4.8 (1) is an a-helpful quantifier for (psEC, $\Leftarrow\!\Leftarrow$). Observe that if $\| (\kappa,\lambda) \ll (\bar{\kappa},\bar{\lambda}) \|_{\underline{M}} = 1$, the models

$$\langle \underline{M}, \|\kappa\|_{\underline{M}}, \|\lambda\|_{\underline{M}} \rangle , \quad \langle \underline{M}, \|\bar{\kappa}\|_{\underline{M}}, \|\bar{\lambda}\|_{\underline{M}} \rangle$$

coincide. $\mathcal{U}$ is defined by the decomposition of $\{0, \times, 1\}^2$ into one-element tufts:

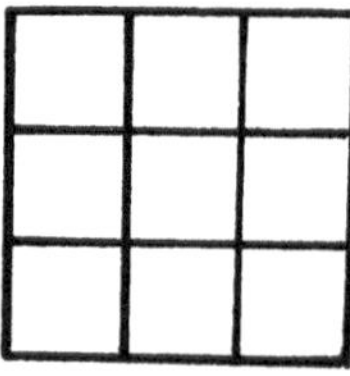

6.4.10 Theorem. Say that a set $\mathcal{U}$ respects equivalence of conjunctions if $\langle u,v,\bar{u},\bar{v}\rangle \in \mathcal{U}$ implies $u \& v = \bar{u} \& \bar{v}$ (cf. 6.4.7).

(1) The largest economical a-helpful set for (psEC, $\Leftarrow\Leftarrow$) respecting equivalence of conjunctions is given by

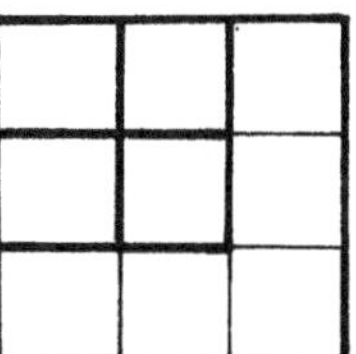

(2) The largest economical i-helpful set for (psEC, $\Leftarrow\Leftarrow$) respecting equivalence of conjunctions is given by

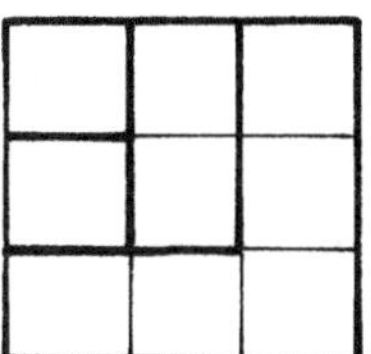

Proof. Remember 6.3.20 and cf. 6.3.21, 6.3.22. The decomposition must be finer than

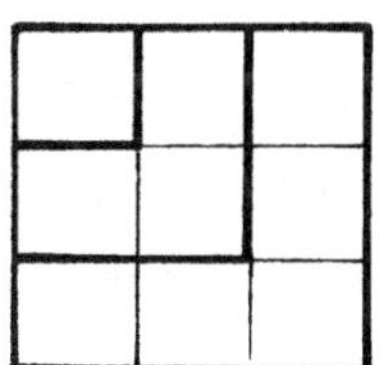

6.4.11 Remark. (1) Note that if an economical a-helpful set $\mathcal{U}$ respects equivalence of conjunctions it satisfies 6.4.6 (i), i.e. $\langle 1,1\rangle$ forms a one-element tuft. The a-helpful set defined by 6.4.10 (1) is the unique a-helpful set $\mathcal{U}$ satisfying both 6.4.6 (i) and (ii) and respecting equivalence of conjunctions.

(2) Note that in the case of qualitative calculus (without incomplete information)

(a) for the quantifiers of 6.4.8 (1), $(\kappa,\lambda) \ll (\bar{\kappa},\bar{\lambda})$ is logically equivalent to $(\kappa \Leftrightarrow \bar{\kappa})$ & $(\lambda \Leftrightarrow \bar{\lambda})$,

(b) for the quantifiers defined by

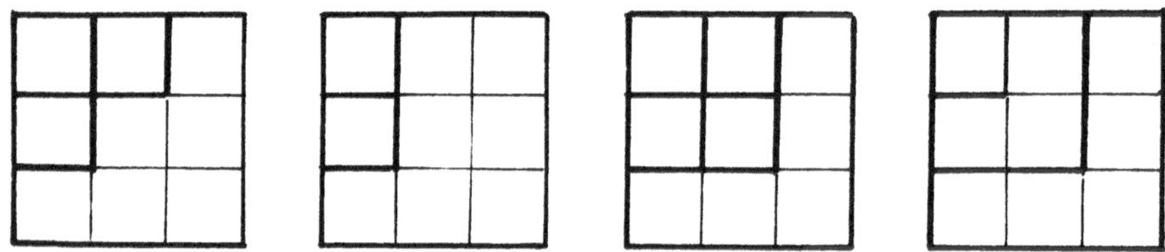

$(\kappa,\lambda) \ll (\bar{\kappa},\bar{\lambda})$ is logically equivalent to $(\kappa \,\&\, \lambda) \Leftrightarrow (\bar{\kappa} \,\&\, \bar{\lambda})$.

6.4.12 Remark and definition. Consider now (pseudo) elementary pairs $\langle \kappa, \delta \rangle$ and implicational quantifiers. Till the end of the present section, $\ll$ will denote the i-helpful quantifier for $(psEP, \leftarrow\!\lhd)$ defined by 6.3.25 (2) (iii), i.e., given by

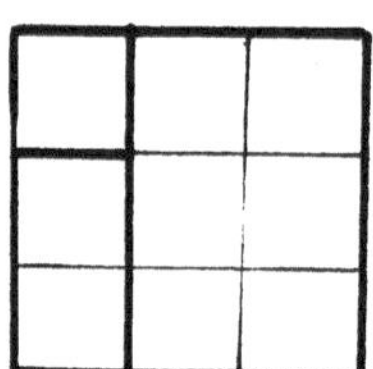

For any disjointed $\langle \kappa, \delta \rangle$, let Regant $\ll_{\underline{M}} (\kappa, \delta)$ be the $\leftarrow$-sup of all $\bar{\kappa} \geq \kappa$ such that $\|(\kappa,\delta) \ll (\bar{\kappa},\delta)\|_{\underline{M}} = 1$. In analogy to the above considerations, we ask whether Regant (κ, δ) has the incompressibility property and whether (under some assumption on $\langle \kappa, \delta \rangle$) κ is a subconjunction of Regant $\ll_{\underline{M}} (\kappa, \delta)$.

6.4.13 Definition. κ is strongly M-incompressible w.r.t. δ if for each κ_o poorer than κ and different from κ there is an $o \in M$ such that $\| \kappa \,\&\, \delta \|_{\underline{M}}[o] = 1$ but $\| \kappa_o \,\&\, \delta \|_{\underline{M}}[o] = 1$.

6.4.14 <u>Remark</u>. If κ is strongly $\underline{M}$-incompressible w.r.t. δ, then κ is obviously $\underline{M}$-incompressible. The intuitive meaning is as follows: The set of all objects having κ is described by κ in an economical way, with particular emphasis to objects having δ. Indeed, let $\underline{M}_\delta = \{ o \in \underline{M}; \|\delta\|_{\underline{M}}[o] = 1\}$. Then making a coefficient in κ poorer we obtain a κ_o such that we find in $\underline{M}_\delta$ objects having κ but not having κ_o.

6.4.15 <u>Lemma</u>. Under the present notation, Regant $\ll_{\underline{M}} (\kappa, \delta)$ is strongly $\underline{M}$-incompressible w.r.t. δ.

<u>Proof</u>. We can see that if $\|(\kappa,\delta) \ll (\bar{\kappa},\delta)\|_{\underline{M}} = 1$ and if $\bar{\kappa}_o \sqsubseteq \bar{\kappa} = \text{Regant} \ll_{\underline{M}} (\kappa,\delta)$ is such that, for each $o \in M$, ($\|\bar{\kappa}\ \&\ \delta\|_{\underline{M}}[o] = 1$ implies $\|\bar{\kappa}_o \&\ \delta\|_{\underline{M}} = 1$), then we have $\|(\bar{\kappa},\delta) \ll (\kappa_o,\delta)\|_{\underline{M}} = 1$ and thus $\|(\kappa,\delta) \ll (\bar{\kappa}_o,\delta)\|_{\underline{M}} = 1$, Hence, $\bar{\kappa}_o \leftarrow \bar{\kappa}$, which implies $\bar{\kappa}_o = \bar{\kappa}$.

6.4.16 <u>Theorem</u>. If κ is strongly $\underline{M}$-incompressible w.r.t. δ, then $\kappa \subseteq \text{Regant} \ll_{\underline{M}} (\kappa,\delta)$.

<u>Proof</u>. Analogous to the proof of 6.4.8.

6.4.17 <u>Remark</u>. Let $\sim$ be an implicational quantifier. Observe that if $\|\kappa \sim \delta\|_{\underline{M}} = 1$ but κ is not strongly $\underline{M}$-incompressible w.r.t. δ, then there is a κ_o poorer than κ, strongly $\underline{M}$-incompressible w.r.t. δ and such that $\|\kappa_o \sim \delta\|_{\underline{M}} = 1$. (Take the subconjunction of Regant $\ll_{\underline{M}} (\kappa,\delta)$ with the function symbols of κ). This suggests that sentences $\kappa \sim \delta$ true in $\underline{M}$ and such that κ is strongly $\underline{M}$-incompressible w.r.t. δ are of particular interest. (See the next chapter .)

6.4.18 <u>Theorem</u>. If there is an $o \in M$ such that $\|\kappa\ \&\ \delta\|_{\underline{M}}[o] = 1$ and if κ is an EC then Regant $\ll_{\underline{M}} (\kappa,\delta)$ is an EC (all coefficients non-empty).

Proof analogous to the proof of 6.3.17 (i) $\Rightarrow$ (ii).

6.4.19 <u>Key words:</u> M-incompressibility, economical helpful sets respecting equivalence of conjunctions, strong M-incompressibility.

PROBLEMS AND SUPPLEMENTS TO CHAPTER VI.

(1) Note the following concerning independence and sufficiency: (a) X is I(X) sufficient; (b) If X is Z-independent then X is Z'-independent for each $Z' \supseteq Z$ (consequently, if X is strongly independent then X is Z-independent for each $Z \supseteq X$, hence it is weakly independent); (c) X is Y-independent iff X is $(I(X) \cap Y)$-independent; (d) if X is Z-independent then X need not be Z'-independent for a proper subset Z' of Z . (Consider Sent = $\{1,2,3\}$, I = $\{\frac{1}{2}, \frac{2}{3}\}$ and X = $\{1,2\}$, then X is weakly independent but not strongly independent.) (e) Z-independence is not hereditary (consider Z = Sent = $\{1,2,3,4,5\}$, I = $\{\frac{1}{4}, \frac{1}{2}, \frac{2,3}{5}\}$, X = $\{1,2,3\}$ and $X \supseteq X' = \{1,2\}$.) (f) Strong independence is hereditary.

(2) We can define (as in Hájek 1973) a <u>linearly ordered syntactic system</u> (ℓ.o. syntactic system) as a triple $\mathcal{L} = \langle Sent, I, S \rangle$, where $\langle Sent, I \rangle$ is a syntactic system and S is a linear ordering of Sent. If $\mathcal{L}$ is a ℓ.o. syntactic system, then a set $X \subseteq$ Sent is <u>increasingly independent</u> if there is no $\varphi \in X$ that would be a conclusion from the preceding elements of X, i.e., for each $\varphi \in X$ we have $\varphi \notin I(X - \{\varphi\}) \cap SEG_S(\varphi)$.

Prove: (a) Any subset of an increasingly independent set is increasingly independent. (b) If Sent is finite then for each $Y \subseteq$ Sent there is a $\subseteq$-minimal $X \subseteq Y$ such that X is increasingly independent and Y-sufficient. (c) Find a condition on S and I implying that, for each $X \subseteq$ Sent , X is increasingly independent iff X is strongly independent. (Note that if Sent is finite then there is an increasingly independent direct solution for each given r-problem.)

(3) (a) <u>Definition</u>. (cf. Hájek 1973) . Let A be a finite set. A **<u>monotone covering</u>** of A is a system H of subsets of A such that (i) H is linearly ordered by the inclusion, (ii) $A \in H$.

Let $\mathcal{S} = \langle$ Sent, $\mathcal{M}$, V, Val $\rangle$ be a semantical system. A <u>hierarchical r-problem</u> in $\mathcal{S}$ is a quadruple $\mathcal{P} = \langle RQ, V_o, I, H \rangle$, where $\langle RQ, V_o, I \rangle$ is an r-problem in $\mathcal{S}$ (denoted by $\mathcal{P}^o$) and H is a monotone covering of Sent. A solution of $\mathcal{P}$, for $\underline{M} \in \mathcal{M}$, is a system $\{X_h, h \in H\}$ such that, for each $h, h' \in H$, $h \subseteq h'$ implies $X_h \subseteq H_{h'}$ and that, for each $h \in H$, $X \cap h$ is a solution of the r-problem $\mathcal{P}^o \upharpoonright h$ in $\mathcal{S} \upharpoonright h$ (obviously, X_{Sent} is then a solution of $\mathcal{P}^o$).

(b) <u>Remark.</u> The definition of a hierarchical problem and of its solution is motivated by two facts: (1) We imagine that the computer will <u>successively construct</u> the sets X_h for increasing h: the program will thus have the form of a loop with parameter h. If it is necessary to break off the computation, and if h is the last processed value of the parameter, then we have a solution of $\mathcal{P}^o \upharpoonright h$.

(2) <u>The interpretation of results</u> is also divided by a hierarchical solution into a set of subtasks, namely the interpretations of the various sets X_h as solutions of r-problems $\mathcal{P}^o \upharpoonright h$.

(c) Let H be a monotone ordering of Sent and let I be an inference rule on Sent. H is <u>I-saturated</u> if the following holds for each $h \in H$: $\frac{e}{\varphi} \in I$ and $\varphi \in h$ implies $e \subseteq h$ for each φ, e.

(d) <u>Theorem.</u> Let $\mathcal{S}$ be a semantic system, let $\mathcal{P} = \langle RQ, V_o, I, H \rangle$ be a hierarchical r-problem in $\mathcal{S}$ and let H be I-saturated. Then for each model $\underline{M}$ there is a locally $\subseteq$-minimal solution $\{X_h; h \in H\}$ of for $\underline{M}$.

(4) Consider observational monadic <u>predicate</u> calculi (two-valued): show that there are exactly four universally definable a-helpful quantifiers for (psCPF, $\subseteq\subseteq$).

(5) There is no quantifier $\ll$ a-helpful w.r.t. (psCPF, $\lhd\lhd$) . (Hint: find CPF's κ_i, λ_i ($i = 1,2,3$) and a model $\underline{M}$ such that (a) $\langle\kappa_1, \lambda_1\rangle \lhd\lhd \langle\kappa_2, \lambda_2\rangle \lhd\lhd \langle\kappa_3, \lambda_3\rangle$ and (b) for each $o \in M$, the $\underline{M}$-value of $\kappa_1, \lambda_1, \kappa_3, \lambda_3, \lambda_2$ is 1 but the $\underline{M}$-value of λ_2 is 0 ; then $\langle M, \|\kappa_1\|_{\underline{M}}, \|\lambda_1\|_{\underline{M}}\rangle = \langle M, \|\kappa_3\|_{\underline{M}}, \|\lambda_3\|_{\underline{M}}\rangle$ so that $\|(\kappa_1, \lambda_1) \ll (\kappa_3, \lambda_3)\|_{\underline{M}} = 1$ hence $\|(\kappa_1, \lambda_1) \ll (\kappa_2, \lambda_2)\|_{\underline{M}} = 1$ but $\langle M, \|\kappa_2\|_{\underline{M}}, \|\lambda_2\|_{\underline{M}}\rangle$ is not a-better than $\langle M, \|\kappa_1\|_{\underline{M}}, \|\lambda_1\|_{\underline{M}}\rangle$:

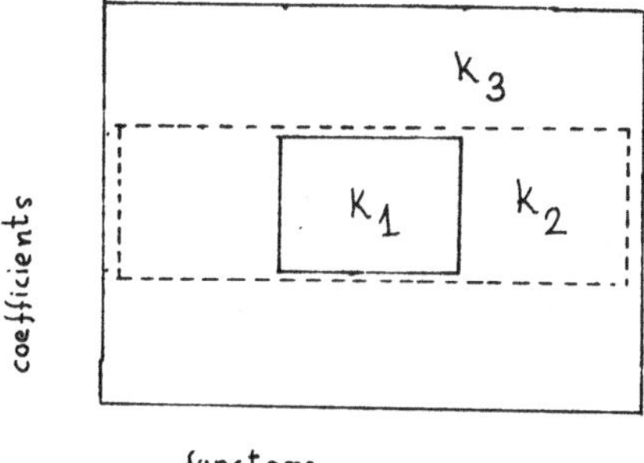

(6) We can define the theoretical notion of incompressibility:

κ is U-incompressible if there is no $\kappa_o \subsetneqq \kappa$ such that $P^{\underline{U}}_{\kappa_o} = P^{\underline{U}}_{\kappa}$.

From theoretical considerations of Chapter IV we obtain the condition:

(*) for each $\langle j_1, \ldots, j_n\rangle$, where $j_i \in V_i$, $P^{\underline{U}}_{(j_1) F_1 \& \ldots \& (j_n) F_n} > 0$.

(Theorem.) If $\underline{U} \models (*)$, then each $\kappa \in EC$ is $\underline{U}$-incompressible.

Chapter VII. A General GUHA-Method with Associational Quantifers

In the present chapter, we use the considerations of Chapter VI for the description and investigation of a particular (rather complex) GUHA-method. The whole chapter can be viewed as an extensive example capable of concrete machine realization (cf. the postscript). Remember the notion of a GUHA-method as a parametrical system $\langle \mathcal{S}(p), \mathcal{P}(p), X(p); p$ parameter$\rangle$ where each $\mathcal{S}(p)$ is a semantic system, $\mathcal{P}(p)$ is an r-problem in $\mathcal{S}(p)$, and $X(p)$ is a function associating with each model $\underline{M}$ of $\mathcal{S}(p)$ a solution of $\mathcal{P}(p)$ in $\underline{M}$. The whole of Section 1 is in fact a single (commented) definition: We successively define the set Par of parameters, and the system $\mathcal{S}(p)$ and the r-problem $\mathcal{P}(p)$ defined by the parameter p. In fact, we do not define a single method since some details remain undecided. First, we neglect some formal questions concerning the particular representation (coding) of things, i.e. Par will not be defined uniquely as a set, and, secondly, we do not discuss questions of the particular bounds for various subparameters since this question is relevant only when one is going to write a program for a particular machine. Hence, the notion we shall define is: $\mathcal{G}$ is a GUHA-method with associational quantifiers. We wish to avoid unnecessary formalism: one can read Section 1 as a list (review) of aspects involved in determining an r-problem with an associational quantifier.

In Section 2, we describe a solution of an r-problem of the form discussed in Section 1 and investigate properties of that solution. For this purpose, we classify r-problems of Section 1 into four classes according to those of their properties expressible without mentioning any structure of sentences (while mentioning the properties of the deduction rule w.r.t. a certain ordering of relevant questions only). In the present context, the reader will always see classes of r-problems of Section 1 and apply our considerations to them; general formulations help to stress relevant features and might perhaps be useful elsewhere.

Section 3 discusses questions of optimized machine realization of the method described. In our opinion, these questions are discussed in enough detail so that the programmer can clearly see his task and, in addition, there are some suggestions as to how to proceed. Moreover, we use our considerations to briefly discuss questions of the complexity of machine computations; we show under what conditions the method is realizable in polynomial time. Some simple strategies (heuristics) for the search of the solution are described in Problem 4.

VII.1 A system of r-problems

We are going to describe successively the set Par of parameters and associate with each parameter p a function calculus $\mathcal{F}(p)$ and an r-problem $\mathcal{P}(p)$. If the reader wishes to simplify the example, he may omit things concerning incompressibility (assuming FORQ to be SIMPLE below) or concerning helpful quantifiers (assuming WHELP to be NO). If the reader makes both restrictions simultaneously, then the example will be rather short (and unnecessarily poor).

7.1.1 Definition (beginning). The set Par of parameters of the GUHA-method with associational quantifiers is supposed to have the following structure: Each parameter p decomposes into three parts, namely (a) the part describing the function calculus in question, (b) the part determining the set of relevant questions, and (c) the part deciding whether and what helpful quantifiers will be taken into consideration. We write $p = \langle CALC, QUEST, HELP \rangle$. (To be continued.)

7.1.2 Remark. Our function calculus will be a cross-qualitative MOFC with an associational quantifier $\sim$, a quantifier $\ll$ of type $\langle 1^4 \rangle$ (helpful for something) and possibly other quantifiers. We must be specific as regards the number and range of our function symbols and as regards the associated functions of our quantifiers. Hence, we continue the definition as follows:

7.1.3 Definition (continued). Let CALC be the calculus-part of a parameter p. Then CALC decomposes into three parts, namely (a) the characteristic CHAR of the calculus in question, (b) the KQUANT part determining the kind of the associational quantifier used, and (c) the PQUANT part reserved for parameters determining uniquely the associated function of the quantifier $\sim$ (in accordance with the declared kind). The caharacteristic determines (aa) the number of function symbols, (ab) for each function symbol F_i its set of regular values $V_i = \{0, 1, \ldots, h_{i-1}\}$, and (ac) information whether we admit models with incomplete information. The possible kinds of associational quantifiers are: IMPL - implicational, SYMNEG - obeying the rules **SYM and NEG** (cf. 3.2.17), and OTHER. We require that the particular associational quantifier defined by PQUANT satisfies the following satisfiability condition: Whenever $\|\varphi \sim \psi\|_{\underline{M}} = 1$, then $\varphi \,\&\, \psi$ is satisfiable in $\underline{M}$, i.e., there is an $o \in M$ such that $\|\varphi \,\&\, \psi\|_{\underline{M}}[o] = 1$ (φ, ψ designated open). (To be continued.)

7.1.4 Remark. (1) We shall not be specific about the form of PQUANT; e.g., if we include $\Rightarrow^{!}_{p,\alpha}$ among the particular quantifiers allowed, then PQUANT could be the triple $< !, p, \alpha >$, where ! indicates that we mean the quantifier of probable implication and p, α are its parameters.

(2) Note that all particular examples of associational quantifiers presented in Chapter IV Section 5 were either implicational (namely, $\Rightarrow^{!}_{p,\alpha}$, $\Rightarrow^{?}_{p,\alpha}$, $\Rightarrow_{p}$) or satisfied SYM and NEG (namely $\sim, \sim_{\alpha}, \sim^{2}_{\alpha}$). All of them satisfied the satisfiability condition.

(3) Our function calculus $\mathcal{F}(p)$ is uniquely determined except for the associated function of $\ll$ (and of the remaining quantifiers, if any). We postpone the definition of that function (those functions) to the time when HELP will be described; then we can easily define $Asf_{\ll}$.

7.1.5 Definition (continued). Let QUEST be the part of a parameter p determining the set of relevant questions. Then QUEST decomposes into three parts: (a) the KRPF part determining the kind of relevant pairs of formulae, (b) the FORQ part determining the form of relevant questions, and (c) the SYNTR part determining syntactic restrictions for the occurrence of literals in relevant questions. The admissible kinds of relevant pairs of formulae are: (aa) CPF - then relevant pairs of formulae are (some) disjointed conjunctive pairs of formulae, and (ab) EPF - then relevant pairs are (some) elementary pairs of formulae. The admissible forms of relevant questions are: (ba) SIMPLE - relevant questions are prenex sentences $\varphi \sim \psi$ where $<\varphi, \psi>$ is a relevant pair of formulae, and (bb) INCOMPR - relevant questions are conjunctions $(\varphi \sim \psi) \,\&\, \ldots$ where $\varphi \sim \psi$ is as above and ... is a sentence expressing a certain incompressibility condition (to be made precise below).

Thus, in all cases the set of relevant questions consists of all sentences $S(\varphi, \psi)$, where $<\varphi, \psi>$ varies over the set of relevant pairs of formulas and $S(-,-)$

is a function such that $S(\varphi,\psi)$ is $\varphi\sim\psi$ either alone or in conjunction with an incompressibility condition.

Each parameter must satisfy the following correctness condition: If the kind of relevant pairs is EPF, then the kind of the associational quantifier is IMPL (i.e., $\sim$ is implicational). (To be continued.)

7.1.6 Remark and Definition. (1) For the choice of CPF's or EPF's and for the requirement that EPF's are to be used only with implicational quantifiers see 6.3.12.

(2) If κ is a (pseudo) EC then the incompressibility of κ is expressible as follows: κ is $\underline{M}$-incompressible iff $\| \bigwedge_{\kappa_0 \subsetneqq \kappa} \neg(\kappa \Leftrightarrow \kappa_0) \|_{\underline{M}} = 1$. Write Incompr$(\kappa)$ for $\bigwedge_{\kappa_0 \subsetneqq \kappa} \neg(\kappa \Leftrightarrow \kappa_0)$. Similarly, let $\| \varphi \Leftrightarrow_1 \psi \|_{\underline{M}} = 1$ iff, for each $o \in M$, ($\|\varphi\|_{\underline{M}}[o] = 1$ iff $\|\psi\|_{\underline{M}}[o] = 1$). Then κ is strongly $\underline{M}$-incompressible w.r.t. δ iff $\|\mathrm{SInc}(\kappa,\delta)\|_{\underline{M}} = 1$, where $\mathrm{SInc}(\kappa,\delta)$ is $\bigwedge_{\kappa_0 \subsetneqq \kappa} (\neg(\kappa_0 \,\&\, \delta) \Leftrightarrow_1 (\kappa \,\&\, \delta))$.

(3) We shall not be specific as regards the syntactic restrictions; but let us assume that KRPF together with SYNTR define uniquely the set of relevant pairs of formulas. (Cf. Problem (2).) SYNTR may postulate that some function symbols may occur only in the antecedents and some only in the succedents, some function symbols can be allowed to have only certain specific arguments, one can impose upper and lower bounds to the number of literals in the antecedents and succedents, etc.

7.1.7 Definition (contd). We make the following economy assumption: If the quantifier satisfies SYM and NEG (i.e., if KQUANT is SYMNEG), then

(a) $\langle\varphi,\psi\rangle \in \mathrm{RPF}(p)$ and $\varphi \neq \psi$ implies $\langle\psi,\varphi\rangle \notin \mathrm{RPF}(p)$;

(b) $\langle\varphi,\psi\rangle \in \mathrm{RPF}(p)$ implies $\langle \underline{\mathrm{neg}}(\varphi), \underline{\mathrm{neg}}(\psi)\rangle \notin \mathrm{RPF}(p)$. (To be continued.)

7.1.8 Remark. (1) First, note that in a predicate calculus each EC is incompressible; hence, in this case (two-valued data), it would **make no sense** to declare the form of relevant questions as INCOMPR.

(2) In the general case, declaring the form of relevant questions as SIMPLE one considers as relevant truths all sentences $\varphi\sim\psi$ true in a given model,

where $<\varphi, \psi> \in$ RPF; declaring INCOMPR one considers as relevant truths - mutatis mutandis - only those true sentences $\varphi \sim \psi$ for which the pair $<\varphi, \psi>$ satisfies a certain incompressibility condition. This means that we restrict the set of relevant truths. What incompressibility condition should be imposed on the relevant pairs ? This depends on the desired sound deduction rules. Remember that for EPF (and implicational quantifiers) we have the despecifying-dereducing rule ; for CPF we have no non-trivial direct rule (without auxiliary formulae), but if we admit helpful quantifiers we have modus ponens for helpful quantifiers. In all these rules we have dealt with prenex formulae $\varphi \sim \psi$; we want to find our incompressibility conditions in such a way that our rules remain sound when $\varphi \sim \psi$ is replaced by $S(\varphi, \psi)$. We observe that our first task is to describe the HELP part of our parameters (determining our helpful quantifiers) ; then the description of the structure of parameters will be completed, and we shall also complete the description of relevant questions and deduction rules (i.e., of our r-problem). We begin with the definition of some particular helpful quantifiers.

7.1.9 Definition (cont). (1) The conservative helpful quantifier is the quantifier $\ll$ universally defined by the set

$$U = \{<u, v, \bar{u}, \bar{v}> ; u = \bar{u} \text{ and } v = \bar{v} , u, v \in \{0, \times , 1\}\}.$$

(2) If the kind of relevant pairs is CPF, then the designated helpful quantifier is universally defined by the economical a-helpful (i-helpful) set determined by the following table:

KQUANT	IMPL	IMPL	SYMNEG or OTHER	SYMNEG or OTHER
FORQ	SIMPLE	INCOMPR	SIMPLE	INCOMPR
	(a)	(b)	(c)	(d)

(3) If the kind of relevant pairs is EPF, then the designated i-helpful quantifier is universally defined by the following economical i-helpful set for (psEPF, $\leftarrow\!\lhd$) :

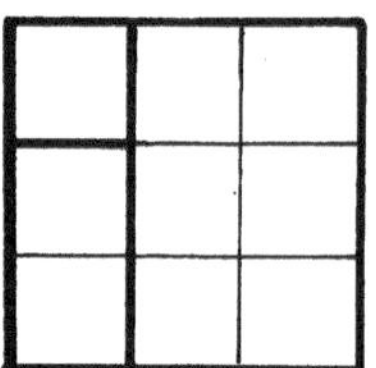

7.1.10 Remark. (1) Hence, if $\ll$ is the conservative helpful quantifier, then (a) $(\varphi,\psi) \ll (\bar{\varphi},\bar{\psi})$ is logically equivalent to $(\varphi \Leftrightarrow \bar{\varphi})$ & $(\psi \Leftrightarrow \bar{\psi})$, (b) $\ll$ is a-helpful w.r.t. (psCPF, $\leftarrow\leftarrow$) and i-helpful w.r.t. (psEPF, $\leftarrow\!\lhd$) . (Cf. 6.4.8 (1).)

(2) The designated quantifier is as strong as possible given some desired properties (first of all, it must be a-helpful or i-helpful respectively). The quantifiers in 7.1.9 (2) are i-helpful and a-helpful for (psCPF, $\leftarrow\leftarrow$) by 6.3.21 and 6.3.22 respectively; 6.4.6 gives reasons for our choice of (a) and (c) among the quantifiers of 6.3.21, 22, while 6.4.10 gives reasons for our choice of (b) and (d). The quantifier in 7.1.9 (3) is i-helpful for (psEPF, $\leftarrow\!\lhd$) by 6.3.26 ; cf. also 6.4.16.

(3) Remember the meaning of the diagrams: e.g., if $\ll$ is defined by 7.1.9 (b), then $\|(\kappa,\lambda) \ll (\bar{\kappa},\bar{\lambda})\|_{\underline{M}} = 1$ (for $\langle\kappa,\lambda\rangle \leftarrow\leftarrow \langle\bar{\kappa},\bar{\lambda}\rangle$) if for each $o \in M$ we have the following: Put

$$\langle \|\kappa\|_{\underline{M}}[o], \|\lambda\|_{\underline{M}}[o], \|\bar{\kappa}\|_{\underline{M}}[o], \|\bar{\lambda}\|_{\underline{M}}[o] \rangle = \langle u, v, \bar{u}, \bar{v} \rangle.$$

Then $[(\langle u,v\rangle = \langle 1,1\rangle$ or $\langle \times, 1\rangle$ or $\langle \times,\times\rangle)$ implies $\langle\bar{u},\bar{v}\rangle = \langle u,v\rangle]$ and $[\langle u,v\rangle = \langle 1,\times\rangle$ implies $\langle\bar{u},\bar{v}\rangle = \langle 1,\times\rangle$ or $\langle\bar{u},\bar{v}\rangle = \langle\times,\times\rangle]$

7.1.11 Definition (contd). The HELP part of a parameter decomposes into two parts: WHELP which indicates whether helpful quantifiers are used or not and can be YES or NO, and KHELP indicating the kind of the helpful quantifier used. If the first part is NO, then the second part is empty; if the first part is YES, then the second part can be either CONSV or DESIGN (conservative or designated helpful quantifiers).

7.1.12 Remark. This completes the definition of the set of parameters and of its structure; in what remains of the present section we shall complete the description of the function calculus and of the problem determined by a parameter, and also define other notions concerning the GUHA-method described.

7.1.13 Definition (contd). We describe the set RQ(p) of relevant questions. It consists of all sentences $S(\varphi,\psi)$ where $\langle\varphi,\psi\rangle$ is a relevant pair of formulae ($\langle\varphi,\psi\rangle \in$ RPF(p)) and S is defined as follows:

(a) If FORQ is SIMPLE, then $S(\varphi,\psi)$ is $\varphi\sim\psi$.

(b) If FORQ is INCOMPR, then $S(\varphi,\psi)$ is as follows:

KRPF	WHELP	KHELP	
CPF	NO	-	$S(\kappa,\lambda)$ is $\kappa\sim\lambda$ & Incompr(κ) & Incompr(λ)
CPF	YES	CONSV	$S(\kappa,\lambda)$ is $\kappa\sim\lambda$ & Incompr(κ) & Incompr(λ)
CPF	YES	DESIGN	$S(\kappa,\lambda)$ is $\kappa\sim\lambda$ & Incompr (con (κ,λ))
EPF	NO	-	$S(\kappa,\delta)$ is $\kappa\sim\delta$ & Incompr (κ)
EPF	YES	CONSV	$S(\kappa,\delta)$ is $\kappa\sim\delta$ & Incompr (κ)
EPF	YES	DESIGN	$S(\kappa,\delta)$ is $\kappa\sim\delta$ & SInc (κ, δ)

7.1.14 Remark. Hence, if we use no helpful quantifier or if we use the conservative helpful quantifier, the sentence expresses both the association and the incompressibility of the conjunctions in question; for the designated helpful quantifier the additional sentence is still stronger. We repeat that the choice is determined by our desire that the rules described below be sound.

7.1.15 Definition (completed).

(1) If relevant pairs are CPF and WHELP is NO, then I(p) is the trivial identity rule:

$$I(p) = \left\{ \frac{S(\kappa,\lambda)}{S(\kappa,\lambda)} \; ; \quad \langle\kappa,\lambda\rangle \text{ a conjunctive pair} \right\}.$$

(2) If relevant pairs are EPF and WHELP is NO, then (KQUANT is necessarily IMPL) I(p) is the despecifiying-dereducing rule:

$$I(p) = \left\{ \frac{S(\kappa,\delta)}{S(\bar{\kappa},\bar{\delta})} \; ; \begin{array}{l} \langle\bar{\kappa},\bar{\delta}\rangle \text{ results from } \langle\kappa,\delta\rangle \\ \text{by successive despecification and dereduction,} \\ \langle\kappa,\delta\rangle, \langle\bar{\kappa},\bar{\delta}\rangle \in \text{EPF} \end{array} \right\}$$

(3) If KRPF is CPF and WHELP is YES, then I(p) is the modified modus ponens:

$$I(p) = \left\{ \frac{S(\kappa,\lambda), (\kappa,\lambda) \ll (\bar{\kappa},\bar{\lambda})}{S(\bar{\kappa},\bar{\lambda})} \; ; \; \begin{array}{l} \langle \kappa,\lambda\rangle \Leftarrow\Leftarrow \langle\bar{\kappa},\bar{\lambda}\rangle \Leftarrow\Leftarrow \langle\bar{\bar{\kappa}},\bar{\bar{\lambda}}\rangle \\ \text{conjuctive pairs} \end{array} \right\}.$$

(4) If KRPF is EPF and WHELP is YES, then I(p) is the set of all pairs

$$\frac{S(\kappa,\delta), (\kappa,\delta) \ll (\bar{\bar{\kappa}},\delta)}{S(\bar{\kappa},\bar{\delta})}$$

where $\kappa \Leftarrow \bar{\bar{\kappa}}$ and $\langle \bar{\kappa},\bar{\delta}\rangle$ is constructed as follows:
(a) One takes a κ_0 such that $\kappa \Leftarrow \kappa_0 \Leftarrow \bar{\bar{\kappa}}$ (improves the antecedent), and (b) one despecifies and dereduces the pair $\langle \kappa_0, \delta\rangle$. Call this I(p) the combined rule.

7.1.16 Remark. Note that we have indeed defined a set of parameters with a certain structure on it and for each p a function calculus $\mathcal{F}(p)$, a set RQ(p) of relevant questions and a rule I(p). To prove that $\mathcal{P}(p) = \langle RQ(p), \{1\}, I(p) \rangle$ is an r-problem it remains to verify the following:

7.1.17 Lemma. For each parameter p, the rule I(p) is sound.

Proof. First, suppose WHELP to be NO (no helpful quantifiers). Then the case of CPF is trivial. If KRPF is EPF and if FORQ is SIMPLE (no incompressibility sentences), we have the usual despecifying-dereducing rule which is sound for each implicational quantifier. (Remember that in the present case KQUANT is IMPL .) If FORQ is INCOMPR, recall that, by 6.4.3., a subconjunction of an incompressible conjunction is incompressible.

Suppose now that we have the conservative helpful quantifier. (KHELP is CONSV.) If FORQ is SIMPLE (no incompressibility), then for CPF we have the modified modus ponens

$$\left\{ \frac{\kappa\sim\lambda, (\kappa,\lambda) \ll (\bar{\bar{\kappa}},\bar{\bar{\lambda}})}{\bar{\kappa}\sim\bar{\lambda}} \; ; \; \langle \kappa,\lambda\rangle \Leftarrow\Leftarrow \langle\bar{\kappa},\bar{\lambda}\rangle \Leftarrow\Leftarrow \langle\bar{\bar{\kappa}},\bar{\bar{\lambda}}\rangle \right\}$$

which is obviously sound; for EPF the soundness also is obvious. If FORQ is INCOMPR, and if KRPF is CPF we have to verify the following: Let $\|\kappa\sim\lambda\|_{\underline{M}} = 1$

and $\|(\kappa,\lambda) \ll (\bar{\bar{\kappa}},\bar{\bar{\lambda}})\|_{\underline{M}} = 1$, let κ,λ be $\underline{M}$-incompressible and let $\langle\kappa,\lambda\rangle \lessdot\lessdot \langle\bar{\kappa},\bar{\lambda}\rangle \lessdot\lessdot \langle\bar{\bar{\kappa}},\bar{\bar{\lambda}}\rangle$. Then $\bar{\kappa},\bar{\lambda}$ are $\underline{M}$-incompressible. But here $\ll$ is the conservative helpful quantifier and hence it follows, by 6.4.8 (1), that $\kappa \subseteq \bar{\kappa} \subseteq \bar{\bar{\kappa}}$ and $\lambda \subseteq \bar{\lambda} \subseteq \bar{\bar{\lambda}}$. Now, $\bar{\bar{\kappa}},\bar{\bar{\lambda}}$ are $\underline{M}$-incompressible by 6.4.5, thus $\bar{\kappa},\bar{\lambda}$ are also $\underline{M}$-incompressible. For EPF one proceeds similarly.

Finally, assume KHELP to be DESIGN. Everything is obvious if FORQ is SIMPLE (cf. 7.1.10 (2)); hence, assume KHELP to be INCOMPR. First let KRPF be CPF. Then relevant questions have the form $\kappa \sim \lambda$ & Incompr$(\underline{\text{con}}(\kappa,\lambda))$; $\ll$ is now the quantifier 7.1.9 (2)(b) or (d). To verify the soundness of the modified modus ponens, assume $\|\kappa \sim \lambda\|_{\underline{M}} = 1$, $\underline{\text{con}}(\kappa,\lambda)$ $\underline{M}$-incompressible, $\|(\kappa,\lambda) \ll (\bar{\bar{\kappa}},\bar{\bar{\lambda}})\|_{\underline{M}} = 1$, $\langle\kappa,\lambda\rangle \lessdot\lessdot \langle\bar{\kappa},\bar{\lambda}\rangle \lessdot\lessdot \langle\bar{\bar{\kappa}},\bar{\bar{\lambda}}\rangle$, We know that $\|\bar{\kappa} \sim \bar{\lambda}\|_{\underline{M}} = 1$, we have to prove that $\underline{\text{con}}(\bar{\kappa},\bar{\lambda})$ is $\underline{M}$-incompressible. We can assume $\langle\bar{\bar{\kappa}},\bar{\bar{\lambda}}\rangle = \text{Reg}\ll_{\underline{M}}(\kappa,\lambda)$ without loss of generality. Now, $\underline{\text{con}}(\bar{\bar{\kappa}},\bar{\bar{\lambda}})$ is $\underline{M}$-incompressible by 6.4.5: by 6.4.8(2), $\underline{\text{con}}(\bar{\kappa},\bar{\lambda})$ is a subconjunction of $\underline{\text{con}}(\bar{\bar{\kappa}},\bar{\bar{\lambda}})$ and, hence, $\underline{\text{con}}(\bar{\kappa},\bar{\lambda})$ is $\underline{M}$-incompressible. Secondly, let KRPF be EPF. Suppose $\|(\kappa \sim \delta)\|_{\underline{M}} = 1$, $\|(\kappa,\delta) \ll (\bar{\bar{\kappa}},\bar{\delta})\|_{\underline{M}} = 1$, $\kappa \lessdot \kappa_0 \lessdot \bar{\bar{\kappa}}$, and let $\langle\bar{\kappa},\bar{\delta}\rangle$ result from $\langle\bar{\bar{\kappa}},\bar{\delta}\rangle$ by despecification and dereduction (i.e., $\langle\bar{\kappa},\bar{\delta}\rangle$ is more acute than $\langle\kappa_0,\bar{\delta}\rangle$). We have to prove that $\bar{\kappa}$ is strongly $\underline{M}$-incompressible w.r.t. $\bar{\delta}$. It follows easily as in the previous paragraph that κ_0 is strongly incompressible w.r.t. δ (use 6.4.15,16). To prove that $\bar{\kappa}$ is strongly $\underline{M}$-incompressible w.r.t. $\bar{\delta}$ it suffices to observe the following two easy facts: (a) If $\bar{\kappa} \subseteq \kappa_0$, then SInc $(\kappa_0,\bar{\delta})$ logically implies SInc $(\bar{\kappa},\bar{\delta})$ (b) If $\delta \lhd \bar{\delta}$, then SInc $(\bar{\kappa},\delta)$ logically implies SInc $(\bar{\kappa},\bar{\delta})$. This completes the proof.

7.1.18 Theorem. For each $p \in$ Par, $\mathcal{F}(p)$ is a cross-qualitative MOFC and $\mathcal{P}(p) = \langle RQ(p), \{1\}, I(p)\rangle$ is an r-problem in the semantic system given by $\mathcal{F}(p)$.

Proof. Immediate from the preceding.

7.1.19 <u>Discussion</u>. First, let us summarize the things determined by a parameter (and determining the calculus and r-problem):

CALC	CHAR	characteristic		
	KQUANT	kind of assoc. quantifier (IMPL, SYMNEG, OTHER)		
	PQUANT	parameters of the assoc. quantifier		
QUEST	KRPF	kind of relev. pairs of formulae (CPF, EPF)		
	FORQ	form of relev. questions (SIMPLE, INCOMPR)		
	SYNTR	syntactic restrictions to literals		
HELP	WHELP	whether	helpful quant.	YES, NO
	KHELP	what		CONSV, DESIGN

Presumably, the choice of particular values of the above parameters <u>except</u> FORQ and HELP will be satisfactorily determined by the extramathematical problem to be solved, but the hypothetical user may be ill at ease when answering the following questions: (1) Whether to make the restriction to incompressible things (choose FORQ to be INCOMPR) and (2) whether and what helpful quantifier should be used (how to choose HELP). Some remarks are in order. In fact, we shall repeat things already stated elsewhere above; we will be able to give some more information in the next section (in dependence on the solutions).

When one changes FORQ from SIMPLE to INCOMPR (keeping other things fixed), one diminishes the set of relevant truths: A true sentence $\varphi \sim \psi$ is relevant only if the pair $\langle\varphi, \psi\rangle$ satisfies the respective incompressibility condition. Hence, if one is afraid that the set of relevant truths will be too large this is a reasonable restriction. (The statistical significance of this restriction is considered below.)

Helpful quantifiers are intended to strengthen our capability of "seeing at a glance" and, in particular, to provide non-trivial deduction rules for KQUANT not being IMPL. This often helps to diminish the solution (see the next section) by replacing the whole set $\{\bar{\kappa} \sim \bar{\lambda}\ ;\ \langle\kappa, \lambda\rangle \twoheadleftarrow \langle\bar{\kappa}, \bar{\lambda}\rangle \twoheadleftarrow \langle\bar{\bar{\kappa}}, \bar{\bar{\lambda}}\rangle\}$ (where $\kappa \sim \lambda$ and $(\kappa, \lambda) \ll (\bar{\bar{\kappa}}, \bar{\bar{\lambda}})$ are true) by $\kappa \sim \lambda$ and $(\kappa, \lambda) \ll (\bar{\bar{\kappa}}, \bar{\bar{\lambda}})$. The stronger the quantifier $\ll$, the larger is the hope for a better (smaller) solution. Even the conservative quantifier can be of considerable help: **the designated** quantifier is the strongest possible (for a given case). On the other hand, the conservative quantifier is called conservative since if $\|(\kappa,\lambda) \ll (\bar{\kappa},\bar{\lambda})\|_{\underline{M}} = 1$ (where $\ll$ is conservative) then $\langle M, \|\kappa\|_{\underline{M}}, \|\lambda\|_{\underline{M}}\rangle$ equals $\langle M, \|\bar{\kappa}\|_{\underline{M}}, \|\bar{\lambda}\|_{\underline{M}}\rangle$, and hence every statistic takes the same value for κ, λ as for $\bar{\kappa}, \bar{\lambda}$ in $\underline{M}$, which might be useful. For the designated quantifier $\ll$, $\langle M, \|\bar{\kappa}\|_{\underline{M}}, \|\bar{\lambda}\|_{\underline{M}}\rangle$ is a-better (i-better) than $\langle M, \|\kappa\|_{\underline{M}}, \|\lambda\|_{\underline{M}}\rangle$ so that for reasonable statistics (defining associational quantifiers) the value for $\bar{\kappa}, \bar{\lambda}$ is better than for κ, λ .

Note that a practical user need not know the particular definitions of designated helpful quantifiers for all the cases; it is sufficient if he knows the notion of an a-(i-) helpful quantifier and knows the fact that the corresponding rule is sound. Neither is he obliged to know the optimality properties as expressed by 6.3.21, 22, 6.4.6 , 6.4.10, etc.

When one uses helpful quantifiers and also restricts oneself to incompressible pairs, then for CPF the designated helpful quantifier is slightly weaker than the corresponding designated quantifier for FORQ being SIMPLE. It is a delicate question which is then better (and in what sense), whether to consider all pairs or only the incompressible ones.

The restriction to incompressible pairs has one more advantage, namely it makes it possible to order relevant questions such that both syntactical simplicity and logical strength is respected. We shall go into details in the next section.

7.1.20 Key words: The set of parameters of a GUHA method with associational quantifiers ; structure of parameters: CALC - description of the function calculus, QUEST - determination of the set of relevant questions, HELP - deciding the usage of helpful quantifiers ;

CALC: characteristic of calculus, kind of associational quantifiers, parameter of associational quantifiers; QUEST: kind of relevant pairs of formulae, form of relevant questions, syntactic restrictions; HELP: no, conservative or designated helpful quantifiers.

VII.2 Solutions

Now our task is to describe, for each parameter $p \in Par$ and each model $\underline{M}$ of the corresponding characteristic , a solution $X(p, \underline{M})$ of the problem $\mathcal{P}(p)$. For this purpose it is reasonable to classify problems into four groups depending on their deduction rules (cf. 7.1.15). (If the reader has disregarded helpful quantifiers and/or incompressibility in Section 1 he also must - and can - disregard them in the present section.)

We shall isolate some general properties of problems in connection with certain orderings on sets of relevant questions. This makes **it possible to have** a uniform definition of the solution as the set of all $\mathcal{P}(p)$-prime sentences of $\underline{M}$ and, in addition, if the solution is indirect, of some auxiliary sentences. The following definition will be useful:

7.2.1 Definition. Let $\mathcal{P} = \langle RQ, V_o, I \rangle$ be an r-problem and let $\emptyset \neq RQ_o \subseteq RQ$. The restriction of $\mathcal{P}$ to RQ_o is the problem $\mathcal{P} \upharpoonright RQ_o = \langle RQ_o, V_o, I \rangle$.

7.2.2 Definition. Let $\mathcal{P} = \langle RQ, V_o, I \rangle$ be an r-problem. $\mathcal{P}$ is deductionless (or: of the first kind) if I consists only of (some) pairs where $\frac{\varphi}{\varphi} \in$ Sent.

7.2.3 Remark. For the problems of Section 1, if relevant pairs are CPF and helpful quantifiers are not used (WHELP is NO), $\mathcal{P}(p)$ is deductionless - cf. 7.1.15.

7.2.4 Lemma. Let $\mathcal{P}$ be deductionless. Then

(a) for each M there is a uniquely determined solution consisting of all relevant truths $(X = RQ \cap Tr_{V_o}(\underline{M}))$;

(b) for each non-empty $RQ_o \subseteq RQ$, $\mathcal{P} \upharpoonright RQ_o$ is deductionless; if X is the solution of $\mathcal{P}$ in $\underline{M}$, then $X \cap RQ_o$ is the solution of $\mathcal{P} \upharpoonright RQ_o$ in $\underline{M}$.

Proof. Obvious, for a generalization see Problem (5).

7.2.5 Definition (G - i.e., concerning the system of Sect. 1 - part 1). If KRPF is CPF and WHELP is NO, then for each $\underline{M}$ we put $X(p,\underline{M}) = RQ(p) \cap Tr_{\{1\}}(\underline{M})$. For the sake of uniformity, in this case call each relevant question Φ true in $\underline{M}$ a $\mathcal{P}(p)$-prime sentence of $\underline{M}$ (or: sentence $\mathcal{P}(p)$-prime in $\underline{M}$); $X(p,\underline{M})$ consists of all $\mathcal{P}(p)$-prime sentences of $\underline{M}$. Observe that $X(p,\underline{M})$ is a solution .

7.2.6 Definition. $\mathcal{P} = \langle RQ, V_o, I \rangle$ is a simple problem (or: problem of the second kind) if there is an ordering $\leq$ on RQ such that I consists exactly of all pairs $\frac{\varphi}{\psi}$ such that $\varphi, \psi \in RQ$ and $\varphi \leq \psi$.

7.2.7 Lemma. Let $\mathcal{P}$ be a simple problem.

(a) For an arbitrary $\underline{M}$ let X be the set of all $\leq$-minimal elements of $RQ \cap Tr_{V_o}(\underline{M})$. Then X is the $\subseteq$-least solution; i.e., X is a solution and is a subset of each solution.

(b) Let RQ_o be a non-empty subset of RQ and let $\underline{M}$ be a model. Let X_o be the set of all $\leq$-minimal elements of $RQ_o \cap Tr_{V_o}(\underline{M})$. Then X_o is the $\subseteq$-least direct solution of $\mathcal{P} \upharpoonright RQ_o$. If RQ_o is a lower $\leq$-segment of RQ (i.e., if $\Phi \leq \Psi \in RQ_o$ implies $\Phi \in RQ_o$), then $X_o = X \cap RQ_o$.

Proof. Obvious.

7.2.8 Remark (G). Let KRPF be EPF and WHELP be NO. Let $\hat{\mathcal{P}}(p)$ be the problem differing from $\mathcal{P}(p)$ only in the fact that relevant pairs of formulae are all EPF's (without any restrictions). Then $\mathcal{P}(p) = \hat{\mathcal{P}}(p) \upharpoonright RQ(p)$ and $\hat{\mathcal{P}}(p)$ is a simple problem; the corresponding ordering is $\propto$ (more acute than, cf. 6.3.27; more precisely, one considers the ordering of relevant questions induced by the ordering of relevant pairs of formulae).

7.2.9 Definition (G - part 2). If KRPF is EPF and WHELP is NO, then call a $\Phi \in$ RQ(p) a $\mathcal{P}$(p)-prime sentence of $\underline{M}$ if $\|\Phi\|_{\underline{M}} = 1$ and there is no Ψ distinct from Φ such that $\Psi \propto \Phi$ and $\|\Psi\|_{\underline{M}} = 1$. Let X(p,$\underline{M}$) be the set of all $\mathcal{P}$(p)-prime **sentences of** $\underline{M}$.

7.2.10 Definition. Let $\mathcal{P}$ be a problem, let $\leq$ be an ordering of RQ, and suppose that I consists of some pairs of the form $\frac{\varphi,\ aux}{\psi}$ where $\varphi, \psi \in$ RQ, $\varphi \leq \psi$ and aux $\notin$ RQ. Call ψ M-obtainable from φ if there is an aux such that $\frac{\varphi,\ aux}{\psi} \in I$ and aux $\in Tr_{V_0}(\underline{M})$. Call $\mathcal{P}$ a tuft problem w.r.t. $\leq$ (or: a problem of the third kind) if, for each $\underline{M}$, RQ $\cap\ Tr_{V_0}(\underline{M})$ is a union of disjoint (upper) tufts $Y_1, \ldots, Y_k$ satisfying the following property: For arbitrary $\varphi \leq \psi$, φ, ψ belong to the same tuft iff $\|\varphi\|_{\underline{M}} = 1$ and ψ is $\underline{M}$-obtainable from φ .

7.2.11 Lemma (G). Let KRPF be CPF and let WHELP be YES. Let $\hat{\mathcal{P}}$(p) be the problem differing from $\mathcal{P}$(p) only in the fact that relevant pairs of **formulae** are all CPF's (without any restrictions). Then $\mathcal{P}(p) = \hat{\mathcal{P}}(p) \upharpoonright$ RQ(p) and $\hat{\mathcal{P}}$(p) is a tuft problem w.r.t. $\lessdot\lessdot$ (more precisely, w.r.t. the ordering induced by the ordering $\lessdot\lessdot$ of relevant pairs of formulae). In fact, if Φ is $S(\kappa_1, \lambda_1)$ and if Ψ is $S(\kappa_2, \lambda_2)$, then Φ, Ψ are in the same tuft iff ($\|\Phi\|_{\underline{M}} = \|\Psi\|_{\underline{M}} = 1$ and $Reg\ll_{\underline{M}}(\kappa_1, \lambda_1) = Reg\ll_{\underline{M}}(\kappa_2, \lambda_2)$.

Proof. Obviously, $S(\kappa_2, \lambda_2)$ is $\underline{M}$-obtainable from $S(\kappa_1, \lambda_1)$ iff $\langle\kappa_1, \lambda_1\rangle \lessdot\lessdot \langle\kappa_2, \lambda_2\rangle$ and $\|(\kappa_1, \lambda_1) \ll (\kappa_2, \lambda_2)\|_{\underline{M}} = 1$. Since $\ll$ is a closure quantifier it follows by 6.2.12 that the set psCPF decomposes **into** pairwise disjoint tufts $Z_1, \ldots, Z_l$ such that $\langle\kappa_1, \lambda_1\rangle$, $\langle\kappa_2, \lambda_2\rangle$ are in the same tuft iff $Reg\ll_{\underline{M}}(\kappa_1, \lambda_1) = Reg\ll_{\underline{M}}(\kappa_2, \lambda_2)$. For each such tuft Z_i, either there is no $\langle\kappa, \lambda\rangle \in Z_i$ such that $\|S(\kappa, \lambda)\|_{\underline{M}} = 1$ or, otherwise, the collection $\{\langle\kappa, \lambda\rangle \in Z_{ij}; \|S(\kappa, \lambda)\|_{\underline{M}} = 1\}$ forms a subtuft Z_i° of Z_i with the same top point (since I(p) is sound). Put $Y_i = \{S(\kappa, \lambda);\ \langle\kappa, \lambda\rangle \in Z_i^{\circ}\}$. Note that for a CPF $\langle\kappa, \lambda\rangle$, $\|S(\kappa, \lambda)\|_{\underline{M}} = 1$ implies that $Reg\ll_{\underline{M}}(\kappa, \lambda)$ is a CPF, not only a psCPF, since then we have $\|\kappa \sim \lambda\|_{\underline{M}} = 1$. Hence, $\kappa \& \lambda$ is satisfiable (by the satisfiability requirement 7.1.3); then $\mathrm{Reg} \ll_{\underline{M}}(\kappa,\lambda)$ is a CPF by 6.3.17. Thus the sets Z_i° are tufts in $(\mathrm{CPF}, \ll)$.

7.2.12 Discussion. Let $\mathcal{P}$ be a tuft problem w.r.t. $\leq$; suppose that for each $\varphi \in \mathrm{RQ}$ and $\underline{M}$ we have a sentence $\mathrm{Reg}_{\underline{M}}(\varphi)$ such that

(*) ψ is $\underline{M}$-obtainable from φ iff $\dfrac{\varphi, \mathrm{Reg}_{\underline{M}}(\varphi)}{\psi} \in$ I. This is satisfied by the problem $\mathcal{P}(p)$ of **7.2.11**; $\mathrm{Reg}_{\underline{M}}(S(\kappa,\lambda))$ is $(\kappa,\lambda) \ll (\bar{\kappa},\bar{\lambda})$ where $\langle\bar{\kappa},\bar{\lambda}\rangle = \mathrm{Reg}\ll_{\underline{M}}(\kappa,\lambda)$.

(a) Let $\underline{M}$ be a model; let $\mathrm{RQ} \cap \mathrm{Tr}_{V_o}(\underline{M}) = Y_1 \ldots Y_k$ where $Y_1,\ldots,Y_k$ are tufts as described in 7.2.10. Call $\varphi \in \mathrm{RQ}$ a $\mathcal{P}$-prime sentence of $\underline{M}$ if φ is a minimal element of a tuft Y_i, i.e., φ is true in $\underline{M}$ and not $\underline{M}$-obtainable from any true $\psi < \varphi$. Let X be the set containing, for each $\mathcal{P}$-prime sentence φ of $\underline{M}$, both φ and $\mathrm{Reg}_{\underline{M}}(\varphi)$. Then X is obviously a solution of $\mathcal{P}$ in $\underline{M}$ (since if $\varphi \in Y_i$ and if ψ is the top point of Y_i, then

$$\frac{\varphi,\ \mathrm{Reg}_{\underline{M}}(\varphi)}{\psi} \in \mathrm{I}).$$

(b) Let $\emptyset \neq \mathrm{RQ}_o \subseteq \mathrm{RQ}$ and put $\mathcal{P}_o = \mathcal{P} \restriction \mathrm{RQ}_o$. Call a $\varphi \in \mathrm{RQ}_o$ a $\mathcal{P}$-prime sentence of $\underline{M}$ if φ is $\underline{M}$-true and is not $\underline{M}$-obtainable from any $\underline{M}$-true $\psi < \varphi$, $\varphi \in \mathrm{RQ}_o$. Let X_o be the set, containing for each $\mathcal{P}_o$-prime sentence φ of $\underline{M}$, **both** φ and $\mathrm{Reg}_{\underline{M}}(\varphi)$. Then X_o is a solution of $\mathcal{P}_o$ in $\underline{M}$.

(Note that $\mathrm{Reg}_{\underline{M}}(\varphi)$ is determined by I and not by RQ_o.) In general, $\mathcal{P} \restriction \mathrm{RQ}_o$ is not a tuft problem since, e.g., the supremum of sentences in RQ_o obtainable from a $\varphi \in \mathrm{RQ}_o$ need not belong to RQ_o.

(c) How good is the solution just described? Would it not be better to take the direct solution $Z = \mathrm{RQ} \cap \mathrm{Tr}_{V_o}(\underline{M})$? Unfortunately, we cannot assert that card(X) is always $\leq$ card(Z) (e.g., if $\mathrm{RQ} \cap \mathrm{Tr}_{V_o}(\underline{M})$ has exactly one element, then X has two: φ and $\mathrm{Reg}_{\underline{M}}(\varphi)$). But we have the following lemma giving satisfactory reasons for our preference of the indirect solution:

(Lemma.) For each tuft problem $\mathcal{P}$ satisfying $(*)$, if X is the indirect solution described in (a) above and if $Z = RQ \cap Tr_{V_o}(\underline{M})$, then $card(X) \leq 2\, card(Z)$. On the other hand, for each natural number m there is a tuft problem satisfying $(*)$ such that $card(Z) > m.card(X)$.

Proof. Let $RQ \cap Tr_{V_o}(\underline{M}) = Y_1 \ldots Y_k$ as above; consider Y_i. Let Y_i have p minimal elements; then Y_1 produces $\leq 2p$ elements of X and $\geq p$ elements of Z. This proves the first part. As concerns the second part, let RQ_m be the tuft of all the non-empty subsets of $\{0,1,\ldots,m-1\}$ ordered by inclusion; if $\underline{M}$ is such that $RQ_m \subseteq Tr_{V_o}(\underline{M})$ and if I_m is such that ψ is $\underline{M}$-obtainable from φ iff $\varphi \subseteq \psi$, then $card(X) = 2m$ and $card(Z) = 2^m - 1$; the ratio $(2^m - 1):(2 \cdot m)$ converges to infinity with m. (It is easy to find a tuft problem with CPF and an associational quantifier simulating the described situation, see Problem(3).)

7.2.13 Definition (G - part 3). If KRPF is CPF and WHELP is YES, then call a sentence $S(\kappa,\lambda) \in RQ(p)$ a $\mathcal{P}(p)$-prime sentence of $\underline{M}$ if $\|S(\kappa,\lambda)\|_{\underline{M}} = 1$ and there is no $\langle\kappa_o,\lambda_o\rangle \Leftarrow\Leftarrow \langle\kappa,\lambda\rangle$, $\langle\kappa_o,\lambda_o\rangle$ different from $\langle\kappa,\lambda\rangle$ and such that $\|S(\kappa_o,\lambda_o)\|_{\underline{M}} = 1$ and $\|(\kappa_o,\lambda_o) \ll (\kappa,\lambda)\|_{\underline{M}} = 1$. We define $X(p,\underline{M})$ to be the set containing, for each $\mathcal{P}(p)$-prime sentence $S(\kappa,\lambda)$, both $S(\kappa,\lambda)$ and $(\kappa,\lambda) \ll (\bar{\kappa},\bar{\lambda})$ where $\langle\bar{\kappa},\bar{\lambda}\rangle = Reg\ll_{\underline{M}}(\kappa,\lambda)$. The sentence $(\kappa,\lambda) \ll (\bar{\kappa},\bar{\lambda})$ is omitted if there is no $S(\kappa',\lambda') \in RQ(p)$ distinct from $S(\kappa,\lambda)$ and such that $\langle\kappa,\lambda\rangle \Leftarrow\Leftarrow \langle\kappa',\lambda'\rangle \Leftarrow\Leftarrow \langle\bar{\kappa},\bar{\lambda}\rangle$. (This is the case e.g. if $\langle\bar{\kappa},\bar{\lambda}\rangle = \langle\kappa,\lambda\rangle$.)

7.2.14 Lemma. In the situation of 7.2.13, if FORQ is INCOMPR, then $S(\kappa,\lambda)$ is a prime sentence of $\underline{M}$ iff $\|S(\kappa,\lambda)\|_{\underline{M}} = 1$ and if there is no $\langle\kappa_o,\lambda_o\rangle \subseteq\subseteq \langle\kappa,\lambda\rangle$ different from $\langle\kappa,\lambda\rangle$ and such that $\|S(\kappa_o,\lambda_o)\|_{\underline{M}} = 1$ and $\|(\kappa_o,\lambda_o) \ll (\kappa,\lambda)\|_{\underline{M}} = 1$.

Proof. $\Rightarrow$ is obvious. We prove $\Leftarrow$. Let $\langle\kappa_o,\lambda_o\rangle \Leftarrow\Leftarrow \langle\kappa,\lambda\rangle$, let $S(\kappa,\lambda)$ and $(\kappa_o,\lambda_o) \ll (\kappa,\lambda)$ be $\underline{M}$-true. We prove $\langle\kappa_o,\lambda_o\rangle \subseteq\subseteq \langle\kappa,\lambda\rangle$. Suppose $\langle\kappa,\lambda\rangle = Reg\ll_{\underline{M}}(\kappa_o,\lambda_o)$ without loss of generality. For the conservative helpful quantifier we obtain from $\|S(\kappa_o,\lambda_o)\|_{\underline{M}} = 1$ the $\underline{M}$-incompressibility of κ and λ and, hence, 6.4.8 (1) yields $\kappa_o \subseteq \kappa$, $\lambda_o \subseteq \lambda$.

For the designated helpful quantifier, we have the M-incompressibility of con(κ,λ) and 6.4.8 (2) gives con$(\kappa_\circ,\lambda_\circ) \subseteq$ con(κ,λ) ; but $<\kappa_\circ,\lambda_\circ>$ is disjointed and by 6.4.7 if a function symbol F occurs both in κ and in λ, then it has the same coefficient in both formulae. Hence $\kappa_\circ \subseteq \kappa$ and $\lambda_\circ \subseteq \lambda$.

7.2.15 Remark. The preceding lemma will be useful when we discuss the order in which the solution is to be generated. The lemma yields an additional argument for the restriction to incompressible things (FORQ taken to be INCOMPR); namely, the solution can be obtained in a more natural ordering. See below.

7.2.16 Definition. A problem $\mathcal{P} = <\text{RQ}, V_0, I>$ is combined (or: of the fourth kind) if there are rules I_1, I_2 and orderings $\leq_1$, $\leq_2$ such that $\mathcal{P}_1 = <\text{RQ}, V_0, I_1>$ is a simple problem w.r.t. $\leq_1$, $\mathcal{P}_2 = <\text{RQ}, V_0, I_2>$ is a tuft problem w.r.t. $\leq_2$ satisfying (*) of 7.2.12, and, moreover, the following holds: $\frac{\varphi,\ \text{aux}}{\psi} \in I$ iff there is a $\bar{\varphi}$ such that $\frac{\varphi,\ \text{aux}}{\bar{\varphi}} \in I_2$ and $\frac{\bar{\varphi}}{\psi} \in I_1$. Finally, we assume the following: For each M, if $\|\varphi\|_M = 1$, $\frac{\varphi}{\psi_\circ} \in I_1$ and ψ is M-obtainable from $\psi_\circ$ using I_2, then there is a $\bar{\varphi}$ M-obtainable from φ such that $\frac{\bar{\varphi}}{\psi} \in I_1$.

7.2.17 Remark and Definition. Let KRPF be EPF and let WHELP be YES. Let $\hat{\mathcal{P}}(p)$ be the problem differing from $\mathcal{P}(p)$ only in the fact that relevant pairs of formulae are all disjointed EPF's (without further restrictions). Then $\mathcal{P}(p) = \hat{\mathcal{P}}(p) \restriction \text{RQ}(p)$ and $\hat{\mathcal{P}}(p)$ is a combined problem in which I_1 is the despecifying-dereducing rule, $\leq_1$ is $\propto$ (more - acute-than), I_2 is the modified modus ponens of the following form:

$$I_2 = \left\{ \frac{\kappa \sim \delta,\ (\kappa,\delta) \ll (\bar{\bar{\kappa}},\delta)}{\bar{\kappa} \sim \delta}\ ;\ \kappa \lessdot \bar{\kappa} \lessdot \bar{\bar{\kappa}} \right\}$$

and $\leq_2$ is $\lessdot\Box$ ($<\kappa_1,\delta_1> \lessdot\Box <\kappa_2,\delta_2>$ iff $\kappa_1 \lessdot \kappa_2$ and $\delta_1 = \delta_2$.

Note that we admit only disjointed EPF's . **Evidently, for each disjointed** EPF $<\kappa,\delta>$ and each M such that $\|S(\kappa,\delta)\|_M = 1$ there is a $\leq_2$-largest disjointed EPF $<\bar{\kappa},\bar{\delta}>$ such that $\|(\kappa,\delta)<<(\bar{\kappa},\bar{\delta})\|_{M} = 1$. (Take Regant $<<_{\underline{M}}(\kappa,\delta)$ and omit from it all literals with function symbols occurring in δ.) Denote this EC by Regant $\overset{D}{<<}_{\underline{M}}(\kappa,\delta)$, the disjointed regularization of the antecedent of $<\kappa,\delta>$. By 7.1.15, I(p) is obtained from I_1 and I_2 exactly as one requires in 7.2.16. As regards the last condition of the definition, cf. 6.3.33.

7.2.18 Discussion. Let $\mathcal{P}$ be a combined problem (notation as in 7.2.16). For each φ, let $\mathrm{Reg}_{\underline{M}}(\varphi)$ be the formula guaranteed by (*) of 7.2.12 w.r.t. I_2. Call $\varphi \in RQ$ $\mathcal{P}$-prime in $\underline{M}$ if φ is both $\mathcal{P}_1$-prime and $\mathcal{P}_2$-prime in $\underline{M}$. Let X be the set containing, for each $\mathcal{P}$-prime sentence φ both φ and $\mathrm{Reg}_{\underline{M}}(\varphi)$. Then X is a solution of $\mathcal{P}$ in $\underline{M}$. (Cf. 6.3.33 again; use the last condition of 7.2.16.) Similarly, if $\emptyset \neq RQ_o \subseteq RQ$ and $\mathcal{P}_o = \mathcal{P} \upharpoonright RQ_o$, then put $\mathcal{P}_1^o = \mathcal{P}_1 \upharpoonright RQ_o$ and $\mathcal{P}_2^o = \mathcal{P}_2 \upharpoonright RQ_o$. Then call φ $\mathcal{P}_o$-prime if φ is both $\mathcal{P}_1^o$-prime and $\mathcal{P}_2^o$-prime (this makes sense by the above). Let X_o be the set containing, for each $\mathcal{P}_o$-prime φ, both φ and $\mathrm{Reg}_{\underline{M}}(\varphi)$ (computed w.r.t $\mathcal{P}_2$!). Then X_o is a solution of $\mathcal{P}_o$. One could make optimality remarks similar to 7.2.12 (c).

7.2.19 Definition (G - part 4). If KRPF is EPF and if WHELP is YES, then call a sentence $S(\kappa,\delta) \in RQ(p)$ a $\mathcal{P}(p)$-prime sentence of $\underline{M}$ if (i) $\|S(\kappa,\delta)\|_{\underline{M}} = 1$, (ii) there is no $<\kappa_o,\delta_o> \propto <\kappa,\delta>$ distinct from $<\kappa,\delta>$ such that $\|S(\kappa_o,\delta_o)\|_{\underline{M}} = 1$ and (iii) there is no $\kappa_o \prec \kappa$, $\kappa_o \neq \kappa$ such that $\|S(\kappa_o,\delta)\|_{\underline{M}} = 1$ and $\|(\kappa_o,\delta)<<(\kappa,\delta)\|_{\underline{M}} = 1$. Define $X(p,\underline{M})$ to be the set containing, for each $\mathcal{P}(p)$-prime sentence $S(\kappa,\delta)$ of $\underline{M}$, both $S(\kappa,\delta)$ and $(\kappa,\delta)<<(\bar{\kappa},\delta)$, where $\bar{\kappa} = \mathrm{Regant}\overset{D}{<<}_{\underline{M}}(\kappa,\delta)$.

The sentence $(\kappa,\delta)<<(\bar{\kappa},\delta)$ is omitted if there is no $S(\kappa',\delta) \in RQ(p)$ distinct from $S(\kappa,\delta)$ and such that $\kappa \prec \kappa' \prec \bar{\kappa}$. (This is the case e.g. if $\kappa = \bar{\kappa}$.)

7.2.20 Lemma. In the situation of 7.2.17, if FORQ is INCOMPR, then $S(\kappa,\delta)$ is $\mathcal{P}(p)$-prime in $\underline{M}$ iff (i), (ii) and (iii'), where (iii') is as follows: (iii') There is no $\kappa_0 \subseteq \kappa$ distinct from κ such that $\|S(\kappa_0, \delta)\|_{\underline{M}} = 1$ and $\|(\kappa_0, \delta) \ll (\kappa, \delta)\|_{\underline{M}} = 1$.

The proof is similar to that of 7.2.14.

7.2.21 Conclusion (G). In all cases, the set $X(p,\underline{M})$ defined by 7.2.5, 7.2.8, 7.2.13 and 7.2.19 is a solution of the problem $\mathcal{P}(p)$, hence the system

$$\langle \mathcal{F}(p), \mathcal{P}(p), X(p,\underline{M});\ p \in \text{Par}, \underline{M} \text{ a } \mathcal{F}(p)\text{- model} \rangle$$

is a GUHA-method called the general GUHA method with associational quantifiers.

(Strictly speaking, in the first place one expects not a function calculus but a semantic system. Hence, let Sent (p) be the set of all sentences involved in I (p), then replace $\mathcal{F}(p)$ by $\mathcal{S}(p)$- the semantic system determined by $\mathcal{F}(p)$ and Sent (p).)

In all cases, moreover, $X(p,\underline{M})$ is an RQ(p)-independent solution and, hence, a $\subseteq$ -minimal solution (cf. 6.1.14).

7.2.22 Remark and Definition. For each p and each $\underline{M}$ we have the notion "a $\mathcal{P}(p)$-prime sentence of $\underline{M}$". If WHELP is NO, then $X(p,\underline{M})$ is a direct solution consisting of all the prime sentences; if WHELP is YES, then $X(p,\underline{M})$ is indirect and contains sentences with helpful quantifiers as well as the prime sentences. For each p, we describe a quasiordering $\leq_{des}$ on Sent p such that " Ψ M-obtainable form Φ " implies $\Phi \leq_{des} \Psi$. The quasiordering will be induced by a corresponding ordering of RPF (p).

7.2.23 Definition. For each $p \in$ Par, we define the designated ordering of RPF(p) as follows:

KRPF	WHELP	FORQ	design. ordering	remark
CPF	NO	arb.	$\sqsubset\!\sqsubset$	identity
EPF	NO	arb.	$\propto$	"more acute than"
CPF	YES	SIMPLE	$\lessdot\!\lessdot$	
CPF	YES	INCOMPR	$\subseteq\!\subseteq$	
EPF	YES	SIMPLE	$(\lessdot \sqsubset) \propto$	composition of $(\lessdot \sqsubset)$ and $\propto$
EPF	YES	INCOMPR	$(\subseteq \sqsubset) \propto$	similarly

The designated ordering extends to a quasiordering of Sent(p) as follows: First, define for each $\Phi \in$ Sent(p) its characteristic pair. If Φ is $S(\varphi,\psi)$ then $cp(\Phi) = \langle\varphi,\psi\rangle$; if Φ is $(\varphi,\psi) \ll (\bar{\varphi},\bar{\psi})$, then $cp(\Phi) = \langle\varphi,\psi\rangle$. Then let $\Phi \leq_{des} \Psi$ iff $cp(\Phi) \leq_{des} cp(\Psi)$, where $\leq_{des}$ is the designated ordering of RPF(p).

7.2.24 Remark. The composition of $\lessdot\Box$ and $\propto$ (in this order) **is an** ordering of the set of all disjointed EPF's by 6. 3. 30. Obviously, $\langle\kappa_1,\delta_1\rangle \subseteq\Box \langle\kappa_2,\delta_2\rangle$ iff $\kappa_1 \subseteq \kappa_2$ and $\delta_1 = \delta_2$; the fact that the composition of $\subseteq\Box$ and $\propto$ is an ordering is proved exactly as 6. 3. 30.

7.2.25 Lemma. For each $p \in$ Par the designated quasiordering of Sent(p) restricted to RQ(p) is an ordering (hence, if WHELP is NO, then $\leq_{des}$ is an ordering).

(2) Let WHELP be NO. If $\Phi, \Psi \in$ RQ(p) and if $\frac{\Phi}{\Psi} \in$ I(p), then $\Phi \leq_{des} \Psi$.

(3) Let WHELP be YES. If $\Phi, \Psi \in$ RQ(p), $\|\Phi\|_{\underline{M}} = 1$, $\|\text{aux}\|_{\underline{M}} = 1$ and $\frac{\Phi,\ \text{aux}}{\Psi} \in$ I(p), then $\text{aux} \equiv_{des} \Phi \leq_{des} \Psi$.

Proof. (1) and (2) are obvious; (3) is obvious if FORQ is SIMPLE; (3) if FORQ is INCOMPR - cf. 7.2.14 and 7.2.20.

7.2.26 Corollary. If H is a hierarchy on Sent (p) such that each $h \in H$ is a lower $\leq_{des}$-segment (i.e., $\Psi \leq_{des} \Phi \in h$ implies $\Psi \in h$), then X (p,$\underline{M}$) is a hierarchical solution of $\mathcal{P}$(p) w.r.t. H. (Obvious; cf. Problem (3) of Chapter VI .)

7.2.27 Remark. For example, if the designated ordering is $\subseteq\subseteq$, then we can somehow linearize the ordering $\lhd\lhd$ of RPF(p) and extend the linearization to Sent(p); the corresponding segments of Sent(p) form a hierarchy "respecting syntactical simplicity". If the designated ordering is $\lessdot\lessdot$ then we can only partly respect syntactical simplicity (we can respect $\subseteq$ but not $\models$). It is reasonable to use a linearization of the designated ordering for successive treatment of relevant pairs in the construction of the solution, cf. the next section.

7.2.28 Key words: r-problems: deductionless (of the first kind), simple (of the second kind), tuft problems (of the third kind) and combined (of the fourth kind); its solutions; prime and auxiliary sentences: application to GUHA-problems from VII.1, designated orderings.

VII.3 Remarks on realization and optimalization

In the present section, we shall discuss three topics: (i) How the solution should be represented on the machine output, and how to find the input corresponding to one prime sentence quickly, (ii) how to verify quickly the truth of $\varphi \sim \psi$ in a model with incomplete information and, finally, (iii) under what conditions the method is realizable in polynomial time.

7.3.1 Discussion. Our first question is uninteresting for the case without helpful quantifiers (WHELP being NO): If the parameter is known, then the solution is fully represented by the list of pairs $\langle\varphi,\psi\rangle$ of relevant formulae such that $S(\varphi,\psi)$ is a $\mathcal{P}$(p)-prime sentence of $\underline{M}$. According to 7.2.25, one goes through RPF(p) in a linear order linearizing the designated ordering of RPF(p). For the case of helpful quantifiers (WHELP is YES) the question is, assuming that $S(\varphi,\psi)$ is prime, how to find (represent) the corresponding auxiliary sentence of the form $(\varphi,\psi)\ll(\bar{\varphi},\bar{\psi})$. (Recall that if relevant pairs are CPF, then $\langle\bar{\varphi},\bar{\psi}\rangle$ is $\mathrm{Reg}\ll_{\underline{M}}(\varphi,\psi)$; if relevant pairs are EPF, then $\bar{\varphi}$ is $\mathrm{Regant}\ll_{\underline{M}}(\varphi,\psi)$ and $\bar{\psi}$ is ψ.) Here we use Theorem 6.3.14 (cf. remark 6.3.15). Recall that $\mathrm{Ant}(\varphi,\psi,\chi)$ is logically equivalent to $(\varphi,\psi)\ll(\varphi\&\chi,\psi)$ and, similarly, $\mathrm{Suc}(\varphi,\psi,\chi)$ is logically equivalent to $(\varphi,\psi)\ll(\varphi,\psi\&\chi)$. We have the following:

7.3.2 Lemma. Let WHELP be YES (helpful quantifiers used). (1) If relevant pairs are CPF and if $S(\kappa,\lambda)$ is a $\mathcal{P}$(p)-prime sentence of $\underline{M}$, then $\mathrm{Reg}\ll_{\underline{M}}(\kappa,\lambda)$ can be constructed as follows: For each function symbol F_i, ask whether there is an $X \subsetneqq V_i$ such that $\|\mathrm{Ant}(\kappa,\lambda,(X)F_i)\|_{\underline{M}} = 1$; if so, let X_i be the least such X and put i into A. Further, ask whether there is a $Y \subsetneqq V_i$ such that $\|\mathrm{Suc}(\kappa,\lambda,(Y)F_i\|_{\underline{M}} = 1$; if so, let Y_i be the least such Y and put i into S. Let $\bar{\kappa} = \bigwedge_{i\in A}(X_i)F_i$ and let $\bar{\lambda} = \bigwedge_{i\in S}(Y_i)F_i$. Then $\mathrm{Reg}\ll_{\underline{M}}(\kappa,\lambda) = \langle\bar{\kappa},\bar{\lambda}\rangle$.

(2) If relevant pairs are EPF and if $S(\kappa,\delta)$ is a $\mathcal{P}(p)$-prime sentence of M, then $\mathrm{Regant}\overset{D}{\ll}_{\underline{M}}(\kappa,\delta)$ can be constructed as follows; For each function symbol F_i not occurring in δ ask whether there is an $X \subsetneqq V_i$ such that $\|\mathrm{Ant}(\kappa,\delta,(X)F_i)\|_{\underline{M}}=1$; if so, let X_i be the least such X and put i into A. Let $\bar{\kappa}$ be $\bigwedge_{i \in A} (X_i)F_i$ then $\bar{\kappa} = \mathrm{Regant}\overset{D}{\ll}_{\underline{M}}(\kappa,\delta)$

Proof. (1) follows immedately from 6.3.14; (2) is proved analogously. Note that we know that each X_i is non-empty (cf. 7.2.10 and 7.2.15).

7.3.3 Remark. (1) We can now answer the question of the desired output in the case with helpful quantifiers. For each prime sentence $S(\varphi,\psi)$, the output contains:

(a) the pair $\langle \varphi,\psi \rangle$

(b) the list of all literals $(X)F_i$ where X is the smallest coefficient $Z \subsetneqq V_i$ such that $(Z)F_i$ improves the antecedent of $S(\varphi,\psi)$ (if KRPF is EPF disregard the function symbols occurring in ψ);

(c) in addition, if KRPF is CPF, then the output contains the list of all literals $(X)F_i$ where X is the smallest coefficient $Z \subsetneqq V_i$ such that $(Z)F_i$ improves the succedent of $S(\varphi,\psi)$.

In dependence on the particular quantifier used (PQUANT), we may also require further information, e.g. the exact value of the statistic used in the definition of the quantifier, etc.

(2) Our next aim is to show how to find X_i $(i \in A)$ and Y_i $(i \in S)$ directly. Let κ, λ, F_i, $\underline{M}$ be given; recall that $\ll$ is given by the parameter p. The quantifier $\ll$ is universally defined by an economical set $\mathcal{U}$. Let $o \in M$ and consider the quadruple

$$\langle \|\kappa\|_{\underline{M}}[o], \|\lambda\|_{\underline{M}}[o], \|\kappa \,\&\, (X)F_i\|_{\underline{M}}[o], \|\lambda\|_{\underline{M}}[o]\rangle = \langle u, v, \bar{u}, v\rangle .$$

We want to choose the least X such that this quadruple is in $\mathcal{U}$, for all $o \in M$. Take note of the following (1) if $u \neq 0$ and $\|F_i\|_{\underline{M}}[o] = \times$, then $\bar{u} = \times$ independently of X. (2) If $u = 1$ and $\|F_i\|_{\underline{M}}[o] \in X$, then $\bar{u} = 1$. (3) If $u = \times$ and $\|F_i\|_{\underline{M}}[o] \in X$ or $\|F_i\|_{\underline{M}}[o] = \times$, then $\bar{u} = \times$.

(4) If $u = 0$, then $\bar{u} = 0$. Hence if, for given u, v, the only w such that $\langle u, v, w, v\rangle \in U$ is $w = u$, then we can force $\bar{u}$ to be equal to u (possibly enlarging the coefficient <u>unless</u> $u = 1$ and $\|F_i\|_{\underline{M}}[o] = \times$). If $\langle u, v, w, v\rangle \in U$ implies $w \neq 0$, then we may always force $\bar{u}$ to be $\neq 0$. We obtain the following definition:

7.3.4 <u>Definition</u>. Let $\ll$ be <u>universally defined by</u> a set $U \subseteq \{0, \times, 1\}^4$.
(1) <u>Strongly A-critical pairs</u> for $\ll$ are pairs $\langle 1, v\rangle$ ($v \in \{0, \times, 1\}$) such that $\langle 1, v, w, v\rangle \in U$ implies $w = u = 1$.

(2) <u>Weakly A-critical pairs for</u> $\ll$ are pairs $\langle u, v\rangle$ with $u \neq 0$ such that $\langle u, v, w, v\rangle \in U$ implies $w \neq 0$.

(3) <u>Strongly S-critical pairs</u> (<u>weakly S-critical pairs</u>) are pairs $\langle u, v\rangle$ where $v = 1$ ($v \neq 0$) and $\langle u, v, u, w\rangle \in U$ implies $w = 1$ ($w \neq 0$).

7.3.5 <u>Lemma</u>. Let WHELP be YES. (1) Let relevant pairs be CPF, let $S(\kappa, \lambda)$ be $\mathcal{P}(p)$-prime in $\underline{M}$, and let F_i be a function symbol. There is an X such that $\|Ant(\kappa, \lambda, (X)F\|_{\underline{M}} = 1$ iff there is no object $o \in M$ such that $\langle \|\kappa\|_{\underline{M}}[o], \|\lambda\|_{\underline{M}}[o]\rangle$ is strongly A-critical and $\|F_i\|_{\underline{M}}[o] = \times$.

If the last condition is satisfied, then the least such X is

$X = \{\|F_i\|_{\underline{M}}[o];\ \|F_i\|_{\underline{M}}[o] \neq \times$ and $\langle \|\kappa\|_{\underline{M}}[o], \|\lambda\|_{\underline{M}}[o]\rangle$ is A-critical.

(2) Analogously if KRPF is CPF and Ant is replaced by Suc, A-replaced by S-. (3) Analogously if KRPF is EPF and λ is replaced by δ.

<u>Proof.</u> By using 7.3.3.

7.3.6 <u>Remark.</u> We make a list of critical pairs for quantifiers involved (first come the strongly critical pairs; they are separated from those not strongly critical by a semicolon).

KHELP	KRPF	FORQ	KQUANT	A-critical	S-critical
CONSV	CPF	arb.	arb.	11,1×,10;×1,××,×0	11,×1,01;1×,××,0×
CONSV	EPF	arb.	arb.	11,1×,10;×1,××,×0	———
DESIGN	CPF	SIMPLE	IMPL	11;×1	11;×1
DESIGN	CPF	SIMPLE	not IMPL	11,1×;×1	11,×1;1×
DESIGN	CPF	INCOMPR	IMPL	11;1×,×1,××	11,×1;1×,××
DESIGN	CPF	INCOMPR	not IMPL	11,1×;×1,××	11,×1;1×,××
DESIGN	EPF	arb.	IMPL	11; —	———

Hence, we see that to decide whether there is an X such that $\| Ant(\kappa, \lambda, (X)F_i \|_{\underline{M}} = 1$ and if so to find the least X one needs only one inspection of the model, object by object. Caution: It is possible that the least coefficient X such that $\| Ant(\kappa, \lambda, (X)F_i \|_{\underline{M}} = 1$ is $X = V_i$, then $(X)F_i$ will not be included in the regularized antecedent.

We see directly why the constructed coefficient cannot be empty: $\langle 1,1 \rangle$ is always a critical pair and, since $S(\varphi, \psi)$ is true, $\varphi \,\&\, \psi$ is satisfied by at least one object (cf. 7.1.3).

7.3.7 Remark. If we have a particular pair $\langle \varphi, \psi \rangle$ (CPF or EPF) and want to evaluate $\varphi \sim \psi$ in a model $\underline{M}$ then everything is determined by $\underline{M}_{\varphi,\psi} = \langle M, \|\varphi\|_{\underline{M}}, \|\psi\|_{\underline{M}} \rangle$, which is a three-valued ($\{0, \times, 1\}$ -valued) model. Should one apply the definition directly, one would have to consider all two-valued completions of the last model, which would be tiresome. It is useful if we can effectively associate with $\underline{M}_{\varphi,\psi}$ one particular two-valued completion $w(\underline{M}_{\varphi,\psi})$ such that $\|\varphi \sim \psi\|_{\underline{M}} = 1$ iff $Asf_{\sim}(w(\underline{M}_{\varphi,\psi})) = 1$. Then we can call $w(\underline{M}_{\varphi,\psi})$ the worst completion of $\underline{M}_{\varphi,\psi}$.

We consider three-valued models of type $\langle 1,1 \rangle$.

7.3.8 Lemma. Let $\underline{M}$ be a three-valued model of type $\langle 1,1 \rangle$ and let $\sim$ be an associational quantifier. Put

$$\underline{N} = \underline{M}\,(\langle 1,\times\rangle : \langle 1,0\rangle)(\langle \times,0\rangle : \langle 1,0\rangle)(\langle \times, 1\rangle : \langle 0,1\rangle)(\langle 0,\times\rangle : \langle 0,1\rangle)$$

(i.e., each card $\langle 1,\times \rangle$ is replaced by $\langle 1,0 \rangle$ etc .). Then $\underline{N}$ and $\underline{M}$ are a-equivalent, i.e., $\underline{M}$ is a-better than $\underline{N}$ and $\underline{N}$ is a-better than $\underline{M}$.

This follows directly from 3.3.24.

7.3.9 Remark (continued). Hence, looking for the worst completion w M we know what to do with all cards except $\langle \times,\times \rangle$. Obviously, for symmetrical quantifiers, the last card must be completed partly to $\langle 1,0 \rangle$ and partly to $\langle 0,1 \rangle$; but how many objects with card $\langle \times,\times \rangle$ should be completed to $\langle 1,0 \rangle$ and how many to $\langle 0,1 \rangle$? We shall show that in the particular cases discussed in Chapter IV we can answer this question.

7.3.10 Definition (1) $\underline{N}$ is a symmetric completion of $\underline{M}$ if all cards except $\langle x, x \rangle$ are completed as in 7.3.8, i.e.,

cards	1×	×0	×1	0×
are completed to	10	1 0	0 1	0 1

and if the cards $\langle \times,\times \rangle$ are completed partly to $\langle 1,0 \rangle$ and partly to $\langle 0,1 \rangle$ in such a way that $|b_{\underline{N}} - c_{\underline{N}}|$ is as small as possible.(Remember that $b_{\underline{N}}$ is the frequency of the card $\langle 1,0 \rangle$ in $\underline{N}$,etc.)

(2) $\underline{N}$ is an implicational completion of $\underline{M}$ if all cards are completed as follows:

cards	1×	×0	××	×1	0×
are completed to	1 0	1 0	1 0	0 1	0 1

7.3.11 Theorem. Let $\sim$ be one of the quantifiers $\sim, \sim_\alpha, \sim_\alpha^2$ (i.e., in words, the simple, Fisher, and χ^2 quantifiers). Then, for each $\underline{M}$ and each symmetric completion $\underline{N}$ of $\underline{M}$, $\underline{N}$ is a-equivalent to $\underline{M}$, i.e., $Asf_\sim(\underline{N}) = 1$ iff $Asf_\sim(\underline{N}') = 1$ for each completion $\underline{N}'$ of $\underline{M}$.

Proof. Our proofs will be easy in all cases except the Fisher quantifier; for the Fisher quantifier the result is due to Rauch.

In all cases, we know by lemma 7.3.8 that we can restrict ourselves to completions of a model M having the following 3×3 table of frequencies:

	1		0	
1	a	0	b	r
	0	n	0	n
0	c	0	d	s
	k	n	l	m

Then a completion $\underline{N}$ has a 2×2 table of the form:

a	$b + n'_{\underline{N}}$
$c + n - n'_{\underline{N}}$	d

(1) First, consider the quantifier of simple association $\sim$: $(r + n'_{\underline{N}})(k + n - n'_{\underline{N}})$ attains its maximum if

$$\delta(\underline{N}) = |(r + n'_{\underline{N}}) - (k + n - n'_{\underline{N}})| = |b_{\underline{N}} - c_{\underline{N}}|$$

attains its minimum. Now we have a completion $\underline{N}$ for which $\delta(\underline{N})$ attains its minimum and $Asf_{\sim}(\underline{N}) = 1$, i.e., $am > (r + n'_{\underline{N}})(k + n - n'_{\underline{N}})$; then clearly $Asf_{\sim}(\underline{N}) = 1$ for each completion $\underline{N}'$ of $\underline{M}$.

(2) Consider now the quantifier of χ^2- association $\sim^2_\alpha$: $Asf_{\sim^2_\alpha}(\underline{N}) = 1$ iff

$$H(\underline{N}) = \frac{(ad - (b + n'_{\underline{N}})(c + n - n'_{\underline{N}}))^2 \; m}{(r + n'_{\underline{N}})(k + n - n'_{\underline{N}})(s + n - n'_{\underline{N}})(1 + n'_{\underline{N}})} \geq \chi^2_\alpha .$$

$H(\underline{N})$ attains its minimum iff $\delta(\underline{N})$ attains its minimum. Thus if $\underline{N}$ is a symmetrical completion, then $H(\underline{N}') \geq H(\underline{N})$ for each completion $\underline{N}$ of $\underline{M}$.

(3) The last case is the quantifier of the Fisher association $\sim_\alpha$: Suppose (without any loss of generality due to the symmetry of $\sim_\alpha$) that $r + n'_{\underline{N}} \geq k + n - n'_{\underline{N}}$. Observe that the completions $\underline{N}_1$, $\underline{N}_2$ with $n'_{\underline{N}_1} = n'_{\underline{N}_2}$ are equivalent for our purposes and thus we shall consider the function $\delta(n') = r - k + 2n' - n \geq 0$. Denote $I(n, r, k) = \{n' ; \delta(n') \geq 0, 0 \leq n' \leq n\}$. $I(n,r,k)$ is an interval in $\mathbb{N}$ and $\delta(n')$ is strictly increasing on $I(n, r, k)$. The least element of $I(n,r,k)$ corresponds to a symmetrical completion.

We know that the associated function of $\sim_\alpha$:

$Asf_{\sim_\alpha}(\underline{N}) = 1$ if $\Delta(a_N, r_N, k_N\; m_N) \leq \alpha$.

Hence, it is sufficient to prove that the function

$H(n') = \Delta(a, r + n', k + n - n', m)$ is decreasing in $n' \in I(n,r,k)$, i.e. that

$H(n') \geq H(n'+1)$. We already know that

$$H(n') = \Delta(a, r+n', k+n-n', m)$$
$$= \Delta(a, r+n, r+n' - \delta(n'), m) =$$
$$\sum_{i=0}^{r+n'-\delta(n')} \sigma(i, r+n', r+n' - \delta(n'), m),$$

where

$$\sigma(i, r+n', r+n' - \delta(n'), m) =$$

$$= \frac{(r+n)!\,(m-(r+n'))!\,(r+n'-\delta(n'))!\,(m-(r+n'-\delta(n')))!}{i!\,(r+n'-i)!\,(r+n'-\delta(n')-i)!\,(m-2(r+n')-\delta(n')+i)!\,m!}$$

Note that $r+(n'+1)-\delta(n'+1) = r+n'-\delta(n')-1$.

We want to prove that, for **arbitrarily given** k,r,m,n, $H(n') \geq H(n'+1)$ for each possible value of a (see 4.4.23.), i.e. for

$a_o = \max(0, 2(r+n') - m - \delta(n') \leq a \leq r+n' - \delta(n') - 1$.

For $a_o = a$, $H(n') = 1$ for each n'.

Thus we shall suppose that $a_o < a < r+n' - \delta(n') - 1$.

Denote $\sigma_i = \sigma(i, r+n', r+n - \delta(n'), m)$

and $\sigma'_i = \sigma(i, r+n'+1, r+(n'+1) - \delta(n'+1), m) =$

$$= \sigma(i, r+n'+1, r+n' - \delta(n') - 1, m).$$

Observe that

$$\frac{\sigma_i}{\sigma'_i} = \frac{(m-(r+n'))(r+n'-\delta(n'))}{(r+n'+1)(m-(r+n')+\delta(n')+1)} \; \frac{r+n'+1-i}{r+n'-\delta(n')-i} =$$

$$= C\,\frac{(r+n'+1-i)}{(r+n'-\delta(n')-i)} = C\,f(i).$$

It is easy to see that $f(i)$ is an increasing function of i. Then we have two cases: First: $\sigma_i < \sigma_i'$ for each i; and second: $\sigma_i \leq \sigma_i'$ for $i \leq i_o$, and

$\sigma_i > \sigma_i'$ for $i > i_o$. In the second case, if $a > i_o$, then clearly

$$H(n') - H(n'+1) = \sum_{i=a_o}^{r+n'-\delta(n')-1} (\sigma_i - \sigma_i') + \sigma_{r+n'-\delta(n')} > 0 .$$

If $a \leq i_o$ or the first case occurs, then we use the following equality:

$$H(n') - H(n'+1) = \sum_{i=a}^{r+n'-\delta(n')-1} (\sigma_i - \sigma_i') + \sigma_{r+n'-\delta(n')} - \sum_{i=a_o}^{a-1} (\sigma_i - \sigma_i') .$$

(We can see that

$2(r+n'+1) - \delta(n'+1) = 2(r+n') - \delta(n').$)

We have $\sum_{i=a_o}^{a-1} (\sigma_i - \sigma_i') \leq 0$, hence

$$H(n') - H(n'+1) \geq \sum_{i=0}^{r+n'-n'-1} (\sigma_i - \sigma_i') + \sigma_{r+n'-\delta(n')} = 0 .$$

7.3.12 Theorem. Let $\sim$ be an implicational quantifier. Then for each $\underline{M}$ and the implicational completion $\underline{N}$ of $\underline{M}$, $\underline{N}$ is i-equivalent to $\underline{M}$, i.e., $Asf_{\sim}(\underline{N}) = 1$ iff $Asf_{\sim}(\underline{N}') = 1$ for each completion $\underline{N}'$ of $\underline{M}$.

Proof. By using 3.3.24, we have $\langle \times \times \rangle \equiv_i \langle 1, \times \rangle \equiv_i \langle 1, 0 \rangle$ and $\langle 0, \times \rangle \equiv_i \langle \times, 1 \rangle \equiv_i \langle 0, 1 \rangle \equiv_i \langle 0, 0 \rangle$, cf. Remark 3.3.25 (Note that, in fact, $\langle 0, \times \rangle$ can be completed to $\langle 0, 0 \rangle$ and/or to $\langle 0, 1 \rangle$.) The proof of the present theorem is rather trivial now, but the theorem is stronger than Theorem 7.3.11; it gives the worst completion for each implicational quantifier.

7.3.13 Discussion. Clearly, one does not need to know the symmetric completion but only its frequencies: they are easily computable from the frequencies of all the 9 cards (elements of $\{0,\times,1\}^2$) in $\underline{M}$. Hence, returning to 7.3.7, to evaluate $\varphi \sim \psi$ one needs only one inspection of the model, object by object. Our next questions are: How does one decide on incompressibility? Granted that $S(\varphi,\psi)$ is true, how does one decide whether it is prime?

The first question is easy to answer: If there is a $\kappa_0 \subsetneqq \kappa$ such that $\|\kappa_0 \Leftrightarrow \kappa\|_{\underline{M}} = 1$, then there is such a κ_0 which differs from κ only in one literal, say, containing F_i, and if $(X)F_i$ is in κ and $(X_0)F_i$ is in κ_0, then we may assume that the difference $X - X_0$ has exactly one element. Hence, κ is $\underline{M}$-incompressible if whenever one takes a literal $(X)\ F_i$ from κ and diminishes X by omitting one element one obtains a conjunction κ_0 not equivalent to κ in $\underline{M}$. Hence, if $\underline{M}$ has n function symbols and for each i, V_i has at most h elements, then to decide incompressibility one needs to inspect the model not more than $n \cdot h$ times.

The situation for primeness is similar. One can summarize the definition of a prime sentence as follows: $S(\varphi,\psi)$ is a $\mathcal{P}(p)$-prime sentence of $\underline{M}$ iff $S(\varphi,\psi)$ is true in $\underline{M}$ and there is no $\langle\varphi_0,\psi_0\rangle$ strictly less than $\langle\varphi,\psi\rangle$ in the designated ordering (given by p) such that $S(\varphi_0,\psi_0)$ is true (and - if helpful quantifiers are used - $(\varphi_0,\psi_0) \ll (\varphi,\psi)$ is true in $\underline{M}$). It is easy to see that the words "strictly less than" can be replaced by "immediate predecessor of" ; and it is easy to verify that the pair $\langle\varphi,\psi\rangle$ has at most $n(h+1)$ immediate predecessors. (For example, consider α : Immediate predecessors result either by removing one element from one coefficient in the succedent - if the coefficient was a one-element set, omit the whole literal- or, otherwise, by transferring a literal from the succedent into the antecedent with the obvious change; this yields at most $n(h+1)$ cases.) We already know that the evaluation of $\varphi_0 \sim \psi_0$ needs one inspection of the model; the evaluation of $(\varphi_0,\psi_0) \ll (\varphi,\psi)$ also clearly needs only one inspection since our $\ll$ is universally definable. And we showed that the decision whether φ,ψ are incompressible ($\varphi \,\&\, \psi$ is incompressible or φ is strongly incompressible w.r.t. ψ respectively) needs at most n.h inspections.

7.3.14 The above considerations are not only useful for the construction of reasonable machine programs but also enable the formulation of some theoretical consequences concerning the complexity of the realizing algorithm. **Suppose that** we have a "natural" syntactically described linear order $\leq_{(p)}$ on RPF(p) linearizing the designated ordering. The algorithm realizing our method can be described by the simple flow-diagram presented in 6.1.22 (for further optimization see Problem (4)). The input consists of the investigated model $\underline{M}$ (represented e.g. as a matrix) and of the parameter p. The complexity of $\underline{M}$ can be measured by three numbers: m - the number of objects in M, n - the number of function symbols, and h - the maximum of cardinalities of sets of regular values of the function symbols (V_i). It is hoped that the above considerations give enough evidence for the claim that each single item of the flow-diagram is realizable in polynomial time (in the three variables m, n, h). Here, n and h are given by the parameter p (in particular, by CHAR); $\underline{M}$ must have the prescribed characteristic (but it is allowed to have, theoretically, any finite non-zero cardinality). Hence we come to the following

Conclusion. If the cardinality of RPF(p) depends polynomially on n and h and if the statement of 7.3.11 holds for each associational quantifier admitted by PQUANT, then the time necessary for the construction of the solution X(p, $\underline{M}$) depends polynomially on m, n, h. (Cf. 6.1.23.)

Hence, let us make the following assumption:

7.3.15 Assumption. The syntactic restrictions SYNTR always imply that RQ(p) consists of some sentences of complexity less than b (in addition, each quantifier allowed by PQUANT satisfies the assertion of 7.3.11).

Here, of course, we wish the complexity of a sentence $S(\varphi, \psi)$ - or, say, of the pair $\langle\varphi, \psi\rangle$ - to be defined in such a way that the number of all disjointed CPF's (EPF's) of complexity at most b is polynomial in n and h.

This can be achieved as follows: First, impose an upper bound on the number of function symbols occurring in φ and ψ - say, b_1.

Second, impose an upper bound - say b_2 - on the number of elements of V_i determining a single coefficient. "Determine" can mean "form" (i.e., determine by listing) but it need not. For example, we may allow only coefficients that are intervals in the set of natural numbers (when our attributes have a more or less comparative and not entirely qualitative character); each interval is determined by two elements - its end-points. It is then elementary to see that the number of possible coefficients is majorized by $\sum_{i=1}^{b_2} \binom{h}{i} = p_2(h)$, which is a polynomial in h

(b_2 being fixed), the number of pairs $<A, S>$ of disjoint sets of function symbols satisfying $card(A) + card(S) \leq b_1$ is $\sum_{i+j \leq b_1} \binom{n}{i}\binom{n-i}{j} = p_1(n)$ and, hence, the cardinality of $RQ(p)$ is majorized by $(p_2(h))^{b_1} p_1(n)$, which is a polynomial in n,h.

7.3.16 Conclusion. Under the assumption 7.3.15 the time needed to construct the solution $X(p, \underline{M})$ depends polynomially on m,n,h. Thus the GUHA method with associational quantifiers is realizable in polynomial time (assuming 7.3.15).

7.3.17 Key words. Representation of the solution, critical pairs, worst completions, intelligibility bound.

VII. 4 Some suggestions concerning GUHA methods based on rank calculi

We now present some suggestions concerning the further development of new particular GUHA methods. In Chapter V we developed a theory of calculi with mixed two-valued, enumerational and rational valued models and generalized quantifiers inspired by statistical rank tests. The next step is to apply these calculi in the logic of suggestion. Such methods could be practically applicable in the whole field of underlying statistical rank tests or tests on enumerational structures in general.

The construction of particular GUHA methods and their machine realization is, at the present stage of research, only just beginning. Many questions in this field are as yet open, therefore we give only some suggestions for their further development. A promising area for further investigation and construction remains open here.

7.4.1 First, we concentrate on calculi with distinctive quantifiers. The reader should have in mind both the notion of distinctive quantifiers (with mixed two-valued and enumeration models) from Chapter V, Section 3, and the notion of distinctive rank quantifiers from 5.4.13 - 5.4.15.

7.4.2 We shall consider some r-problems. We put $V_o = \{1\}$ and $RQ = \{q_\alpha(\varphi, F)\}_{\alpha \in A,\ \varphi \in B,\ F \in C}$, where $A \subseteq (0, 0.5] \cap Q$ and B, C are non-empty sets of designated open formulae of the appropriate sort. Suppose that $\{q_\alpha\}_{\alpha \in A}$ is a monotone class of d-executive quantifiers. The following deduction rules could be used in such a situation:

$$\text{M:} \left\{ \frac{q_\alpha(\varphi, F)}{q_{\alpha'}(\varphi, F)} ;\ \alpha' > \alpha \right\}_{\langle \varphi, F \rangle \in B \times C} ,$$

$$\text{E:} \left\{ \frac{q_\alpha(\varphi, F) ,\ \varphi \Leftrightarrow \varphi'}{q_\alpha(\varphi', F)} \right\}_{\varphi, \varphi' \in B,\ F \in C} .$$

The usefulness of rule M is clear. The algorithmic usage of E needs a particular form of designated open formulae from B. Let B be the set of all EC's built up from some function symbols $F_1, \ldots, F_k$. Suppose that B is ordered by $\subseteq$ ($\kappa \subseteq \lambda$ means that κ is included in λ ; the generalization to calculi with incomplete information using x is straightforward). Here relevant pairs of formulae (RPF) are $\langle \varphi, F \rangle$, where φ is an EC and F is a function symbol of sort b. Define $\langle \kappa, F \rangle \subseteq \langle \lambda, G \rangle$ if $\kappa \subseteq \lambda$ and $F = G$.

Let $RQ = \{q_\alpha(\kappa, F)\}_{\langle \kappa, F \rangle \in RPF}$ (α fixed) and consider the r-problem $\mathcal{P} = \langle RQ, E', \{1\} \rangle$, where

$$E' = \left\{ \frac{q_\alpha(\kappa_2, F),\ \kappa_2 \Leftrightarrow \kappa_2}{q_\alpha(\kappa_3, F)} \ ;\ \kappa_1 \subseteq \kappa_3 \subseteq \kappa_2,\ \kappa_1, \kappa_2, \kappa_3 \in EC \right\}.$$

A <u>prime sentence</u> of $\underline{M}$ is each $q_\alpha(\varphi, F)$ such that $q_\alpha(\varphi, F) \in Tr(\underline{M})$ and there is no $\kappa' \subsetneq \kappa$ such that $\|\kappa' \Leftrightarrow \kappa\|_{\underline{M}} = 1$. Let $X_{\underline{M}}$ be the set containing, for each prime sentence $q_\alpha(\varphi, F)$, both the formula $q_\alpha(\varphi, F)$ and the formula $\kappa \Leftrightarrow \lambda$ where λ is the maximal conjunction, $\kappa \subseteq \lambda$, $\underline{M}$ - equivalent to κ. Then $X_{\underline{M}}$ is a solution of $\mathcal{P}$.

Define a quantifier $\square$ of type $\langle a, b, b \rangle$ as follows:

$Asf_\square(\langle M, f_1, f_2, f_3 \rangle) = 1$, if $\langle M, f_1, f_2 \rangle \preceq_d \langle M, f_1, f_3 \rangle$,

($Asf_\square(\langle M, f_1, f_2, f_3 \rangle) = 0$ otherwise), then we have the following deduction rule

$$C: \left\{ \frac{q_\alpha(\varphi, F),\ \square(\varphi, F, G)}{q_\alpha(\varphi, G)} \right\} \quad \alpha \in A,\ \langle \varphi, F, G \rangle \in B \times C \times C .$$

More generally, we can define, in analogy to Chapter VI, <u>d-improving quantifiers</u>. For example, $\boxtimes$ of type $\langle a, a, b, b \rangle$ with

$Asf_\boxtimes(\langle M, f_1, f_2, f_3, f_4 \rangle) = 1$ iff $\langle M, f_1, f_3 \rangle \preceq_d \langle M, f_2, f_4 \rangle$.

We have a good algorithm to decide whether $\|\boxtimes(\varphi_1, \varphi_2, F, G)\|_{\underline{M}} = 1$. But if we want to use sentences with auxiliary quantifiers in solutions of r-problems and if we want to know whether the use of such sentences is reasonable (at least as in VII.2), then we need more: We have said in Section 2 of the present chapter that one prime sentence and one auxiliary sentence with a helpful quantifier has determined the whole (syntactically defined and easily comprehensible) tuft of true sentences. The only analogy known in the present situation is the quantifier $\Leftrightarrow$ of equivalence. Further development could push the analogy further.

7.4.3 The reader can derive from the considerations of 5.4.13 - 5.4.15 the way in which distinctive quantifiers could be used in the construction of r-problems with <u>distinctive rank quantifiers</u>. For this situation (r-problems concerning mixed binary and rational valued models) two further deduction rules can be introduced:

$$R = \left\{ \frac{q_\alpha(\varphi_1, \varphi_2),\ \varphi_2 \equiv_R \varphi_3}{q_\alpha(\varphi_1, \varphi_3)} \right\}$$

and the analogue to PE (cf. 7.4.6). It seems to be clear how to combine them with that of 7.4.2. If we have a monotone class of distinctive rank quantifiers $\{q_\alpha\}_\alpha$, we can define a r-problem with

$$RQ = \{ q_\alpha(\varphi_1, \varphi_2); \alpha, \varphi_1, \varphi_2 \}$$

where φ_1 are, e.g., EC's based on $P_1, \ldots, P_{k_1}$ and φ_2 can be some designated open formulae of sort c. $V_o = \{1\}$ and I is based on deduction rules from 7.4.2 and the above mentioned.

Our knowledge concerning executability (and best ways; see Problem (8)) can be used for the optimalization of algorithms.

7.4.4 Second, we turn our attention to correlational quantifiers. Before we discuss rank correlational quantifiers, we shall show a way in which some r-problems can be stated in calculi with mixed two-valued and enumeration models. These r-problems can be generalized for rational-valued models in the usual way; cf. 5.1.13 and further discussions in the present section.

We shall make a slight generalization of correlational quantifiers. Let q_o be a correlational quantifier and define a new quantifier q of type $\langle a, b, b\rangle$ as follows: If $\underline{M} = \langle M, f_1, f_2, f_3\rangle$ is of type $\langle a,b,b\rangle$, let $M_{f_1} = \{o \in M ; f_1(o) = 1\}$ and $\underline{M}_{f_1} = \langle M_{f_1}, f_2 \restriction M_{f_1}, f_3 \restriction M_{f_1}\rangle$ (the submodel of all objects satisfying f_1). Put

$$\mathrm{Asf}_q(\underline{M}) = \begin{cases} 0 & \text{if } M_{f_1} = \emptyset, \\ \mathrm{Asf}_{q_o}(\underline{M}_{f_1}) & \text{otherwise}. \end{cases}$$

Let A be a class of designated open formulae of sort a and let $F_1, \ldots, F_k$ be functors of type b. Put

$$RQ = \{ q(\varphi, F_i, F_j) \}_{\varphi \in A}, \quad V_o = \{1\}.$$

In fact, this is an old idea of Metoděj Chytil called ELICO - Elimination of nuisance objects in correlation. The question of reasonable deduction rules and solutions is open. We can see that if φ are EC's, then our remarks from 5.4.11 could be very useful for the algorithmic solution of the above mentioned problem.

7.4.5 We now turn our attention to rank correlational quantifiers (in calculi with rational valued models) . Remember that items that are to be interpreted in our calculi as real valued are said to be of sort c. Now we shall be more specific as to the form of designated open formulae of sort c. Define elementary unijunctions as follows: Let Jct_1 be a finite (but possibly big) set of unary junctors with $\mathrm{Asf}_2 : \mathbb{Q} \longrightarrow \mathbb{Q}$. Then elementary unijunctions (EU) are formulae built up from designated atomic formulae $F_1(x), \ldots F_{k_2}(x)$ (or simply $F_1, \ldots, F_{k_2}$) of sort c with the help of junctors from Jcf_1.

Note that each EU has the form τF_i where τ is a sequence of junctors from Jcf_1 (possibly empty). Put $\mathrm{Asf}_{\emptyset}(u) = u$ (for $u \in \mathbb{Q}$) and, if Asf_{τ}

is defined and $\iota \in Jct_1$, put $Asf_{\iota\tau}(u) = Asf_{\iota}(Asf_{\tau}\ u))$. Note that for $\iota \in Jct_1$, Asf_{ι} is defined by the function calculus in question.
In the function calculi we have in mind, the unary junctors - elements of Jct_1 are names of some standard functions (for example, particular polynomials in one variable or some rational-valued approximations of sin, log, ... etc.). We may assume that junctors from Jct_1 together with their associated functions are fixed in the sequel ; other components of function calculi (type, quantifiers, binary junctors) may vary.

Write $\varphi \subseteq \psi$ if φ is a subformula of ψ .

7.4.6 A relation $\triangleleft$ on EU^2 is called a <u>relation of positive expansion</u> if (1) $\varphi_1 \triangleleft \varphi_2$ implies $\varphi_1 \subseteq \varphi_2$, i.e., $\varphi_2 = \tau\varphi_1$, and (2) $\varphi_1 \triangleleft \varphi_2$ implies that Asf_{τ} is increasing. If $\triangleleft$ is such a relation, then

$$PE = \left\{ \frac{q(\varphi_1, \varphi_2)}{q(\varphi_1, \varphi_3)} \ ; \ \varphi_2 \triangleleft \varphi_3 \right\}$$

is a sound deduction rule for each strong rank quantifier.

7.4.7 <u>Lemma.</u> Let R be a recursive relation on EU^2 such that $\varphi_1\, R\, \varphi_2$ implies $\varphi_1 \subseteq \varphi_2$. If q is an executive rank correlational quantifier and if a rule

$$\left\{ \frac{q(\varphi_1, \varphi_2)}{q(\varphi_1, \varphi_3)} \ ; \ \varphi_2\, R\, \varphi_3 \right\}$$

is sound , then R is a relation of positive expansion.

<u>Proof</u>. Consider $q(\varphi_1, \varphi_1)$; if $\varphi_1\, R\, \varphi_2$ $(\varphi_2 = \tau\varphi_1)$, then $\underline{M} \models q(\varphi_1, \tau\varphi_1)$ for each model $\underline{M}$ such that $m_{\underline{M}} > m_{min}(q)$. But this is the case iff Asf_{τ} is increasing.

7.4.8 We can introduce the binary junctor + with the usual associated function. Then we can consider further deduction rules, e.g.,

$$\left\{ \frac{q(\varphi, \varphi_1 + \varphi_2), \varphi_2 \equiv_R \varphi_1 + \varphi_2}{q(\varphi, \varphi_1 + \varphi_2 + \varphi_3)} ; \varphi_1 \lhd \varphi_3 \right\} .$$

In particular,

$$AD = \left\{ \frac{q(\varphi, \varphi_1)}{q(\varphi, \varphi_1 + \varphi_3)} ; \varphi_1 \lhd \varphi_3 \right\} .$$

For example, consider junctors $(\ )^3$ (third power) and B, where $B > 0$ (multiplication by the positive number B), with the usual semantics. Let φ_1 be a unijunction of sort C; then $\varphi_3 = B(\varphi_1)^3$ is a unijunction of sort c and $\varphi_1 \lhd \varphi_3$, hence from $q(\varphi, \varphi_1)$ we can infer $q(\varphi, \varphi_1 + B(\varphi_1)^3)$. Further deduction rules can be based on the relation $\preceq_c$ between $\langle b,b \rangle$ models. But the question of helpful deduction rules of such kinds is an open one.

7.4.9 Consider a monotone class $\{q_\alpha\}_\alpha$ of correlational quantifiers; for each α, let q^*_α be the extension of q_α to all models of type $\langle c,c \rangle$ described in 5.4.16. Let φ_0 be a fixed designated open formula and let A be a class consisting of some designated open formulae built up from $F_1, \ldots, F_{k_2}$ using junctors from Jct_1 and also +. Put $V_0 = \{1\}$, $I = \{PE\} \cup \{AD\} \cup \{M\}$, where

$$M = \left\{ \frac{q^*_\alpha(\varphi_1, \varphi_2)}{q^*_{\alpha'}(\varphi_1, \varphi_2)} ; \alpha' \geq \alpha, \varphi_1, \varphi_2 \right\} .$$

The aim is to find all "good" approximations (or correlates) of a given form to φ_1. (Cf. Bendová, Havránek.)

7.4.10 A class of rank quantifiers which cannot be treated on enumeration models are <u>regression rank quantifiers</u>. They are of type $\langle c,c \rangle$, and they are rank quantifiers. I.e., for each model of type $\langle c, c \rangle$, $Asf_q(\underline{M}) = Asf_q(Rk(\underline{M}))$. They have the following basic property: Let

$\underline{M}_1 = \langle M, f, g_1 \rangle$, $\underline{M}_2 = \langle M, f, g_2 \rangle$ be two models of type $\langle c, b \rangle$. Suppose that $f(o_1) < f(o_2)$ and $g_1(o_1) > g_1(o_2)$ implies $g_2(o_1) - g_2(o_2) < g_1(o_1) - g_1(o_2)$. Then $Asf_q(\underline{M}_1) = 1$ implies $Asf_q(\underline{M}_2) = 1$.

Such quantifiers can be treated similarly as in the previous cases, see Problem (1) of Chapter V.

7.4.11 There remains an important open question and a crucial point for quick application: In the present case can one construct deduction rules based on analogues of helpful quantifiers from Chapter VI?

7.4.12 Key words: r-problems with distinctive quantifiers; r-problems with correlational quantifiers; regression rank quantifiers.

PROBLEMS AND SUPPLEMENTS TO CHAPTER VII.

(1) Define a natural linearization of the designated ordering of relevant questions. This linearization is necessary for the successive generation of the solution and also for the interpretation of results.

(2) Define the parameter SYNTR in more details. Suggestion:
(a) One declares four sets of function symbols: IMPTA, REMNA, IMPTS, REMNS - important and remaining antecedent (succedent) function symbols. Say that an open formula φ respects a set B of function symbols if either $B = \emptyset$ or at least one function symbol from B occurs in φ. A sentence $\varphi \sim \psi$ satisfies the conditions on function symbols if (i) contains only function symbols from IMPTA $\cup$ REMNA and φ respects IMPTA and (ii) ψ contains only function symbols from IMPTS $\cup$ REMNS and ψ respects IMPTS.

(b) One declares maximal allowed length of antecedents and maximal allowed length of succedents.

(c) Conditions concerning coefficients: for each function symbol F_i in IMPTA $\cup$ REMNA one declares a set $H_i^A \subseteq V_i$ and a number c_i^A. Then F_i can only occur in an antecedent with arguments $X \subseteq H_i^A$ such that $\mathrm{card}(X) \leq c_i^A$. Similarly for H_i^S, c_i^S.

This is the realization from the textbook [Hájek at al.].Think of other possibilities.

(3) Specify parameters of the GUHA-method with associational quantifiers and find a model $\underline{M}$ in such a way that the conditions of 7.2.11(c) (end of the discussion) are satisfied.

(4) The "jump" principle: Given the linearization $<$ from Problem (1), the program realizing the GUHA-method with associational quantifiers can do the following: having processed a relevant question Φ which is not p-prime in the input model, ask whether a whole interval of relevant questions (w.r.t. $<$) beginning with Φ and ending with a Ψ can be omitted from the consideration (since either all elements of that interval are true but not prime or all are false).

This question is reasonable if we have a jump function $j(\Phi, \underline{M})$ determining ψ such that the computation of $j(\Phi, \underline{M})$ is, in many cases, quicker than the successive investigation of all members of the respective interval.

Consider the following possibilities of jumping:

I - for CPF: Φ is $\kappa \sim \lambda$. Assume that $<$ has the following property: relevant questions of any fixed length with the same antecedent form an interval. Denote the $\underline{M}$-frequency of κ by r, the $\underline{M}$ - frequency of λ by k, the $\underline{M}$-frequency of κ & λ by a and the cardinality of M by m. We know that $\|\kappa \sim \lambda\|_{\underline{M}}$ is determined by a, r, k, m, say, $\|\kappa \sim \lambda\|_{\underline{M}} = Asf_{\underset{\sim}{}}(a,r,k,m)$. Say that the $\underline{M}$-frequency of λ is too low w.r.t. κ if the following holds: For each a, $Asf_{\underset{\sim}{}}(a,r,k,m) = 0$. If the $\underline{M}$-frequency of λ is too low w.r.t. κ then look for an appropriate subconjunction $\lambda_0 \subseteq \lambda$ such that the $\underline{M}$-frequency of λ_0 is too low w.r.t. κ. Then for any λ' such that $\lambda_0 \subseteq \lambda'$ the $\underline{M}$- frequency of λ' is too low w.r.t. κ. This can be used to find $j(\Phi, \underline{M})$.

II - for EPF: (a) define (and use) the notions "the $\underline{M}$-frequency of δ w.r.t. κ is too high", "the $\underline{M}$-frequency of κ & δ w.r.t. κ is too low".
(b) Iff $\kappa \sim \delta$ is true but not prime then look for an appropriate $\delta_0 \subseteq \delta$ such that $\kappa \sim \delta_0$ is true but not prime; then, for any $\delta' \supseteq \delta_0$, $\kappa \sim \delta$ is true but not prime.

(5) Consider deduction rules I consisting of some pairs $\frac{\varphi}{\psi}$. Suppose that I is an equivalence on RQ. Each class E w.r.t. I can be represented by a minimal element of E w.r.t. an ordering on RQ. Thus we obtain a simple problem. (Cf. 7.2.6.) In this way, other kinds of problem may be generalized.

(6) In all cases, the associated functions of associational (or implicational) quantifiers from Theorem 7.3.11 can be transformed into the form $Asf(\underset{\sim}{N}) = 1$ iff $f(\underset{\sim}{N}) > c_{\alpha\sim}$ where $f_{\sim}$ is a statistic and $c_{\alpha\sim}$ a (critical) value ($\underline{N}$ means a regular model). Now we can apply the general principle of the "least favourable value", i.e., we look for a completion for which $f_{\sim}(\underline{N})$ attains the minimum value. Note hat our proofs of Theorem 7. 3. 11 and 7. 3. 12 in fact show that symmetric (implicational) completions (as defined in 7.3.10) are models in which the least favourable value is attained (for respective quantifiers). For further applications of this principle see V.2 and VII.4.

(7) For $f_1, f_2 : M \rightarrow \{0,1\}$ put $f_1 \leq f_2$ if $f_1(o) = 1$ implies $f_2(o) = 1$ for each $o \in M$. A distinctive quantifier q is <u>interpolable</u> if the following holds: Let $f_1 \geq f_2 \geq f_3$, $f_i : M \rightarrow \{0,1\}$ and $g \in \mathcal{R}_M$ be such that $Asf_q(\langle M, f_1, g\rangle) = Asf_q(\langle M, f_3, g\rangle) = 1$. Then there is a $g \in \mathcal{R}_M$ coinciding with g for each $o \in M$ such that $f_1(o) = 0$ or $f_3(o) = 1$ and such that $Asf_q(\langle M, f_2, g\rangle) = 1$.

Obviously, we may impose the following additional condition on g: If $f_1(o) = f_1(o') = 0$ and $f_3(o) = f_3(o') = 1$, and if $f_2(o) = 1$ and $f_2(o) = 0$, then $\hat{g}(o) > g(o')$. A sequence $f_1 > \ldots > f_k$ of mappings of M into $\{0, 1\}$ is a <u>path</u> if, for each $i = 1, \ldots, k-1$, f_i and f_{i+1} differ in exactly one object $o \in M$. Let g be a distinctive quantifier and let $Asf_q(\langle M, f_k, g\rangle) = Asf_q(\langle M, f_1, g\rangle) = 1$. The path $f_1, \ldots, f_k$ is <u>admissible</u> if $Asf_q(\langle M, f_i, g\rangle) = 1$ for each $i = 1, \ldots, k$. Prove that q is interpolable iff for each $\langle M, f_1, g\rangle, \langle M, f_3, g\rangle$ as above there is an admissible path from f_1 to f_2 w.r.t. g.

(8) Consider $\langle M, f_1, g\rangle$, $\langle M, f_k, g\rangle$, $f_k < f_1$. The <u>worst path</u> from f_1 to f_k w.r.t. g is defined as follows: f_i and f_{i+1} differs in o, for which

$$f_2(o) = \max\{f_2(o');\ f_i(o') = 1 \text{ and } f_k(o') = 0\}.$$

Analogously, we can define the best path. Prove the following:

(a) If the best path is not admissible, then no path is admissible.

(b) If the worst path is admissible, then each path is admissible.

Apply this fact to the deduction for each distinctive quantifier.

(9) A d-executive quantifier with

$$Asf_q(\langle M, f_1, g\rangle) = 1 \text{ if } \sum f_1(o)\,\underline{a}(f_2(o)) \geq c(m,r)$$

is interpolable if, for each $m_{min}(q)$ and each r such that $r-1, r+1 \in (r_{min}(m,q), r_{max}(m,q))$ we have $c(m,r) \leq 1/2\,(c(m,r-1) + c(m,r+1))$.

Are quantifiers based on asymptotical forms of the Wilcoxon and median statistics interpolable? (Hint: use a weaker inequality than the one mentioned abolve.)

Chapter VIII. Further Statistical Problems of the Logic of Discovery

As we have already mentioned in Section 1 of Chapter VI, there are some questions of a statistical nature related to the interpretation and exact understanding of the results obtained by methods of discovery, particularly by GUHA-methods. Roughly speaking, we have to answer the following two questions: given an r-problem $\mathcal{P}$, a model $\underline{M}$ and a solution X of $\mathcal{P}$ in $\underline{M}$: (1) What is the exact statistical meaning of a sentence belonging to X? (2) What is the exact meaning of X as a whole?

In Sections 1 and 2 of the present chapter we answer these questions in a general form, hence our results can be applied to other methods of a similar nature. It is clear that for particular methods one can obtain better results by using specific properties of the methods considered. Some of the ways of looking for such results are explained in Section 3.

Furthermore, in the same section we formulate some motivation for the further development of Mathematical Statistics; this motivation is one of the results of our investigations of the logic of discovery.

VIII.1 Local interpretation

In the present section, we shall investigate some local properties of statistical quantifiers which are important from the point of view of the interpretation of results obtained by GUHA-methods. First, we have to formulate some global frame assumptions guaranteeing the validity of some local assumptions concerning the statistical meaning of the fact that a pure prenex sentence is true in the model.

We shall investigate some particular cases of such assumptions, concerning the tests described in Chapters IV and V (and used in Chapter VII).

In the second part of the present section, the preservation of some statistical properties of hypothesis testing with the aid of GUHA-methods will be investigated. It will be shown that the local properties of hypothesis testing with the help of a GUHA-method are the same as in the case of single testing.

8.1.1 Discussion. In Section 4.4, we considered random structures related to monadic predicate calculi. If we have a fixed type $<1^n>$ and the corresponding predicate calculus $\mathcal{F}$, then a given random $\{0,1\}$- structure $\underline{U} = \langle U, Q_1, \ldots, Q_n \rangle$ and a designated open formula φ of $\mathcal{F}$ determines a random structure, denoted by $\underline{U}_\varphi$ (cf. 2.4.6 and 4.4.0). The generalization for more designated open formulae is natural.

Our question reads: Under what general conditions is it true that if $\underline{U}$ is a regular random V-structure and $\varphi_1, \ldots, \varphi_n$ are designated open formulae then $\underline{U}_{\varphi_1, \ldots, \varphi_n}$ is also a regular random structure?

The question is important as one of the adequacy questions for the methods described above: If we evaluate , e.g., various sentences $\varphi \sim \psi$ in a model $\underline{M}_{\sigma}$, a sample from a random universe $\underline{U}$, then, in fact, we are testing some hypotheses concerning the structures $\underline{U}_{\varphi, \psi}$ and we should know that the assumptions of the respective tests are satisfied (cf. Discussion 8.1.11).

We shall consider general random V-structures and the respective function calculi.

8.1.2 Lemma. Let $\underline{U} = \langle U, Q_1, \ldots, Q_n \rangle$ be a regular $\underline{\Sigma}$-random V-structure, let g be a Borel measurable function from V^k ($k \leq n$) into V.

Let Q_g be defined as follows (composition):

$$Q_g(o,\sigma) = g(Q_1(o,\sigma), \ldots, Q_n(o,\sigma)) .$$

Then $\underline{U}_g = \langle U, Q_g, Q_{k+1}, \ldots, Q_n \rangle$

is a regular $\underline{\Sigma}$-random V-structure.

Proof. Remember conditions (0) - (2) from 4.2.1. Condition (0) is directly satisfied. For (1), we have to prove that Q_g is a random quantity, i.e., that $Q(o, .)$ is a random variate for each $o \in U$. But $Q_1(o,.), \ldots, Q_n(o, .)$ are random variates (measurable functions) and g is a Borel function; so the composition is measurable. For (2), we use the following fact (we restrict ourselves to two n-dimensional variates; but the proof is similar for more variates): Let $\underline{\Sigma} = \langle \Sigma, \mathcal{R}, P \rangle$. Keep in mind the notation

$\underline{\vartheta}_o = \langle Q_1(o,.), \ldots, Q_n(o,.) \rangle$, and denote

$$\mathcal{R}(\underline{\vartheta}_o) = \{A \in \mathcal{R} ; \underline{\vartheta}_o(A) \in \mathcal{B}_n\},$$

where $\mathcal{B}_n$ is the n-dimensional Borel σ-algebra (cf. 4.1.9); $\mathcal{R}(\underline{\vartheta}_o)$ is the σ-algebra induced by the random variate $\underline{\vartheta}_o$. Similarly, introduce $\underline{\vartheta}^g_o$ and $\mathcal{R}(\underline{\vartheta}^g_o)$.

Now, $\underline{\vartheta}_{o_1}$ and $\underline{\vartheta}_{o_2}$ are stochastically independent (in $\underline{U}$) iff for each $A \in \mathcal{R}(\underline{\vartheta}_{o_1})$ and $B \in \mathcal{R}(\underline{\vartheta}_{o_2})$ we have

$$P(A \cap B) = P(A) . P(B).$$

Use the fact that

$$\mathcal{R}(\underline{\vartheta}^g_{o_1}) \subseteq \mathcal{R}(\underline{\vartheta}_{o_1}) \text{ and } \mathcal{R}(\underline{\vartheta}^g_{o_2}) \subseteq \mathcal{R}(\underline{\vartheta}_{o_2}).$$

8.1.3 Definition. Let V be a regular set of abstract values and let $\mathcal{F}$ be an MOFC of type $\langle 1^n \rangle$; assume that the models of $\mathcal{F}$ are all the finite

$(V \cap \mathbb{Q})$ - structures. Call $\mathcal{F}$ <u>continuously generated</u> if for each junctor ι of $\mathcal{F}$, of arity k, there is a function $g_\iota: V^k \longrightarrow V$ continuous on V^k such that Asf_ι is the restriction of g_ι to $(V \cap \mathbb{Q})^k$. Note that if such a g_ι exists then it is determined uniquely (since V is regular). If there is no danger of confusion, we write Asf_ι instead of g_ι.

8.1.4 <u>Remark.</u> (1) Let $\underline{\Sigma}$ be a probability space. Observe that $\mathcal{F}$ is continuously generated iff there is a theoretical calculus $\mathcal{F}^*$ satisfying the following: (i) Models of $\mathcal{F}^*$ are all (regular) $\underline{\Sigma}$ -random V-structures, (ii) $\mathcal{F}$ and $\mathcal{F}^*$ have the same function symbols and junctors; (iii) associated functions of the junctors in $\mathcal{F}^*$ are continuous and (iv) for each junctor ι of arity k,

$$Asf_\iota^{\mathcal{F}} = Asf_\iota^{\mathcal{F}^*} \restriction (V \cap \mathbb{Q})^k .$$

Note that $\mathcal{F}$ and $\mathcal{F}^*$ have the same open formulae (via the identification described in 4.4.0), and the semantics of open formulae on Σ - random structures is determined uniquely by $\mathcal{F}$. Hence e.g. for each $\underline{\Sigma}$ -random V-structure $\underline{U}$ and each tuple $\varphi_1, \ldots, \varphi_r$ of designated open formulae, the derived structure $\underline{U}_{\varphi_1, \ldots, \varphi_r} = \langle U, \|\varphi_1\|_{\underline{U}}, \ldots, \|\varphi_r\|_{\underline{U}} \rangle$ is uniquely determined by $\mathcal{F}$. If the structure $\underline{U}$ is fixed in a consideration, we write Q_φ instead of $\|\varphi\|_{\underline{U}}$.

(2) The generalization to $\mathbb{V}$ -structures and to calculi with more sorts of function symbols is obvious. In the case of the calculi described in Chapter V, we have junctors applicable either to sort a or to sort c. For the first sort, we need no conditions, to the second we apply the continuity condition.

8.1.5 <u>Definition.</u> Let a continuously generated calculus be given. Let PF be a set of pairs of designated open formulae. We say that a distributional statement Φ is <u>global</u> w.r.t. Ψ and PF if the following holds: $\underline{U} \models \Phi$ implies $\underline{U}_{\varphi_1, \varphi_2} \models \Psi$ for each $\langle \varphi_1, \varphi_2 \rangle \in$ PF.

In the following lemmas we apply some cases of global frame assumptions.

8.1.6 <u>Lemma</u>. Let $\mathcal{F}$ be continuously generated. If $\underline{U}$ is d-homogeneous, then, for each pair φ_1, φ_2 of designated open formulae, $\underline{U}_{\varphi_1, \varphi_2}$ is d-homogeneous.

Proof. Let $\underline{\Sigma} = \langle \Sigma, \mathcal{R}, P \rangle$ and consider $\underline{\Sigma}$-random structures: such a structure is d-homogeneous if the joint distribution function is independent of $o \in U$. One can easily prove that $\underline{U}$ is d-homogeneous iff, for each Borel set $A \subseteq \mathbb{R}^n$, $P(\{\sigma; \langle Q_1(o,\sigma), \ldots, Q_n(o,\sigma)\rangle \in A\}$ does not depend on o. It follows by an easy induction that for each tuple $\varphi_1, \ldots, \varphi_k$ of designated open formulae, each $o \in U$ and each Borel set $B \subseteq \mathbb{R}^k$, there is a Borel set $A \subseteq \mathbb{R}^n$ such that

$$\langle Q_1(o,\sigma), \ldots, Q_n(o,\sigma)\rangle \in A \quad \text{iff} \quad \langle Q_{\varphi_1}(o,\sigma), \ldots, Q_{\varphi_k}(o,\sigma)\rangle \in B .$$

Hence, $P(\{\sigma; \langle Q_{\varphi_1}(o,\sigma), \ldots, Q_{\varphi_k}(o,\sigma)\rangle \in B\})$ is independent of o.

8.1.7 Lemma. Consider d-homogeneous $\{0,1\}$ - structures and MOPC's of a given type $\langle 1^n \rangle$. Let Φ be the following distributional statement: For each $\varepsilon \in \{0,1\}^n$, $P(\{\sigma; \mathcal{V}_o(\sigma) = \varepsilon\}) > 0$. If Φ then for each pair of independent designated open formulae φ_1, φ_2 we have $p_{\varphi_1 \& \varphi_2} > 0$,

$p_{\neg\varphi_1 \& \varphi_2} > 0$, $p_{\varphi_1 \& \neg\varphi_2} > 0$, $p_{\neg\varphi_1 \& \neg\varphi_2} > 0$.

The proof is left to the reader.

8.1.8 Discussion. Thus we have global assumptions guaranteeing the satisfaction of the frame assumptions for each pair φ_1, φ_2 w.r.t. the tests based on the quantifiers $\sim_\alpha, \sim_\alpha^2, \sim_\alpha^3$ or $\Rightarrow^!_{p,\alpha}$ ($\Rightarrow^?_{p,\alpha}$) respectively.

The following lemma gives, similarly, the frame assumption for the tests of H_2^- and H_2. In H_2^- we consider pairs of formulae: for the first formula in each pair we have to guarantee the positivity condition by the global assumption of 8.1.7.

8.1.9 Lemma. Let $\mathcal{F}$ be continuously generated. Assume, moreover, that all the functions g_1 are one-to-one (more generally, it suffices to assume that the pre-image of each $u \in V$ is at most countable). Let Φ be the following distributional statement: "For each $o \in U$, $D_{\underline{\mathcal{V}}_o}$ is continuous". Then $\underline{U} \vdash \Phi$ implies for each designated open formula φ that the distribution function D_{Q_φ} of Q_φ is continuous.

<u>Proof.</u> The distribution function D_ϑ of a variate ϑ on $\langle \Sigma, \mathcal{R}, P \rangle$ is continuous iff, for each $u \in \mathbb{R}$, $P(\{\sigma ; \vartheta(\sigma) = u\}) = 0$. Let the random structures we consider be n-dimensional. Under our assumptions, the function g such that $Q_\varphi(o,\sigma) = g(Q_1(o,\sigma), \ldots, Q_n(o,\sigma))$ for each o and σ is one-to-one. Hence, for each $u \in V$, $g^{-1}(u)$ is a point in $\mathbb{R}^n$ and, for each $o \in U$, $P(\{\sigma; \langle Q_1(o,\sigma), \ldots, Q_n(o,\sigma)\rangle = g^{-1}(u)\}) = 0$.

8.1.10 <u>Theorem</u> (global null hypothesis of independence). Let $\mathcal{F}$ be continuously generated and let Φ be the following distributional statement:

For each $o \in U$, $\vartheta_{1,o}, \ldots, \vartheta_{n,o}$

are stochastically independent.
Moreover, let PF be a set of pairs of disjointed designated open formulae. Then $\underline{U} \models \Phi$ implies that, for each $o \in U$ and $\varphi_1, \varphi_2 \in$ PF, the variates $\vartheta_{\varphi_1,o}$ and $\vartheta_{\varphi_2,o}$ are stochastically independent.

<u>Proof</u>. Supoose that φ_1 contains function symbols $F_1, \ldots, F_{k_1}$ while φ_2 contains function symbols $F_{k_2}, \ldots, F_n$, $k_1 < k_2$. For each $A_1 \in \mathcal{R}(\vartheta_{\varphi_1,o})$ and $A_2 \in \mathcal{R}(\vartheta_{\varphi_2,o})$, we have to consider $P(A_1 \cap A_2)$. But

$$\mathcal{R}(\vartheta_{\varphi_1,o}) \subseteq \mathcal{R}(\vartheta_{1,o}, \ldots, \vartheta_{k_1,o}) \text{ and } \mathcal{R}(\vartheta_{\varphi_2,o}) \subseteq \mathcal{R}(\vartheta_{k_2,o}, \ldots, \vartheta_{m,o}).$$

Use inductively the measurability of $g_\imath$ for each junctor.

8.1.11 <u>Discussion.</u> (1) If we consider the null hypothesis of the stochastical independence of the two quantities Q_{φ_1}, Q_{φ_2} corresponding to **logically** independent designated open formulae φ_1, φ_2, then rejecting this null hypothesis (i.e., inferring, on the basis of some data $\underline{M}$, the alternative hypothesis) means rejecting the global hypothesis of independence too.

(2) Consider now, under some frame assumptions Φ, a null hypothesis Φ_0 and an alternative hypothesis Φ_A, each of these distributional statements

concerning random structures of type $\langle 1, 1\rangle$. Let f be a test statistic (i.e., $P^{\underline{U}}(\{\sigma ; f(\underline{M}_{\sigma}) \in V_0\}) \leq \alpha$ whenever $\underline{U} \models \Phi$ & Φ_0). Now, if we have a simultaneous frame assumption Ψ w.r.t. Φ and a set of pairs of formula PF, and if $\underline{U} \models \Psi$, then we can use this test for testing samples obtained from $\underline{U}_{\varphi_1, \varphi_2}$ for each $\langle \varphi_1, \varphi_2\rangle \in$ PF.
Note that the assertion "Φ_0 is true in $\underline{U}_{\varphi_1, \varphi_2}$" expresses a particular property of the original structure $\underline{U}$ since $\underline{U}_{\varphi_1, \varphi_2}$ is derived from $\underline{U}$; hence, let $\Phi_0[\varphi_1, \varphi_2]$ be a sentence such that

$$\underline{U} \models \Phi_0[\varphi_1, \varphi_2] \quad \text{iff} \quad \underline{U}_{\varphi_1, \varphi_2} \models \Phi_0 ;$$

similarly for Φ_A, etc. .

(3) We can now be more specific as to the structure of statistical inference rules considered in 4.3.3. Let q be the quantifier defined by f. We have the rule

$$\left\{ \frac{\Psi, q(\varphi_1, \varphi_2)}{\Phi_A[\varphi_1, \varphi_2]} ; \langle \varphi_1, \varphi_2\rangle \in \text{PF} \right\} .$$

Cf. again 1.1.6 (L3).

(4) Moreover, the same conclusion can be made for other cases of statistical inference, e.g. for point estimation.

8.1.12 Example. (1) Consider random $\{0, 1\}$-structures and the corresponding MOPC's. Under the global assumption Φ from 8.1.7, the sentences $\varphi_1 \sim \varphi_2$, where $\langle \varphi_1, \varphi_2\rangle \in$ CPF (or $\langle \varphi_1, \varphi_2\rangle \in$ EPF) and $\sim$ is $\sim_\alpha, \sim_\alpha^2, \sim_\alpha^3$ or $\Rightarrow^!_{p,\alpha}$ (or $\Rightarrow^!_{p,\alpha}$) , can serve as observational tests of null hypotheses $Asc_0[\varphi_1, \varphi_2]$ and alternative hypotheses $Asc_A[\varphi_1, \varphi_2]$, where, for CPF and $\sim_\alpha, \sim_\alpha^2, \sim_\alpha^3$ Asc_0 is the hypothesis of independence and Asc_A is the alternative of positive association (cf. 4.4.27), while for CPF or EPF and $\Rightarrow^!_{p,\alpha}$ they are respectively the null and alternative hypotheses specified in 4.4.16 (equivalent to $p_{\varphi_2/\varphi_1} \leq p$ and $p_{\varphi_2/\varphi_1} > p$).

Hence, we have the inference rule

$$\left\{\frac{\Phi, \varphi_1 \sim \varphi_2}{\mathrm{Asc}_A[\varphi_1, \varphi_2]} ; \langle \varphi_1, \varphi_2 \rangle \in \mathrm{PF}\right\},$$

where PF and $\sim$ are specified above.

(2) Similarly if PF is CPF or EPF and $\Rightarrow^?_{p,\alpha}$, but here the rule is

$$\left\{\frac{\Phi, \varphi_1 \Rightarrow^?_{p,\alpha} \varphi_2}{\mathrm{Imp}_o[\varphi_1, \varphi_2]} ; \langle \varphi_1, \varphi_2 \rangle \in \mathrm{PF}\right\},$$

where $\mathrm{Imp}_o[\varphi_1, \varphi_2]$ means $p_{\varphi_2/\varphi_1} \geq p$.
The situation is slightly different for $\sim$ and $\Rightarrow_p$ (simple quantifiers).

Here, we have

$$\left\{\frac{\Phi, \varphi_1 \sim \varphi_2}{\mathrm{Asc}_A[\varphi_1, \varphi_2]} ; \langle \varphi_1, \varphi_2 \rangle \in \mathrm{PF}\right\}$$

and $\left\{\frac{\Phi, \varphi_1 \Rightarrow_p \varphi_2}{\mathrm{Imp}_o[\varphi_1, \varphi_2]} ; \langle \varphi_1, \varphi_2 \rangle \in \mathrm{PF}\right\}$,

but our criteria for these inferences are the criteria of point estimation only.

8.1.13 Discussion. Our aim now is to investigate random V-structures of a given type, where $V \subseteq \mathbb{Q}$. Let us then have a MOPC $\mathcal{F}$ of the appropriate type, with models which are V - structures. Applying a GUHA-method, we consider, moreover, an r-problem $\langle \mathrm{RQ}, \mathrm{I}, V_o \rangle$. Let a sentence φ from RQ be an observational test of the null hypothesis Φ_o and of an alternative hypothesis Φ_A on the significance level α (under some frame assumptions). We have now a universe of discourse $\underline{U}$ (regular random V-structure).

Instead of evaluating $\|\varphi\|_{\underline{M}_\sigma}$ for obtained $\underline{M}_\sigma$, i.e. instead of testing Φ_o and Φ_A directly, we can use a GUHA-method. In fact, we use a procedure $\mathcal{X}$

giving, for each $\underline{M}_\sigma$, a solution $X_{\underline{M}_\sigma}$; we accept Φ_A if φ is an immediate conclusion from $X_{\underline{M}_\sigma}$. What are the properties of this inference based on $X_{\underline{M}_\sigma}$?

8.1.14 Theorem. For each $\underline{U}$, for each $\varphi \in RQ$ and for each sample $M \subseteq U$, we have

$$P^{\underline{U}}(\{\sigma ; \varphi \in Tr_{V_o}(\underline{M}_\sigma)\}) = P^{\underline{U}}(\{\sigma ; \varphi \in X_{\underline{M}_\sigma}\} \cup \bigcup_{\{B; \frac{B}{\varphi} \in I\}} \{\sigma; B \subseteq X_{\underline{M}_\sigma}\}.$$

Proof. This follows immediately from the obvious fact that for each finite sample

$$\{\sigma; \varphi \in Tr_{V_o}(\underline{M}_\sigma)\} = \{\sigma : \varphi \in X_{\underline{M}_\sigma}\} \cup \bigcup_{\{B; \frac{B}{\varphi} \in I\}} \{\sigma; B \subseteq X_{\underline{M}_\sigma}\}.$$

The inclusion $\supseteq$ follows from the soundness of I, and $\subseteq$ follows from the following basic property of solutions: If for some $\varphi \in RQ$ and $\underline{M}_\sigma$ we have $\varphi \in Tr_{V_o}(\underline{M}_\sigma)$, then either $\varphi \in X_{\underline{M}_\sigma}$ or there is a $B \subseteq X_{\underline{M}_\sigma} \subseteq RQ \cup AQ$ such that $\frac{B}{\varphi} \in I$ (where AQ is a set of auxiliary questions ; see 6.2).

8.1.15 Remark. In 8.1.13 and 8.1.14, we supposed that the random V-structures concerned were such that $V \subseteq \mathbb{Q}$. If $V \nsubseteq \mathbb{Q}$, then the above must be reformulated in the obvious way using the notion of a. c. c. statistics (cf. 5.1.2).

8.1.16 Discussion and Definition. (1) Now, if a $\varphi \in RQ$ is an observational test of Φ_o and Φ_A (on the level α) , then the testing of Φ_o against Φ_A with the aid of the procedure $\mathcal{X}$ has the same characteristics, i.e., significance level and power, as the single test φ . We have similar results for the other kinds of statistical inference (for example for inferences based on the quantifier $\Rightarrow^{?}_{p,\alpha}$) .

(2) Consider observational tests of a null hypothesis Φ_o and an alternative hypothesis Φ_A. Such tests can be considered on different significance levels from the interval $(0,0.5]$. In fact, using a computer, we can consider significance levels α belonging only to a finite ε- net on $(0,0.5]$, i.e., to a finite subset T of $(0,0.5]$ such that for each $\alpha \in (0,0.5]$ there is a $\beta \in T$ such that $|\beta - \alpha| \leq \varepsilon$.

(3) Now let an ε-net T be given. We say that $\{\varphi(\alpha)\}_{\alpha \in T}$ forms a <u>full monotone class</u> of tests (of Φ_o against Φ_A) w.r.t. the net T if

(i) each $\varphi(\alpha)$ is a test of Φ_o on the significance level α ,

(ii) if $\alpha' > \alpha$ and $\alpha, \alpha' \in T$, then $\varphi(\alpha)$ logically implies $\varphi(\alpha')$.

(4) In the following we shall assume an ε -net T to be given.

(5) Note that for T and for all open formulae φ_1, φ_2 of the appropriate sorts the strictly monotone class of quantifiers $\{q_\alpha\}_{\alpha \in T}$ (cf. 5.3.22) defines a full monotone class of tests $\{q_\alpha(\varphi_1, \varphi_2)\}_{\alpha \in T}$, possessing a certain optimality property w.r.t. the power of tests (this last fact is due to the definition of the quantifier being of the level α) .

(6) In the following , we shall consider monotone classes of tests. This means that when we speak about a test of a given Φ_o against Φ_A we shall mean a full monotone class of tests of Φ_o and Φ_A.

(7) Now let all $\varphi \in$ RQ be tests on a given level (instead of RQ write RQ (α)); then the obtained solution is called the <u>solution on the level α</u> (and denoted by $X(\alpha)_{\underline{M}_\sigma}$) . Under our assumptions, $X(\alpha')_{\underline{M}_\sigma} \subseteq X(\alpha)_{\underline{M}_\sigma}$ for $\alpha' \leq \alpha$ if the solutions are direct .

Suppose now that if $\frac{B}{\varphi} \in$ I , then $B = \{\psi\}$ for some ψ such that $\varphi, \psi \in$ RQ for some $\alpha \in$ T. Such a deduction rule is called <u>invariant</u> if the following holds:

If, for some $\alpha \in$ T , $\frac{\psi(\alpha)}{\varphi(\alpha)} \in$ I , then, for each $\alpha \in$ T , $\frac{\psi(\alpha)}{\varphi(\alpha)} \in$ I.

8.1.17 <u>Theorem</u>. Let an r-problem $\mathcal{P} = \langle$ RQ(α) , V_o, I $\rangle$. be given. Let I be invariant and let $X(\alpha)_{\underline{M}_\sigma}$ be a solution of $\mathcal{P}$. Let $0<\alpha'<\alpha$, $\alpha, \alpha' \in$ T. Then

$X'_{M} = \{ \varphi(\alpha') \in Tr(\underline{M}_\sigma) \; ; \; \varphi(\alpha) \in X_{\underline{M}_\sigma} \}$ is a solution of $< RQ(\alpha), V_o, I >$.

Proof. Remember that we have full monotone classes of tests. Then $\varphi(\alpha) \in RQ(\alpha)$ implies that $\varphi(\alpha')$ is a test of the same Φ_o, Φ_A as $\varphi(\alpha)$. If $\psi(\alpha') \in X'_{\underline{M}_\sigma}$, then $\frac{\psi(\alpha)}{\varphi(\alpha)} \in I$ iff $\frac{\psi(\alpha')}{\varphi(\alpha')} \in I$ and $I(\psi(\alpha)) \subseteq Tr_{V_o}(\underline{M}_\sigma)$; on the other hand, if a $\varphi(\alpha') \in Tr_{V_o}(\underline{M}_\sigma)$, then $\varphi(\alpha) \in Tr_{V_o}(\underline{M}_\sigma)$; hence, $\varphi(\alpha) \in I(X(\alpha)_{\underline{M}_\sigma})$.

8.1.18 Discussion. (1) The result $\varphi(\alpha') \in Tr_{V_o}(\underline{M}_\sigma)$ for $\alpha' < \alpha$ is stronger in the statistical sense than $\varphi(\alpha) \in Tr_{V_o}(\underline{M}_\sigma)$. By the previous theorem, having a solution on a level α, a solution on a level $\alpha' < \alpha$ can be found as a subset of $\{\varphi(\alpha') \; ; \; \varphi(\alpha) \in X_{M\sigma}\}$.

(2) We can consider a more general case:

$B = \{\psi(\alpha)\} \cup B_1$, where $B_1 \subseteq AQ$. Then $X'_{\underline{M}_\sigma} = Y_{\underline{M}_\sigma} \cup (AQ \cap X_{\underline{M}_\sigma})$, where

$$Y_{\underline{M}_\sigma} = \{ \varphi(\alpha') \in Tr_{V_o}(\underline{M}_\sigma) \; ; \; \varphi(\alpha) \in X_{\underline{M}_\sigma} \cap RQ \}.$$

Invariance of the rule means here that if $\frac{\{\psi(\alpha)\} \cup B_1}{\varphi(\alpha)} \in I$ for an $\alpha \in T$, then $\frac{\{\psi(\alpha)\} \cup B_1}{\varphi(\alpha)} \in I$ for each $\alpha \in T$.

Note that our associational quantifiers of the test type (i.e., $\sim_\alpha, \sim^2_\alpha, \sim^3_\alpha, \Rightarrow^!_{p,\alpha}$) lead to full monotone classes and our deduction rules (cf. Chapt. VI and VII) are of the above mentioned type (SpRd is of the simpler type from 8.1.17).

(3) Let $\varphi(\alpha) \in X_{\underline{M}_\sigma}(\alpha)$ for some $\underline{M}_\sigma$ and $\alpha \in T$. We shall consider the critical level $\alpha(\varphi, \underline{M}_\sigma) = \min_{\alpha \in T} \{\alpha ; \varphi(\alpha) \in Tr_{V_o}(\underline{M}_\sigma)\}$. If I is invariant, then we know that $\alpha(\varphi', \underline{M}_\sigma) \leq \alpha(\varphi, \underline{M}_\sigma)$ for each $\varphi' \in I(\varphi(\alpha) \cup (AQ \cap X_{\underline{M}_\sigma})) \cap RQ$.

The procedures described in [Hájek 1969] and [Hájek, Bendová, Renc] and considered here in Chapter VII give, in fact, (i) a solution $X(\alpha)_{\underline{M}_\sigma}$ for a given level α (parameter of the method) and (ii) the critical levels for sentences of $X(\alpha)_{\underline{M}_\sigma} \cap RQ$.

This will be useful for considerations of the following section; cf. also point (1) of the present discussion.

8.1.19 Key words: global frame assumptions, continuously generated function calculi, preservation of regularity, d-homogeneity and independence, global assumption of positivity; form of statistical inference rules; local properties of GUHA-methods, invariance of deduction rules, full monotone classes of tests.

VIII.2 Global interpretation

The present section is devoted to the problem of global statistical interpretation of the results obtained by GUHA-methods. Thus, we investigate errors of statistical inference based on sets of observational sentences (tests) true in some given data. This situation is related to the problem of simultaneous statistical inference as formulated in the literature. But our situation requires the investigation of cases differing from cases usual for the application of simultaneous inference. Most of the results we obtain are independent of the structure of sentences and of particular relations in the set of relevant questions. Our results are not too advanced, but they solve completely the problem of the global interpretation of the particular methods considered in this book and indicate the possibilities of easy statistical interpretation of further GUHA-methods.

8.2.1 Discussion and definition. In the following, we shall suppose that all sentences from RQ are (names of) tests of a pair of null and alternative hypotheses and

(i) each $\varphi \in$ RQ is from a full monotone class (w.r.t. a given ε-net T),

(ii) all $\varphi \in$ RQ are of the same significance level $\alpha \in$ T, i.e., we have RQ(α) (cf. 8.1.16).

Thus, we shall consider pairs of theoretical sentences of the form $\underline{\Phi} = \langle \Phi_0, \Phi_A \rangle$, and assume that we have a one-to-one mapping τ associating with each $\varphi \in$ RQ a pair $\underline{\Phi} = \langle \Phi_0, \Phi_A \rangle$ such that φ (under some global frame assumptions)is (names) an observational test of the null hypothesis Φ_0 against the alternative hypothesis Φ_A on the significance level α . Put TQ = τ(RQ).(Note that under our assumptions TQ = τ(RQ(α)) for any $\alpha \in$ T .)So if we have a $\varphi \in$ RQ , it is, in fact, a $\varphi(\alpha)$ from a monotone class. But if there is no danger of a misunderstanding we shall write only φ instead of $\varphi(\alpha)$.

Assume TQ to be finite, card TQ = t. Hence, for each $\alpha \in$ T , the corresponding RQ(α) is finite and of the same cardinality. On the other hand, put $RQ^* = \bigcup_{\alpha \in T} RQ(\alpha)$; the cardinality of RQ^* can be much larger than that of TQ.

Say that a sentence Ψ belongs to $<\Phi_o, \Phi_A>$ if Ψ is either the sentence Φ_o or the sentence Φ_A. A set Z of theoretical sentences is a component of TQ if (i) for each $\Psi \in Z$ there is a $\underline{\Phi} \in TQ$ such that Ψ belongs to $\underline{\Phi}$, and (ii) the conjunction $\wedge Z$ of all members of Z is consistent (i.e., there is a random structure $\underline{U}$ satisfying the global conditions in which $\wedge Z$ is true).

8.2.2 Remark (1) It follows from our finiteness assumption that each component is contained in a maximal component.

(2) It follows from the condition (ii) above that there is no $<\Phi_o, \Phi_A> \in TQ$ such that both Φ_o and Φ_A are in Z.

8.2.3 Discussion. Consider random V-structures satisfying some global frame assumptions. Moreover, consider an r-problem $\mathcal{P} = < RQ, V_o, I >$ under the above conditions on RQ. We shall assume, in accordance with the cases described in Section 8.1, that $\mathcal{H}_o = \{\Phi_o ; \Phi \in TQ\}$ is a maximal component. As measures of the statistical quality of an obtained solution $X = X_{\underline{N}}$, where $\underline{N} \in \mathfrak{M}^V_M$, we can use the following probabilities:

(1) $P_I(X) = P(\{\sigma ; I(X) \cap RQ \cap Tr(\underline{M}_\sigma) \neq \emptyset\} | \wedge \mathcal{H}_o)$

and (2) $P_{II}(X) = P(\{\sigma : I(X) \cap RQ \cap D \subseteq Tr(\underline{M}_\sigma)\} | \wedge \mathcal{H})$,

where $\mathcal{H} \neq \mathcal{H}_o$ is a maximal component and $D = \{\tau^{-1}(\underline{\Phi}) ; \Phi_A \in \mathcal{H}\}$.

Probability (1) is the probability of the global error of the first kind. (2) corresponds to the global power of the procedure giving X .

We shall restrict ourselves to the more substantial case, i.e., to the probability of the global error of the first kind.

8.2.4 Lemma. Let $\alpha \in T$ be given. If $P_I(X) \leq \alpha$ for each solution X of the given r-problem, then $P(\{\sigma ; \tau^{-1}(\underline{\Phi}) \in Tr(\underline{M}_\sigma)\} | \Phi_o) \leq \alpha$ for each $\Phi_o \in \mathcal{H}_o$.

The proof is obvious, for each $\varphi \in RQ \cap I(X)$ we have

$$P_I(X) \geq P(\{\sigma ; \varphi \quad Tr(\underline{M}_\sigma)\} \mid \Phi_o).$$

8.2.5 Definition. Let $\varphi(\alpha)$ be a sentence from $RQ \cap Tr(\underline{N})$ for a given model $\underline{N}$. The sentence $\varphi^{crit}_{\underline{N}} = \varphi(\alpha')$ where $\alpha' = \alpha(\varphi, \underline{N})$, is called the critical strengthening of $\varphi(\alpha)$. If Y is a set of sentences from $RQ(\alpha) \cap Tr(\underline{N})$, then we define the critical strengthening of Y in $\underline{N}$ as

$$Y^{crit}_{\underline{N}} = \{\varphi^{crit}_{\underline{N}} ; \varphi(\alpha) \in Y\}$$

(for $\alpha(\varphi, \underline{N})$ see 8.1.18).

8.2.6 Remark. We can now consider, instead of the given solution $X_{\underline{N}}$ its strengthening $X^{crit}_{\underline{N}} = (X_{\underline{N}} \cap RQ)^{crit}_{N} \cup (X_{\underline{N}} \cap AQ)$.

If I is invariant, then we have the following:

(1) $\tau(I(X_{\underline{N}}) \cap RQ) = \tau(I(X^{crit}_{\underline{N}}) \cap RQ^*)$;

we make the same inferences from $I(X^{crit}_{\underline{N}}) \cap RQ$ as from $I(X_N) \cap RQ$.

(2) All observational tests from $I(X^{crit}_{\underline{N}}) \cap RQ$ are on the level less than or equal to

$$\alpha_{max} = \max\{\alpha(\varphi, \underline{N}) ; \varphi \in RQ \cap X_{\underline{N}}\}.$$

8.2.7 Lemma. Let $X_{\underline{N}}$ be a direct **solution. Then:**

(1) If $X_{\underline{N}}$ is a solution on the level less than or equal to $\frac{\alpha}{k}$, where $k = \text{card}(I(X_{\underline{N}}) \cap RQ)$, then $P_I(X_{\underline{N}}) \leq \alpha$.

(2) Let $I \subseteq RQ \times RQ$ be an invariant deduction rule (with one-element antecedents). Put, for each $\varphi \in I(X_{\underline{N}}) \cap RQ$,

$$\alpha_\varphi = \min\{\alpha(\psi, \underline{N}) ; \psi \in X_{\underline{N}} \text{ and } \frac{\psi}{\varphi} \in I\}.$$

Then $\sum_{\varphi \in I(X_{\underline{N}}) \cap RQ} \alpha_\varphi \leq \alpha$ implies $P_I(X^{crit}_{\underline{N}}) \leq \alpha$.

Proof. Both assertions follow from the fact that

$$\{\sigma;\ I(X_{\underline{N}}) \cap RQ \cap Tr(\underline{M}_{\sigma}) \neq \emptyset\} = \bigcup_{\varphi \in I(X_{\underline{N}}) \cap RQ} \{\sigma; \varphi \in Tr(\underline{M}_{\sigma})\}.$$

Then $P_I(X_{\underline{N}}) \leq \sum_{\varphi \in I(X_{\underline{N}}) \cap RQ} P(\{\sigma; \varphi \in Tr(\underline{M}_{\sigma})\} \mid \mathcal{H}_0)$.

For $X_{\underline{N}}^{crit}$ use $\alpha(\varphi, \underline{N}) \leq \alpha_{\varphi}$ and instead of RQ use the set RQ^*.

8.2.8 Remark. Consider deduction rules of the form $\left\{\frac{\{\psi\}, B}{\varphi}\right\}$, where $\varphi, \psi \in RQ$ (thus they are tests) and $B \subseteq AQ$; hence, a solution can be indirect. Then " $X_{\underline{N}}$ is a solution on the level α " means that the tests $X_{\underline{N}} \cap RQ$ are of the significance level α; in fact, we have all tests from RQ on the significance level α (cf. 8.1.16(7)). Then,(1) of 8.2.7 holds. Moreover, let I be of the above form and invariant (in the sense of 8.1.18 (2)).

If $\frac{\{\psi\} \cup B_1}{\varphi} \in I$, then $\alpha(\varphi, \underline{N}) \leq \alpha(\psi, \underline{N})$. Put

$$\alpha_{\varphi} = \min\{\alpha(\psi, \underline{N});\ \psi \in X_{\underline{N}} \cap RQ \text{ and } \frac{\{\psi\} \cup B}{\varphi} \in I \text{ for a } B \subseteq X_{\underline{N}} \cap AQ\}.$$

Then, (2) of 8.2.7 holds.

8.2.9 Discussion. We shall consider the global properties of solutions in connection with error rates used in multiple comparison problems (see e.g., [Balaam, Federer],[O'Neill, Wetheril].

The first error rate which we are going to consider is the inferencewise error rate:

$$(\text{i.e.r.}) = \frac{\text{number of erroneous inferences}}{\text{number of inferences}}$$

Remember that we are interested in hypothesis testing and our inference rules are then of the form $\left\{\frac{\Phi, \varphi}{\tau(\varphi)_A}\right\}$, i.e., we infer alternative hypotheses.

Such an inference is, naturally, erroneous if $\tau(\varphi)_0$ is true (i.e., if $\underline{U} \models \tau(\varphi)_0$ for the investigated universe $\underline{U}$).

8.2.10 Theorem. Consider an r-problem $\mathcal{P} = \langle RQ, I, \{1\}\rangle$ such that under some frame assumptions the conditions of 8.2.1 are satisfied . Let I be invariant and let $X_{\underline{N}}$ be a solution of $\mathcal{P}$. Then for $X_{\underline{N}}^{crit}$ and each maximal component $\mathcal{H}$ of TQ we have

$$E((i.e.r.) \mid \wedge\mathcal{H}) \leq \alpha_{max}.$$

Proof. Recall that $\alpha_{max} = \{\alpha(\varphi, \underline{N});\ \varphi \in X_{\underline{N}} \cap RQ\}$ and that sentences from RQ attain only the values 0 or 1. Let $I(X_{N}^{crit}) \cap RQ^{*} = \{\varphi_1, \dots, \varphi_k\}$ and let $\underline{\Phi}_1, \dots, \underline{\Phi}_k$ be the corresponding theoretical pairs ($\underline{\Phi}_i = \langle \Phi_{oi}, \Phi_{Ai}\rangle$). Assuming $\wedge\mathcal{H}$ we have

$$(i.e.r.)(\underline{M}_{\sigma}) = \frac{1}{k}\sum_{\Phi_{oi}\in\mathcal{H}} \|\varphi_i\|_{\underline{M}_{\sigma}} .$$

Hence, for each $\underline{M}_{\sigma}$, i.e.r. attains **its maximum under the assumption** $\wedge\mathcal{H}_o$. Then, for each sample,

$$E((i.e.r) \mid \wedge\mathcal{H}_o) = E\left(\frac{1}{k}\sum_{i=1}^{k} \|\varphi_i\| \;\Big|\; \wedge\mathcal{H}_o\right) \leq$$

$$\frac{1}{k}\sum_{i=1}^{k} P(\{\sigma;\ \|\varphi_i\|_{\underline{M}_{\sigma}} = 1\} \mid \Phi_{oi}) \leq \frac{1}{k}\, k\, \alpha_{max} .$$

8.2.11 Discussion. If our aim is to obtain information on the given data which is as complete as possible, and if we do not intend to make reliable conclusions depending directly on the simultaneous correctness of inferences based on $RQ \cap Tr(\underline{N})$, then it is appropriate to consider inferencewise error rates (cf. [Cox]). The original aim of GUHA-methods was to give such complete information (see Postcript), and thus a possibility of the choice of some interesting hypotheses for further testing.

On the other hand, if we point out the reliability of the conclusions based on the whole or on part of $RQ \cap Tr(\underline{N})$, we can use the most rigorous error rate which is the following first experimentwise error rate:

$$(I.\ e.e.r.) = \frac{\text{number of erroneous inferences}}{\text{number of experiments}} .$$

We have a model $\underline{M}$, i.e., the output of one experiment (in the sense of which the above "experiment" is to be interpreted). Then, under $\wedge\mathcal{H}_\circ$, we have

$$(\text{I. e.e.r})(\underline{M}_\sigma) = \sum_{i=1}^{k} \|\varphi_i\| \, \underline{M}_\sigma .$$

8.2.12 Theorem. Under the assumptions of 8.2.10, if $\alpha_\varphi \leq \frac{\alpha}{k}$ for each $\varphi \in I(X_{\underline{N}}) \cap RQ$, then for $X_{\underline{N}}^{crit}$ and each maximal component we have

$$E((\text{I.e.e.r}) \mid \wedge \mathcal{H}) \leq \alpha$$

Proof. Apply Lemma 8.2.7 (1) to $X_{\underline{N}}^{crit}$ and follow the proof of Theorem 8.2.10.

8.2.13 Discussion and conclusions. The present concept of error rate corresponds to R.A.Fisher's concept of error rate in multiple comparison tests.

It seems to be quite impossible in real situations to satisfy the condition $\alpha_\varphi \leq \frac{\alpha}{k}$ for each $\varphi \in I(X_{\underline{N}}) \cap RQ$ for the simple reason that $I(X_{\underline{N}}) \cap RQ$ is supposed to be large. In such situations, we can restrict and strengthen the result of Theorem 8.2.12 in the following way. Note that this way is general and does not depend on the structure of TQ and RQ.

If we have a solution $X_{\underline{N}}$, we see immediately $I(X_{\underline{N}}) \cap RQ$. We can choose a set of "very important" statements $S \subseteq I(X_{\underline{N}}) \cap RQ$ and infer from these. From the point of view of error rates, this is the same as if we had tested $\{\underline{\Phi} ; \tau^{-1}(\underline{\Phi}) \in S\}$ only. Then:

(1) If each $\varphi \in S$ is on the level $\leq \frac{\alpha}{k}$, then $E((\text{I.e.e.r}) \mid \wedge \mathcal{H}) \leq \alpha$ (where (I. e.e.r) and $\mathcal{H}$ correspond to S, i.e., they are restricted).

(2) Moreover, we can use the following consideration:

Let $X_N \downarrow S$ be a minimal subset of X_N such that $S \subseteq I(X_N \downarrow S)$. Then we apply 8.2.12 to $X \downarrow S_{\underline{N}}^{crit}$.

(3) Moreover, we can use Lemma 8.2.7 (2) and Remark 8.2.8, i.e. we require $\sum_{\varphi \in S} \alpha_\varphi \leq \alpha$.

Note that this is a weaker condition than $\alpha_\varphi \leq \frac{\alpha}{k}$ for each $\varphi \in S$ (k = card S), which was required in (2) above. Then we prove by 8.2.7 (2) that, for $I((X_N \downarrow S)_N^{crit})$,

$$E((\text{I.e.e.r}) \mid \wedge \mathcal{H}) \leq \alpha .$$

On the other hand, the structure of RQ (i.e., some relations, e.g. deduction rules on RQ) can be used for the investigation of the second experimentwise error rate:

$$\text{(II. e.e.r.)} = \frac{\text{number of experiments with one or more erroneous inferences}}{\text{number of experiments}}$$

(In many cases, moreover, use can be made of the deductive structure of TQ, cf. [Gabriel],[Miller].)

Considering the expectation of (II. e.e.r.) under $\wedge \mathcal{H}_o$, we in fact consider $P_I(X_{\underline{N}})$.

For the sake of simplicity, assume first that our deduction rules are transitive and simple (i.e. that they consist only of pairs $\frac{\varphi}{\psi}$, where $\varphi, \psi \in RQ$); hence, we can obtain direct solutions only. Now let such a solution $X_{\underline{N}}$ be given.

A set of sentences C from $RQ \cap I(X_{\underline{N}})$ is called <u>coverable</u> if there is sentence $\varphi_o \in I(X_{\underline{N}}) \cap RQ$ such that

$$C \subseteq \{\varphi \in RQ \cap I(X_{\underline{N}});\ \frac{\varphi}{\varphi_o} \in I\}$$

and $I(C \cap X_{\underline{N}}) \supseteq C$.

Then we say that C is <u>coverable with the help</u> of φ_o.

The set $BC = C \cap X_{\underline{N}}$ is called the <u>base</u> of C (w.r.t. $X_{\underline{N}}$) and φ_o. (If there is any danger of a misunderstanding, we shall write $C(\varphi_o, X_{\underline{N}})$ and $BC(\varphi_o, X_{\underline{N}})$). Denote $\mathcal{H}_o(C) = \{\tau(\varphi)_o ; \varphi \in C\}$.

We can illustrate this notion by the following graph ; oriented edges denote the relation I, nodes denote the sentences.

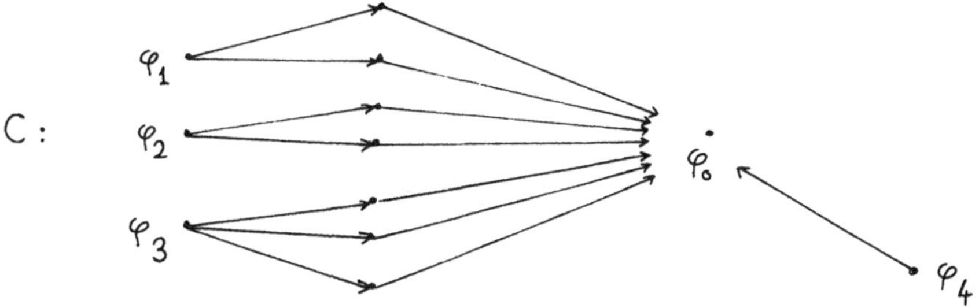

The whole of the above graph is a coverable set C, $BC = \{\varphi_1, \varphi_2, \varphi_3, \varphi_4\}$.

8.2.14 <u>Theorem.</u> Under the above conditions, let $X_{\underline{N}}$ be a direct solution of a given r-problem $\mathcal{P} = \langle RQ, V_o, I \rangle$ and let I be invariant.

Consider a set $C \subseteq RQ \cap I(X_{\underline{N}})$, coverable with the help of a sentence φ_o. Denote $\alpha^1 = \max \{ \alpha(\varphi, \underline{N}) ; \varphi \in BC \}$ and
$\bar{C} = \{ \varphi(\alpha) ; \varphi \in C$ and $\alpha = \min \{ \alpha' \in T ; \varphi(\alpha') \in I (BC_{\underline{N}}^{crit}) \}\}$.

Then $P(\{\sigma ; \bar{C} \cap Tr(\underline{M}_\sigma) \neq \emptyset\} \mid \wedge \mathcal{H}_o(C)) \leq \alpha^1$.

<u>Proof.</u> (Note that $\bar{C} \neq C_{\underline{N}}^{crit}$.)
For each $\underline{M}$ and each $\varphi \in \bar{C}$, $\varphi \in Tr(\underline{M}_\sigma)$ implies $\varphi_o(\alpha^1) \in Tr(\underline{M})$.
(Use the full monotonicity of classes of tests and the invariance property of I.)

Hence, for each given sample M,

$$\bigcup_{\varphi \in C} \{\sigma ; \varphi \in Tr(\underline{M}_\sigma)\} \subseteq \{\sigma ; \varphi_o(\alpha^1) \in Tr(\underline{M}_\sigma)\}.$$

Then
$P(\{\sigma ; \bar{C} \cap Tr(\underline{M}_\sigma) \neq \emptyset\} \mid \wedge \mathcal{H}_o(C)) \leq$
$P(\{\sigma ; \varphi_o(\alpha^1) \in Tr(\underline{M}_\sigma) \mid \wedge \mathcal{H}_o(C)\})$.

But, since $\varphi_o(\alpha^1)$ is a test on the significance level α^1, the left-hand side of the above inequality is less than or equal to α^1.

8.2.15 <u>Discussion.</u> (1) The assertion of the previous theorem signifies that the probability that $I(X_{\underline{N}}) \cap RQ$ rejects one or more true hypotheses from
$\mathcal{H}_o(C) = \{ \tau(\varphi)_o ; \varphi \in \bar{C} \}$ is less than or equal to α^1.

(2) Apply the result of the above theorem to simple problems (cf. 7.2.6). Remember that then $\frac{\varphi}{\psi} \in I$ iff $\varphi \leq \psi$. Consider now the simplest case of coverable sets.

For each $\varphi \in RQ$, define a <u>$\leq$ - interval</u> $[\varphi, \varphi']$ with the least element φ as follows:

$$[\varphi, \varphi'] = \{\psi ; \varphi \leq \psi \leq \varphi'\}.$$

We know that X_N is the set of all $\leq$-minimal elements of $RQ \cap Tr(\underline{N})$. $I(X_{\underline{N}})$ can be thought of as a union of $\leq$ - intervals with least elements in $X_{\underline{N}}$.

Note that each reasonable coverable set C is of the form $C(\varphi_o, X_{\underline{N}}) = \bigcup_{\psi \in BC(\varphi_o, X_{\underline{N}})} C(\varphi_o, \{\psi\})$ where $C(\varphi_o, \{\psi\})$ is a $\leq$-interval. By reasonable we mean the following: If a set C is coverable, then it is a part of a coverable set of the above form; hence, it is reasonable to consider coverable sets of the above form only.

We may apply 8.2.14 to such sets: in particular, let $C(\varphi_o, \{\psi\}) = [\psi, \varphi_o] = \{\psi, \varphi_1, \ldots, \varphi_{k-1}, \varphi_o\}$ (of course, for each $i = 1, \ldots, k-1$ we have $\psi \leq \varphi_i \leq \varphi_o$), then $\alpha^1 = \alpha(\psi, \underline{N})$ and

$$\bar{C}(\varphi_o, \{\psi\}) = \{\psi(\alpha^1), \varphi_1(\alpha^1), \ldots, \varphi_{k-1}(\alpha^1), \varphi_o(\alpha^1)\}.$$

(Note that, for each $\varphi \in \bar{C}(\varphi_o, \{\psi\})$, $\alpha(\varphi, \underline{N}) \leq \alpha^1$; this is useful for local interpretation.)

The reader can easily consider more complicated cases of coverable sets, as mentioned above, i.e., cases in which one has to consider $\leq$-intervals with various least elements and some common members.

(3) Try to generalize Theorem 8.2.14 now to the case of r-problems with indirect solutions. The notion of an invariant rule for such a case was described in 8.1.18 (2). Consider, particularly, a tuft problem w.r.t. an ordering $\leq$ (cf. 7.2.9). Then $RQ \cap Tr_{V_o}(\underline{N})$ is a union of tufts $Y_1, \ldots, Y_k$. Consider one of these tufts. Let φ_o be the largest element and $\varphi_1, \ldots, \varphi_k$ minimal elements; denote this tuft by $Y(\varphi_o)$. Then $\varphi_1, \ldots, \varphi_k \in X_{\underline{N}}$ and φ_o is $\underline{N}$-obtainable from each φ_i, $i=1, \ldots, k$. Hence, using the invariance of I, we have $\varphi_o(\alpha^1) \in Tr(\underline{N})$, where $\alpha^1 = \max\{\alpha(\varphi_i, \underline{N});\ i=1, \ldots, k\}$.
Define

$$\bar{Y}(\varphi_o) = \{\varphi(\alpha);\ \varphi \in Y(\varphi_o) \text{ and } \alpha = \min\{\alpha' \in T;\ \varphi(\alpha') \text{ is } \underline{N}\text{-obtainable from } X^{crit}_{\underline{N}}\}\}.$$

But, generally,

$$\bigcup_{\varphi \in \bar{Y}(\varphi_o)} \{\sigma;\ \varphi \in Tr(\underline{M}_\sigma)\} \nsubseteq \{\sigma;\ \varphi_o(\alpha^1) \in Tr(\underline{M}_\sigma)\}.$$

Note that if ψ is $\underline{N}$-obtainable from φ then we can have

$$\{\sigma;\ \varphi \in Tr(\underline{M}_\sigma)\} \nsubseteq \{\sigma;\ \psi \in Tr(\underline{M}_\sigma)\}.$$

It may happen that $\varphi \in \mathrm{Tr}(\underline{M}_\sigma)$, but $\psi \in \mathrm{Tr}(\underline{M}_\sigma)$ and, hence, aux $\notin \mathrm{Tr}(\underline{M}_\sigma)$. Note that

$$\{\sigma;\ \varphi \in \mathrm{Tr}(\underline{M}_\sigma)\ \&\ \mathrm{aux} \in \mathrm{Tr}(\underline{M}_\sigma)\} \subseteq \{\sigma;\ \psi \in \mathrm{Tr}(\underline{M}_\sigma)\},$$

but this is of little use if aux is of a non-statistical nature, and we know that there are good reasons to use auxiliary sentences of non-statistical nature.

8.2.16 Remark. If I is not transitive but the other assumptions of 8.2.14 hold, one can generalize 8.2.14 in the following manner. A set $C \subseteq RQ \cap I(X_{\underline{N}})$ is coverable if there is a $\varphi_0 \in RQ \cap I(X_{\underline{N}})$ such that $C \subseteq \{\varphi;\ \varphi \vdash_I \varphi_0\}$ and $C \cap X_{\underline{N}} \vdash_I C, \alpha^1$ and $\bar{C}$ can be defined as above and we obtain $P(\{\sigma;\ \bar{C} \cap \mathrm{Tr}(\underline{M}_\sigma) \neq \emptyset\} \mid \wedge \mathcal{H}_0(C)) \leq \alpha^1$ again. For $C \subseteq \{\varphi;\ \frac{\varphi}{\varphi_0} \in I\}$, we then have our assertion as a particular case. See Problem (3) and (4).

8.2.17 Discussion. It is usual, in statistics, to investigate an asymptotical consistency of the procedures used. Consider now regular random structures with domain U which satisfy a frame assumption. Consider a sequence $\{M_n\}_n$ of disjoint samples from U.

Note that for any statistic f the sequence of variates $f_{M_1}, f_{M_2}, \ldots$ is stochastically independent. Now let $\{X_n\}$ be a sequence of solutions based on these samples and $(\mathrm{i.e.r})_n$ the corresponding error rates under $\wedge \mathcal{H}_0$.

We can assume that the number of inferences made for each model $\underline{M}_\sigma$ is finite and bounded (by a number $\hat{C}$).

Now, we formulate a theorem concerning this error rate (cf. [Miller] and, independently and perhaps slightly more precisely, [Havránek 1974]). For other error rates such results are even more trivial.

8.2.18 Theorem. Let $E((\mathrm{i.e.r})_n \mid \wedge \mathcal{H}_0) \leq \alpha_n$ and let there be an $\alpha \in (0, 0.5]$ such that $\alpha_n \leq \alpha$ for each $n \in \mathbb{N}$. Consider the global error rate

$$gl\,(i.e.r.)_n = \frac{\sum_{j=1}^{n} m_j}{\sum_{j=1}^{n} n_j} ,$$

where n_j and m_j are the numbers of inferences and errorneous inferences from M_j respectively.

Then (a.s.) $\lim_{n \to +\infty} gl(i.e.r.)_n \leq \alpha$.

Proof. There is a number C such that $VAR(i.e.r)_j \leq C$ for each $j \in \mathbb{N}$.

We have

$$\frac{\sum_{j=1}^{n} m_j}{\sum_{j=1}^{n} n_j} = \frac{\sum_{j=1}^{n} n_j \frac{m_j}{n_j}}{\sum_{j=1}^{n} n_j} = \frac{\sum_{j=1}^{n} n_j Y_j}{\sum_{j=1}^{n} n_j} .$$

By the Kolmogorov inequality we obtain

$$P\left(\max_{1 \leq j \leq n} \left| \frac{\sum_{j=1}^{n} n_j Y_j}{\sum_{j=1}^{n} n_j} - \frac{\sum_{j=1}^{n} n_j \alpha_j}{\sum_{j=1}^{n} n_j} \right| \geq \varepsilon \right) \leq \frac{C \sum_{j=1}^{n} n_j^2}{\left(\sum_{j=1}^{n} n_j\right)^2} \frac{1}{\varepsilon^2}$$

We know that $\lim_{n \to +\infty} \frac{\sum_{j=1}^{n} n_j^2}{\left(\sum_{j=1}^{n} n_j\right)^2} = 0$ and

$$\lim_{n \to +\infty} \frac{\sum_{j=1}^{n} n_j \alpha_j}{\sum_{j=1}^{n} n_j} \leq \alpha .$$

8.2.19 Key words: Global interpretation ; global error of the first kind ; error rates.

VIII.3 Some questions for statistics

What significance can the methods described in the present book have for statistics ?

It is clear that they cannot substitute statistics in data processing. But they can multiply its power; they can help, on the one hand, to find relevant and reliable statistical statements about data, and, on the other hand, they can help in the orientation in immense empirical data and suggest a large number of hypotheses for further investigations on newly obtained experimental data, on new experimental evidence.

Hypotheses obtained by our methods can be utilized, in this sense, for further statistical testing and/or as an impulse for deeper factual analysis, i.e. for investigations of real processes that caused the inference of **a certain group of** statements.

What other practical questions arise from the application of suggested AI-methods for statistics, besides those mentioned in the preceding chapters? We shall try to expalin this in the discussion and examples in the following three sections.

In the present section we assume a deeper knowledge of statistical theory.

8.3.1 Discussion. In statistics, one frequently considers transformations of a sample space; these transformations are usually required to satisfy some conditions (cf. [Lehmann], [Fergusson]):

Consider a V-structure. Let M be a finite set of objects and denote by $\mathfrak{M}_M$ the set of all V-structures (of the given type) with domain M. Let $\underline{U}$ be a random structure such that $M \subseteq U$. $\underline{U}$ defines a distribution of probabilities on $\mathfrak{M}_M$. Consider one-one mappings of $\mathfrak{M}_M$ as admissible transformations.

For the sake of convenience, we restrict ourselves to the case of d-homogeneous structures of a given type $\langle 1^n \rangle$. As we saw above, usually a frame assumption - distributional statement - is specified in a statistical consideration (in a statistical task). Such a statement Φ can be defined by declaring ($\underline{U} \models \Phi$ iff $D_{\underline{U}} \in \mathfrak{D}_T$), where $\mathfrak{D}_T = \{D_t\}_{t \in T}$ is a system of distribution functions.

Null and alternative hypotheses are described by sets T_o and T_A ($T_o \cap T_A = \emptyset$; usually $T_o \cup T_A = T$ is assumed).

Denote now by $P_t(\{\sigma ; \underline{M}_\sigma \in B\})$ the probability $P^{\underline{U}}(\{\sigma ; \underline{M}_\sigma \in B\})$ under the condition $D_{\underline{U}} = D_t$.

Consider random structures satisfying a frame assumption described by D_T. Let g be a one-to-one mapping on $\mathfrak{M}_M$. $\mathcal{D}_T$ is invariant w.r.t. g if there is a uniquely determined mapping $\bar{g}$ of T into itself such that $P_t(\{\sigma; g(\underline{M}_\sigma) \in B\}) = P_{\bar{g}(t)}(\{\sigma; \underline{M}_\sigma \in B)$ for each Borel set B. No we can say that T is invariant w.r.t. g if $\bar{g}(T) = T$. It is easy to see (cf. [Lehmann], VI.1) that the following holds:

If T is **invariant** w.r.t g and g', then T is invariant w.r.t $g' \circ g$ and g^{-1} and we have $\overline{g' \circ g} = \bar{g}' \circ \bar{g}$ and $\overline{(g^{-1})} = (\bar{g})^{-1}$ (o denotes composition). Moreover, we say that a testing problem is invariant (w.r.t. g) if

$$\bar{g}(T) = T \text{ and } \bar{g}(T_o) = T_o . \qquad (*)$$

If $\mathcal{G}$ is a class of functions satisfying (*) then it is natural to take into account the least group $\mathcal{G}^*$ containing $\mathcal{G}$. From the preceding considerations we know that if a testing problem is invariant w.r.t. elements of $\mathcal{G}$ then it is invariant w.r.t. the whole of $\mathcal{G}^*$. Hence, naturally, having a problem invariant w.r.t. $\mathcal{G}^*$ we define a test to be invariant w.r.t. $\mathcal{G}^*$ iff for the test statistic f the following holds:

for each $g \in \mathcal{G}^*$ and each $\underline{M}$,
$f(g(\underline{M})) \in V_o$ iff $f(\underline{M}) \in V_o$.

In this sense tests of the hypothesis H_o against ASL are invariant w.r.t. strictly increasing transformations of the second functions in models (cf. 5.4.13).

Such a concept of invariance leads to deduction rules on observational sentences of a very specific kind.

Assuming that the test in question is of type $\langle 1, 1\rangle$ we obtain a quantifier of type $\langle 1, 1\rangle$. Let φ_1, φ_2 and ψ_1, ψ_2 be appropriate designated open formulae such that there is a $g \in \mathcal{G}^*$ for which the following holds: for each model $\underline{M}$,

$$\langle M, \|\varphi_1\|_{\underline{M}}, \|\varphi_2\|_{\underline{M}}\rangle = g(\langle M, \|\psi_1\|_M, \|\psi_2\|_M\rangle).$$

Then one could deduce in both directions:

$$\frac{q(\varphi_1, \varphi_2)}{q(\psi_1, \psi_2)} \quad \text{and} \quad \frac{q(\psi_1, \psi_2)}{q(\varphi_1, \varphi_2)} \quad .$$

Hence we obtain deduction rules having the form of an equivalence.

The set of all models can be partitioned to equivalence classes w.r.t. $\mathcal{G}^*$:

$\underline{M}_1 \approx \underline{M}_2$ if $(\exists\, g \in \mathcal{G}^*)(g(\underline{M}_1) = \underline{M}_2)$

($\approx$ denotes the equivalence relation).

Let us compare this notion of invariance with our considerations.

One has a quasiordering on the class of all models of the given type; denote such an ordering by $\leq$ and a test statistic in question by t. Then the following condition is required to hold:

if $\underline{M}_1 \leq \underline{M}_2$ then $t(\underline{M}_1) \in V_o$ implies $t(\underline{M}_2) \in V_o$.

This property can be called the <u>one-sided invariance of the test</u> t w.r.t $\leq$. (If the classical notion of invariance is applied to cases considered in our previous chapters then $g(M_1) = M_2$ always implies $M_1 \leq M_2$ and $M_2 \leq M_1$.).

It is evident, on the one hand, that many statistical procedures, for example rank correlation coefficients and other rank methods, were inspired by intuitive notions of invariance. On the other hand, such properties have not yet been investigated for many tests, for example, for tests of independence in contingency tables.

In general, the question here is the description of a class of tests by their properties (or behaviour) on observational data.

Where does the ordering $\leq$ in particular cases come from?

It can be derived:

(1) from intuitive considerations of the behaviour of "reasonable" statistics on data (e.g., rank correlation coefficients) and/or

(2) from theoretical considerations of an alternative hypothesis in question (i.e., from probabilistic considerations concerning the behaviour of "reasonable" statistics under the alternative hypothesis).

The method of obtaining this ordering can be less straightforward than the one in the case of usual invariance; for example, for the associational contingency tests the ordering of alternative hypotheses determined by theoretical interactions is natural. The ordering of models defined as follows corresponds to the above-

mentioned theoretical ordering:

$$\underline{M}_1 \leq \underline{M}_2 \qquad \text{iff} \qquad \frac{{}^a\underline{M}_1 \ {}^d\underline{M}_1}{{}^b\underline{M}_1 \ {}^c\underline{M}_1} \leq \frac{{}^a\underline{M}_2 \ {}^d\underline{M}_2}{{}^b\underline{M}_2 \ {}^c\underline{M}_2} .$$

We can see that this ordering does not have satisfactory properties for the most frequently used tests.

What are the questions for Mathematical Statistics here?

(1) If one has such an ordering, is there a known test invariant with respect to this ordering?
(Such questions are solved frequently in the present book, cf. 4.5.2, 4.5.3, 5.3.6 and 5.3.2 .)

(2) If one has such an ordering, what is the relation to the alternative hypothesis in question? The relation to the null hypothesis in question is given by the invariance condition.

For example, for contingency tests the following has to hold:

If $\underline{M}_2$ is a-better than $\underline{M}_1$ then

$$P(\{\sigma ; \ \underline{M}_\sigma = \underline{M}_1\} / H_o) \geq P(\{\sigma ; \ \underline{M}_\sigma = \underline{M}_2\} / H_o) .$$

(3) If one has a class of tests for a test problem, find an appropriate ordering with respect to which the tests are invariant. (This question is closely related to the first one, cf. our considerations concerning the associativity and rank tests in Chapters IV and V respectively.)

(4) If one has an ordering, construct one-sided invariant tests with respect to this ordering, i.e. **construct tests** for which non-trivial deduction could be used.

8. 3. 2 Example. Frequently, the investigation of the properties of tests discussed above, for example one-sided invariance, could be less simple. Moreover, these properties can not hold for all models or all $\alpha \in (0,0.5]$ (or $\alpha \in T$, where T is a net, cf. 8.1.16). As an example, the interaction quantifier $\sim_\alpha^3$ can be used (cf. Problem(11) of Chapter IV):

(Theorem) Let $\alpha \in T$ and consider the quantifier $\sim_\alpha^3$. Then there is a number $D(\alpha)$ such that the following holds:

Let $\underline{M}_1, \underline{M}_2$ be two models of type $\langle 1, 1\rangle$ such that $\underline{M}_1$ is a-better than $\underline{M}_2$ and $b_{\underline{M}_2} \geq D(\alpha)$, $c_{\underline{M}_2} \geq D(\alpha)$. Then $Asf_{\sim_\alpha^3}(\underline{M}_2) = 1$ implies $Asf_{\sim_\alpha^3}(\underline{M}_1) = 1$.

Remark. (1) For $\alpha = 0.05$ we have $D(\alpha) = 0$ (Convention: if one of the numbers a, b, c, d is 0 we replace it by 1/2 in the definition of Asf.) For a, b, c, d > 0 we have the following result

α	$D(\alpha)$
0.01	1
0.001	2
0.0005	3
0.00005	4
0.000005	5

For the method of obtaining the numbers $D(\alpha)$ see the proof of the theorem.

(2) Note that we need no conditions on a and d.

(3) The test corresponding to $\sim_\alpha^3$ is asymptotical. It is recommended to be applied if frequences a,b,c,d are not small e.g. a,b,c,d $\geq$ 5. It is easy to see that $\sim_\alpha^3$ is "practically associational" in this field of applicability. (We can say that $\sim_\alpha^3$ is asymptotically associational; but we have the exact bounds of its domain of associationality for each α.)

Proof. [D. Pokorný]. First, it is easy to see that it suffices to restrict ourselves to consider whether the inequality

$$\frac{\log ad/bc}{\sqrt{\frac{1}{a}+\frac{1}{b}+\frac{1}{c}+\frac{1}{d}}} \geq \eta_\alpha \qquad (+)$$

still holds if we substitute a+1 for a or b-1 for b. In the first case it is clear that the answer is positive. In the second case it is easy to see that the following inequality need not hold

$$\frac{\log ad/bc}{\sqrt{\frac{1}{a}+\frac{1}{b}+\frac{1}{c}+\frac{1}{d}}} \leq \frac{\log ad/\ b\text{-}1\ c}{\sqrt{\frac{1}{a}+\frac{1}{b-1}+\frac{1}{c}+\frac{1}{d}}}.$$

The idea of the following proof is that monotonicity is violated for those numbers a, b-1, c,d for which both sides of the previous inequality are **greater** than or equal to $\mathcal{N}_\alpha$ (for some values of α) .

For the sake of convenience, we shall study the preservation of the inequality (+) if b+1 is changed to b. Denote

$$S(a,b,c,d) = (\log ad/bc)\left(\frac{1}{a}+\frac{1}{b}+\frac{1}{c}+\frac{1}{d}\right)^{-1/2} \text{ and } \mathbb{N}^+ = \mathbb{N} - \{0\}.$$

We shall consider the set

$K = \{\langle a,b,c,d\rangle ;\ 0 < S(a,b,c,d) \le S(a, b+1, c,d)\}$. For a given value b we put $K_b = \{\langle a,c,d\rangle ;\ \langle a,b,c,d\rangle \in K\}$ and for given b and q we put $K^q_b = \{\langle a,c,d\rangle ;\ \langle a,c,d\rangle \in K_b \text{ and } \frac{1}{a}+\frac{1}{c}+\frac{1}{d} = q\}$.

The aim now is to find the infimum of the set $S_b = \{S(a,b,c,d) ;\ \langle a,c,d\rangle \in K_b\}$ and to find a number α such that $\inf(S_b) \ge \mathcal{N}_\alpha$.

Note that trivially $S_b = \bigcup_{q\in Q} S^q_b$, where $S^q_b = \{S(a,b,c,d) ;\ \langle a,c,b\rangle \in K^q_b\}$ and Q is the set of possible values of $q = \frac{1}{a}+\frac{1}{c}+\frac{1}{d}$ for $\langle a,c,d\rangle \in \mathbb{N}^+$.

Then $\inf(S_b) = \inf_{q\in Q}(\inf(S^q_b))$. It can be seen that

$$S^q_b = \left\{\frac{\log A/b}{\sqrt{q+\frac{1}{b}}} ;\ \frac{\log A/b}{\sqrt{q+\frac{1}{b}}} \ge \frac{\log(b+1)-\log b}{\sqrt{q+\frac{1}{b}}-\sqrt{q\frac{1}{b+1}}} \text{ and } A \in A_q\right\}$$

where $A_q = \{A ;\ A > 1 \text{ and } (\exists\, a,c,d \in \mathbb{N}^+)(\frac{1}{a}+\frac{1}{c}+\frac{1}{d} = q \,\&\, ad/c = A)\}$.

Now we have immediately that

$$\inf(S^q_b) \ge \frac{\log(b+1) - \log b}{\sqrt{q+\frac{1}{b}}-\sqrt{q+\frac{1}{b+1}}}$$

. An easy **calculation then gives**

$$\inf(S_b) \geq \frac{\sqrt{b(b+1)}}{\sqrt{b+1} - b} \log \frac{b+1}{b} .$$

The left hand side of the previous inequality is denoted LB(b). Note that for $b_1 \geq b_2$ we have $LB(b_1) \geq LB(b_2)$. It remains to compare the values of LB(b) with quantiles η_α. For the further generalizations see Problem (6).

8.3.3 Discussion. The next question to be answered in connection with applicability of statistical tests is the computability (or decidability) problem. It asks whether, for a statistic t in question and a critical region V_o, $t(\underline{M}) \in V_o$ is effectively decidable for each possible model $\underline{M}$.

The first step is to introduce here the notion of computability based on recursive functions (cf. Chapter IV); hence we restrict ourselves to recursive functions on recursive sets $V \cap \mathbb{Q}$.

Usually in statistics one considers a very broad class of tests, namely (in non-randomized case) all measurable functions from the sample space to $\{0, 1\}$ (satisfying conditions on the probability of an error of the first kind). In such a class one then looks for an "optimal" test w.r.t. some rationality criteria.

Obviously if the optimal test is computable in the above sense then it is optimal in the subclass of all computable tests for the test problem in question. Hence we can restrict ourselves to the investigation whether the particular test obtained by the usual statistical methods is computable.

Such problems can occur in applying tests of $\{0, 1\}$ - structures too. It is clear that computability is only one of the properties of realy applicable tests: for practical reasons, in mechanized discovery one needs to take into account a large number of further questions related to the accuracy of the computations (round-off errors), with the occurrence of under-flowing and over-flowing (remember our tests with binomial coefficients) etc. It would seem to be useful in the future to investigate some complexity hierarchies of tests and to take such a hierarchy into the considerations of the optimality of tests.

8.3.4 Example. For large classes of tests we can prove computability trivially (for example, for the tests based on the Neymann-Pearson lemma concerning distributions of the exponential class with the monotone likelihood ratio we can avoid the usage of the exponential function, cf. [Lehmann]).

In other cases the solution is not so simple. As an example we use the interaction quantifier again.

(Theorem) The interaction quantifier is observational.

Proof. We have to decide whether, for given $a,b,c,d \in \mathbb{N}^+$

$$\frac{\log ad/bc}{\sqrt{\frac{1}{a}+\frac{1}{b}+\frac{1}{c}+\frac{1}{d}}} \geq \eta_\alpha$$

assuming η_α is rational. This means deciding whether

(*) $\log A/\sqrt{B} \geq C$ for $A,B,C \in \mathbb{Q}$ and $C > 0$ is a recursive relation. Note that (*) is equivalent to $A > 1$ and $(\log A)^2 \geq C^2 B$. The question remains whether $(\log A)^2 \geq D$ is a recursive relation on $\mathbb{Q}^2$.

We use two known facts:

(1) if A is rational and $A > 1$ then $\log A$ is a transcendental number ;

(2) the function $\log A$ for $A > 1$ can be expressed by the following expansion:

$$\log A = 2\sum_{n=1}^{\infty} \frac{\xi(A)^{2n-1}}{2n-1} \quad \text{where} \quad \xi(A) = \frac{A+1}{A-1}$$

Moreover, the series converges in such a way that (a) the sequence

$$S_k(A) = \sum_{n=1}^{k} 2\,\frac{\xi(A)^{2n-1}}{2n-1}$$

is strictly increasing and (b) the residual error for $S_k(A)$ is less than $R_k = 2\,\dfrac{\xi(A)^{2k+1}}{2k+1}\,\dfrac{1}{1-\xi(A)}$.

Hence if $(\log A)^2 > D$ then $(\exists k)\,((S_k(A))^2 > D)$ and if $(\log A)^2 < D$ then $(\exists k)((S_k(A)+R_k)^2 < D)$. From

(1) we know that equality cannot occur. Hence the relation $(\log A)^2 \geq D$ is recursive.

8.3.5 Discussion. The last question that we are going to discuss concerns the simultaneous inference problem. On the one hand, the reader could use GUHA and similar methods as methods for hypothesis formation only and hence he need not worry about simultaneous inference. He can understand results as working hypotheses which are to be investigated further.

On the other hand, if the reader wants to draw, from a number of obtained results, some general and reliable conclusions, he has to consider the reliability of such inferences, he has to consider the probabilities of global errors and hence he has to take in account problems of simultaneous inference in the sense of the previous section.

Our tests are not stochastically independent, i.e. if $\varphi_1, \ldots, \varphi_k$ are corresponding observational sentences, in general we have

$$P(\{\sigma; \|\varphi_1\|_{\underline{M}_\sigma} \in V_0\} \cap \ldots \cap \{\sigma; \|\varphi_k\|_{\underline{M}_\sigma} \in V_0\}) \neq$$

$$P(\{\sigma; \|\varphi_1\|_{\underline{M}_\sigma} \in V_0\}) \ \ldots \ P(\{\sigma; \|\varphi_k\|_{\underline{M}_\sigma} \in V_0\}).$$

In the considerations of the previous section we did not use this fact. They are applicable without the knowledge as to whether such dependence occurs and what form it really has and without any specific assumptions on the stochastic behaviour of the tests used.

The tasks for statisticians are, hence, the following:

(1) If one hase some test which is used in practical and is applicable in mechanized discovery, to investigate whether it is possible to describe the form of stochastical dependence and to use this dependence or some other probabilistic properties of tests to improve (at least asymptotically) the simultaneous inference properties of the test.

(2) **Constructing** new **tests** having useful properties in the discussed direction.

8.3.6 Example. A great deal of work has been done in this direction for a multidimensional contingency table and generalized interaction tests (see Problem (7)) by J. Anděl [1973]. For χ^2 - test in 2X2 tables derived from 2XC contingency tables, a similar result has been obtained by Sugiura and Otake. Both results are based on Šidák's inequality [Šidák 1962].

We can define a "two-sided" associational quantifier $\approx$ of type $\langle 1,1 \rangle$ with the associated function

$$\mathrm{Asf}_{\approx_\alpha}(\underline{M}) = 1 \quad \text{iff} \quad \frac{|\log(a_{\underline{M}} d_{\underline{M}} / b_{\underline{M}} c_{\underline{M}})|}{\sqrt{\frac{1}{a_{\underline{M}}} + \frac{1}{b_{\underline{M}}} + \frac{1}{c_{\underline{M}}} + \frac{1}{d_{\underline{M}}}}} \geq n_{\alpha/2} .$$

This quantifier corresponds to a test of independence $(\delta = 0)$ against two-sided alternative $(\delta \neq 0)$.

This is an example of quantifiers that could be called "two-sided" associational quantifiers, namely it satisfies the following conditions:

$\mathrm{Asf}_{\approx}(\underline{M}) = 1$, $\underline{M} \leq_a \underline{M}'$ and $a_{\underline{M}} d_{\underline{M}} \geq b_{\underline{M}} c_{\underline{M}}$ implies $\mathrm{Asf}_{\approx}(\underline{M}') = 1$;

$\mathrm{Asf}_{\approx}(\underline{M}) = 1$, $\underline{M}' \leq_a \underline{M}$ and $a_{\underline{M}} d_{\underline{M}} \leq b_{\underline{M}} c_{\underline{M}}$ implies $\mathrm{Asf}_{\approx}(\underline{M}') = 1$.

Such quantifiers should be studied in a systematic way. GUHA methods using two sided associational quantifiers would be very useful since many two-sided associational quantifiers correspond to various two-sided tests of independence in contingency tables.

We now prove a modification of Anděl's Theorem 2 for the quantifier $\approx$.

First, some preliminary considerations and notation.

Consider pairs $<\varphi_i, \psi_i>$ (i=1,...k) of designated open formulae. Now let a d-homogeneous random $\{0,1\}$ - structure $\underline{U} = <U, Q_1, \ldots, Q_n>$ be given. We denote

$$p_{i_1,\ldots,i_n} = P^{\underline{U}}(\{\sigma ; Q_1(o,\sigma) = i_1,\ldots,Q_n(o,\sigma) = i_n\})$$

for each $<i_1,\ldots,i_n> \in \{0,1\}^n$. So we obtain a 2^n - tuple of probabilities $\bar{p} = <p_{1,\ldots,1}, p_{1,\ldots,1,0}, \ldots, p_{0,\ldots,0}>$ which corresponds to 2^n -dimensional multinomial distribution with possible events $<i_1,\ldots,i_n>$.
Similarly, for a model $\underline{M}$ we denote $\bar{m}_{\underline{M}} = <m_{1,\ldots,1},\ldots,m_{0,\ldots,0}>$ where $m_{i_1,\ldots i_n}$ is the number of cards $C_{\underline{M}}(o) = <i_1,\ldots,i_n>$.

Instead of $\bar{m}_{\underline{M}}$ we shall write $\bar{m}$ only.
For a designated open formula φ we put

$$\sum_{\varphi}(\bar{m}) = \sum_{\{<i_1,\ldots,i_n> ;\ \varphi^*(i_1,\ldots,i_n) = 1\}} m_{i_1,\ldots,i_n},$$

where φ^* is the boolean function corresponding to φ.

$$(\varphi^*(C_{\underline{M}}(o)) = 1 \text{ iff } \|\varphi\|_{\underline{M}}[o] = 1.)$$

We see immediately that, for a pair of designated open formulae

$$\|\varphi_i \approx_\alpha \psi_i\| = 1 \text{ iff } \frac{g_i(\bar{m})}{s_i(\bar{m})} \geq \eta_{\alpha/2}, \text{ where}$$

$$g_i(\bar{m}) = \log \sum_{\varphi_i \& \psi_i}(\bar{m}) - \log \sum_{\varphi_i \& \neg\psi_i}(\bar{m}) - \log \sum_{\neg\varphi_i \& \psi_i}(\bar{m}) + \log \sum_{\neg\varphi_i \& \neg\psi_i}(\bar{m})$$

and

$$s_i(\bar{m}) = \sqrt{\frac{1}{\sum_{\varphi_i \& \psi_i}(\bar{m})} + \frac{1}{\sum_{\varphi_i \& \neg\psi_i}(\bar{m})} + \frac{1}{\sum_{\neg\varphi_i \& \psi_i}(\bar{m})} + \frac{1}{\sum_{\neg\varphi_i \& \neg\psi_i}(\bar{m})}}$$

As we mentioned in Section 4.1 the probability $P^{\underline{U}}(\{\sigma ;\ t(\underline{M}_\sigma) \in V_o\})$, where t is a statistic and V_0 a regular set, is the same for all samples M of equal cardinality.

Hence we can consider

$\lim_{m \to +\infty} P^{\underline{U}}(\{\sigma ; t(\underline{M}_\sigma) \in V_0\})$, independently of the particular choice of the sample M. If $\lim_{m \to +\infty} P^{\underline{U}}(\{\sigma ; t(\underline{M}_\sigma) \in V_0\}) \leq \alpha$

we say that the probability of this event is asymptotically less than or equal to α .

Now we can formulate and prove the desired theorem.

(Theorem.) Consider an r-problem with

$RQ(\alpha) = \{\varphi \approx_\alpha \psi ; \langle \varphi, \psi \rangle \in PF\}$, where PF is a set of disjointed designated open formulae. Let a model $\underline{N}$ be given. Consider a subset $S \subseteq I(X_{\underline{N}}) \cap RQ(\alpha)$

Suppose $S = \{\bar{\varphi}_1(\alpha), \ldots, \bar{\varphi}_k(\alpha)\}$ where $\bar{\varphi}_i(\alpha) = \varphi_i \approx_\alpha \psi_i$.

Put $\alpha_S = 1 - \prod_{i=1}^{k} (1 - \alpha(\bar{\varphi}_i, \underline{N}))$.

Then the probability of a global error of the first kind is asymptotically (in the cardinality of samples) less than or equal to α_S.

We have now to formulate and prove a lemma. Notation is from Rao 's book.

(Lemma .) Let $\Theta = \langle \Theta_1, \ldots, \Theta_k \rangle$ be a real vector. Let $\mathbb{T}_n$ be a k-dimensional statistic $\langle T_{1n}, \ldots, T_{kn} \rangle$ such that the asymptotical distribution of $\langle \sqrt{n}(T_{1n} - \Theta_1), \ldots, \sqrt{n}(T_{kn} - \Theta_k) \rangle$ is k-variate normal with mean zero and dispersion matrix $\Sigma = (\sigma_{ij}(\Theta))$

Let $g_1, \ldots, g_q$ be totally differentiable functions. Denote by $\mathbb{G}$ the matrix $\left(\frac{\partial g_i}{\partial \Theta_j}\right)_{i,j}$ and denote by $\vartheta_i(\Theta)$ the diagonal elements of $\mathbb{G}\Sigma\mathbb{G}'$, i.e.

$$\vartheta_i(\Theta) = \sum_r \sum_s \sigma_{rs}(\Theta) \frac{\partial g_i}{\partial \Theta_r} \frac{\partial g_i}{\partial \Theta_s} .$$

(Then $\vartheta_i(\mathbb{T}_n)$ is the value of $\vartheta_i(\Theta)$ if we put $\Theta = \mathbb{T}_n$.)

Let Σ and $\mathbb{G}$ be continuous functions of Θ .

Then asymptotical distribution of

$$\frac{\sqrt{n}\,(g_i(\mathbb{T}_n) - g_i(\Theta)}{\sqrt{\vartheta_i(\mathbb{T}_n)}} ,$$

$i = 1, \ldots, q$, is q-variate normal with a dispersion matrix having diagonal elements equal to 1.

Proof of the lemma. By 6.a.2 (iii) in [Rao] we have that $\sqrt{n}\, u_{1n}, \ldots, \sqrt{n}\, u_{qn}$, where $u_{in} = (g_i(\mathbb{T}_n) - g_i(\Theta))$, has an asymptotical distribution q-variate normal with zero mean and dispersion matrix

Hence $$\left\langle \frac{\sqrt{n}\, u_{1n}}{\sqrt{\vartheta_1(\Theta)}}, \ldots, \frac{\sqrt{n}\, u_{qn}}{\sqrt{\vartheta_q(\Theta)}} \right\rangle$$

has asymptotically q-variate normal distribution with zero mean and a dispersion matrix having diagonal elements equal to 1. Using the continuity of Σ and $\mathbb{G}$ we obtain

$$\left| \frac{\sqrt{n}\, u_{in}}{\sqrt{\vartheta_i(\Theta)}} - \frac{\sqrt{n}\, u_{in}}{\sqrt{\vartheta_i(\mathbb{T}_n)}} \right| \xrightarrow{P} 0$$

for $i = 1, \ldots, q$. Hence, by 2.c.4 (ix) from [Rao] for vector variates, both vector variates have the same asymptotical distribution function.

Proof of the theorem.

(1) Note that m is a statistic having 2^n-dimensional multinomial distribution. Hence

$$\langle \sqrt{m}\,(m_{1,\ldots,1}/m - p_{1,\ldots,1}), \ldots, \sqrt{m}\,(m_{0,\ldots,0}/m - p_{0,\ldots,0}) \rangle$$

has asymptotically a 2^n-variate normal distribution with zero mean and dispersion matrix

$$\mathbb{V} = \begin{pmatrix} p_{1,\ldots,1}(1-p_{1,\ldots,1}), & -p_{1,\ldots,1}\,p_{1,\ldots,0}, & \ldots \\ -p_{1,\ldots,0}\,p_{1,\ldots,1} & , p_{1,\ldots,0}\,p_{1,\ldots,0}, & \ldots \\ \vdots & & \\ & & \vdots \\ & & \ldots, p_{0,\ldots,0}\,p_{0,\ldots,0} \end{pmatrix}$$

(2) To $\langle \sqrt{m}\, g_1(\frac{\bar{m}}{m}) - g_1(\bar{p}), \ldots, \sqrt{m}\, g_k(\frac{\bar{m}}{m}) - g_k(\bar{p})\rangle$ we apply the lemma. Here, we have

$$v_i(\bar{p}) = \sum_{\langle i_1,\ldots,i_n\rangle} \sum_{\langle j_1,\ldots,j_n\rangle} v_{\langle i_1,\ldots,i_n\rangle\langle j_1,\ldots,j_n\rangle} \frac{\partial g_i(\bar{p})}{\partial p_{i_1,\ldots,i_n}} \frac{\partial g_i(\bar{p})}{\partial p_{j_1,\ldots,j_n}}$$

where $v_{\langle i_1,\ldots,j_n\rangle \langle j_1,\ldots,j_n\rangle}$ are elements of the variance matrix $\mathbb{V}$ (by $\frac{\bar{m}}{m}$ we mean $\langle \frac{m_{1,\ldots 1}}{m}, \ldots, \frac{m_{o,\ldots,o}}{m} \rangle$).

(3) For g_i corresponding to $\langle \varphi_i, \psi_i\rangle$ we have $v_i(\bar{p}) = s_i^2(\bar{p})$ i.e.,

$$v_i(\bar{p}) = \frac{1}{\sum_{\varphi_i \& \psi_i}(\bar{p})} + \frac{1}{\sum_{\neg\varphi_i \& \psi_i}(\bar{p})} + \frac{1}{\sum_{\varphi_i \& \neg\psi_i}(\bar{p})} + \frac{1}{\sum_{\neg\varphi_i \& \neg\psi_i}(\bar{p})}$$

(the proof is an essentially elementary but rather cumbersome algebraic exercise).

Note that under our null hypothesis we have $v_i(\bar{p}) = 0$ for $i = 1, \ldots, k$. Moreover, we have $s_i^2\left(\frac{\bar{m}}{m}\right) = m\, s_i^2(\bar{m})$ and $g_i\left(\frac{\bar{m}}{m}\right) = g_i(\bar{m})$. Hence (by the lemma)

$$\left\langle \frac{\sqrt{m}\, g_1(\bar{m}/m)}{s_1(\bar{m}/m)}, \ldots, \frac{\sqrt{m}\, g_k(\bar{m}/m)}{s_k(\bar{m}/m)} \right\rangle =$$

$$= \left\langle \frac{g_1(\bar{m})}{s_1(\bar{m})}, \ldots, \frac{g_k(\bar{m})}{s_k(\bar{m})} \right\rangle$$

has an asymptotically q-variate normal distribution with expectation $0 = \langle 0, \ldots, 0\rangle$ and diagonal elements of the dispersion matrix equal to 1.

(4) Lemma [Šidák]. If the variates $\vartheta_1, \ldots, \vartheta_k$ have k-variate normal distribution, then

$$P(|\vartheta_1| < c_1, \ldots, |\vartheta_k| < c_k) \geq P(|\vartheta_1| > c_1) \ldots P(|\vartheta_k| > c_k).$$

Hence $$P\left(\bigcup_{i=1}^{k} |\vartheta_i| \geq c_i\right) \leq 1 - \prod_{i=1}^{k}\left(1 - P(|\vartheta_i| \geq c_i)\right). \qquad (X)$$

(4) If $\alpha_1, \ldots, \alpha_k$ are the desired critical levels then, if we consider multinormal variates $\vartheta_1, \ldots, \vartheta_k$ with VAR $(\vartheta_i) = 1$ we obtain, by (X),

$$P\left(\bigcup_{i=1}^{k} |\vartheta_i| \geq n_{\alpha_i/2}\right) \leq 1 - \prod_{i=1}^{k}(1 - \alpha_i).$$

(5) Applying pont (3) we have

$$\lim_{m \to +\infty} \left| P\left(\bigcup_{i=1}^{k} |g_i(\bar{m})| \geq n_{\alpha_i/2}\, s_i(\bar{m})\right) - P\left(\bigcup_{j=1}^{k} |\vartheta_i| \geq n_{\alpha_{j/2}}\right)\right| = 0$$

This completes the proof.

(Remark.) Note that we incidentally proved that $\approx_\alpha$ is an asymptotical observational test of the null hypothesis of independence.

8.3.9 Let us make some concluding remarks on the relation of statistics to mechanized hypothesis formation. It is well known that "statistical theory is poor in such suggestion (i.e. suggesting hypotheses); hypotheses are usually assumed to be formulated before statistical theory is invoked. This is a weakness in statistical theory, regarded as a part of scientific method, consequently some new results in this direction should be of interest". This is due to Good [1963]. In this paper, Good made an important contribution concerning the apriori formation of null hypotheses using a theoretical principle of maximum entropy . In mechanized hypothesis formation, we are interested mostly in the formation of alternative hypotheses on the basis of observational data suitable for further statistical analysis.

As far as data of a statistical nature is concerned, many of the methods presented in this book apply to analysis of multidimensional contingency tables. The reader should observe that we mean tables having indeed many dimensions, say thirty. Complete and correct statistical analysis of such a table needs the computation of all frequencies up to the 30th order saturated model in the statistical meaning; cf. Bishop[1974]. This is obviously practically impossible and one has to confine oneself to some simple information derived from the table.

And this is indeed what the GUHA method described in Chapter VII (Sect. 1-3) does; it does it in a way optimalized from the logical and computational point of view. Needless to say, in the present versions of the GUHA method we could not apply everything from the statistics of contingency tables. There are new important results in the statistics of contingency tables, published only after this book was written. But we hope that our approach in connection with some deeper statistical methods may prove to be yet more efficient. The development of exact tests of higher order dependencies in contingency tables, as initiated by Zelen [1971], seems to be particularly promising. Further development of Hypothesis Formation in this direction may bring essentially new impulses for interdisciplinary studies in logic, computer science and statistics.

8.3.10 Key words: Invariant tests, one-sided invariance, simultaneous procedures using probabilistic properties of tests; relation of GUHA-methods to statistics.

PROBLEMS AND SUPPLEMENTS TO CHAPTER VIII.

(1) Consider d-homogeneous random $\langle V_1, \ldots, V_n \rangle$ - structures and appropriate MOPC's as in Sect. 3.2. Then the following distributional statements are equivalent:

(i) For each $\kappa \in EC$, $p_\kappa^{\underline{U}} > 0$.

(ii) For each $\langle j_1, \ldots, j_n \rangle$, where $j_i \in V_i$,

$$p^{\underline{U}}_{(j_1)F_1 \,\&\, \ldots \,\&\, (j_n)F_n} > 0$$

(i.e., all the joint multinomial probabilities are non-zero).
Prove this.

(2) We present here some remarks concerning our particular MOPC's with associational and helpful quantifiers as described in Chapter VI. Consider corresponding random structures satisfying the global condition of 8.1.7. We can define, for each $\langle \kappa, \lambda \rangle \in CPF$,

$$\Delta(\kappa,\lambda) = \frac{p_{\kappa \& \lambda}\, p_{\neg\kappa \& \neg\lambda}}{p_{\kappa \& \neg\lambda}\, p_{\neg\kappa \& \lambda}}$$ as in 4.4.18.

Note that we can define a theoretical sentence $(\kappa,\lambda) \leq^{th}_{a} (\kappa',\lambda')$ such that $\underline{U} \models (\kappa,\lambda) \leq^{th}_{a} (\kappa',\lambda')$ iff $\underline{U} \models \Delta(\kappa,\lambda) \leq \Delta(\kappa',\lambda')$. Similarly for $\leq^{th}_{i}$ and $p_{\delta/\kappa} \leq p_{\delta'/\kappa'}$.

<u>Lemma.</u> (i) Consider $(X)F, (Y)F$; then $(X \subseteq Y)$ implies $p_{(X)F} \leq p_{(Y)F}$,

(ii) $\kappa \leq \lambda$ implies $p_\kappa \geq p_\lambda$, (iii) $\kappa \sqsubseteq \lambda$ implies $p_\kappa \geq p_\lambda$,

(iv) $\kappa \lessdot \lambda$ implies $p_\kappa \geq p_\lambda$, and (v) $\delta_1 \lhd \delta_2$ implies $p_{\delta_1} \leq p_{\delta_2}$.

Remember the ordering $\leq_c$ on $\{0, 1\}^2$ and define similarly the ordering $\leq_{c'}$ on $[0, 1]^2$. Then $\langle \kappa,\lambda \rangle \lessdot\lessdot \langle \kappa',\lambda' \rangle$ implies $\langle p_\kappa, p_\lambda \rangle \geq_{c'} \langle p_{\kappa'}, p_{\lambda'} \rangle$.

The obvious proof is left to the reader.

Discussion. One could investigate $\leq_a^{th}$ as a helpful quantifier on the theoretical level, but the obstacle is that one cannot verify whether $\underline{U} \models (\kappa,\lambda) \leq_a^{th} (\kappa',\lambda')$ or not. One could use an inference rule of the form

$$\left\{ \frac{(\kappa,\lambda) \ll (\kappa',\lambda')}{(\kappa,\lambda) \leq_a^{th} (\kappa',\lambda')} ; \; \langle\kappa,\lambda\rangle \ll \langle\kappa',\lambda'\rangle \right\},$$

where $\ll$ is a helpful quantifier in the sense of Chapters VI; but, statistically, this is an inference based only on point estimation. Hence, this course is inappropriate for our concept based on hypothesis testing.

Nevertheless, we can modify an a-helpful quantifier $\ll$ to be a test (on the level α) with the alternative hypothesis $(\kappa,\lambda) \leq_a^{th} (\kappa',\lambda')$. (Then, $\|(\kappa,\lambda) \ll_\alpha (\kappa',\lambda')\|_{\underline{M}} \in V_0$ implies $\|\langle\kappa,\lambda\rangle \ll \langle\kappa',\lambda'\rangle\|_{\underline{M}} \in V_0$.)

It is easy to see that a procedure for testing $\Delta(\kappa',\lambda') > 1$ using such a quantifier $\ll_\alpha$ will be less powerful than our procedures. In fact, we have to obtain the same set $RQ \cap Tr_{V_0}(\underline{M}_\sigma)$ but the solutions will be greater.

Hence, we use as auxiliary questions only sentences of a non-statistical, i.e., observational, nature.

(3) Remember Section 3.4. Define a theoretical notion of incompressibility: κ is $\underline{U}$ - incompressible if there is no $\kappa_0 \subsetneqq \kappa$ such that $p_{\kappa_0}^{\underline{U}} = p_{\kappa}^{\underline{U}}$. If $\underline{U}$ satisfies (i), then each $\kappa \in$ EC is incompressible. (Prove this.)

(4) Prove the assertion of Remark 8.2.16.

(5) In [Havránek 1974], one finds a slightly stronger form of the assertion of Remark 8.2.16. There, coverable sets $C(\varphi_0, X_{\underline{N}})$ for which $\tau(\varphi_0)_0$ implies $\tau(\varphi)_0$ for each $\varphi \in C(\varphi_0, X_{\underline{N}})$ were considered. Then we have

$$P(\{\sigma ; \overline{C} \cap Tr(\underline{M}_\sigma) \neq \emptyset\} | \tau(\varphi_0)_0) \leq \alpha^1 .$$

Prove this assertion. It can be applied to many situations; remember the global hypothesis of independence.

(6) Let $\vartheta_1, \vartheta_2, \ldots$ be a sequence of variates and E a number. We say that $\{\vartheta_n\}$ converges almost surely to E ((a.s.) $\lim_{n\to+\infty} \vartheta_n = E$) if $P(\{\sigma ; \lim_{n\to+\infty} \vartheta_n(\sigma) = E\}) = 1$.(Note that if σ is an elementary random event then $\{\vartheta_i(\sigma); i \in \mathbb{N}\}$ is a sequence of real numbers ; observe that for each E the set of all σ for which the above sequence converges to E is in the σ-field R.

(2) it has the probability 1.)

(7) Consider d-homogeneous random $\langle V_1, V_2\rangle$ -structures with two quantities such that $V_1 = \{0, \ldots, h_1 - 1\}$, $V_2 = \{0, \ldots, h_2 - 1\}$, We can consider the hypothesis of independence of Q_1 and Q_2 against alternative hypotheses described in the following way:

Let A be a $h_1 \times h_2$ matrix such that for each row and each column its sum is zero, i.e. $\sum_i a_{ij} = 0$ and $\sum_j a_{ij} = 0$ for each i,j . Let $p_{ij} = P(Q_1 = i \cap Q_2 = j)$. We define logarithmic interaction of Q_1 and Q_2 w.r.t. A as follows:

$$\delta(A) = \log \prod_i \prod_j p_{ij}^{a_{ij}} .$$

Then the null hypothesis is $\delta(A) = 0$ and the alternative hypothesis is $\delta(A) > 0$. A test can be defined as follows:

$$t(\underline{M}) = 1 \text{ if } \frac{\log \prod_i \prod_j m_{ij}^{a_{ij}}}{\sqrt{\sum_i \sum_j a_{ij}^2 / m_{ij}}} \geq n_\alpha .$$

Let $\underline{M}_1, \underline{M}_2$ be two models of the corresponding type. We say that $\underline{M}_1$ is A-better than $\underline{M}_2$ if

$m_{ij}(\underline{M}_1) \geq m_{ij}(\underline{M}_2)$ for $a_{ij} \geq 0$

and $m_{ij}(\underline{M}_1) \leq m_{ij}(\underline{M}_2)$ for $a_{ij} \leq 0$.

(a) Prove that the test defined above is a one-sided invariant test w.r.t. the $\mathcal{A}$-better ordering, i.e. prove generalized associativity. Hint: the proof is a generalization of the proof of Theorem 8.3.2; cf. [Pokorný].

(b) Construct appropriate observational calculi and r-problems (open).

(8) We can define conditional associational quantifiers; they are ternary. Consider models $(\{0, 1\})$ of type $\langle 1,1,1 \rangle$; we define conditional associational quantifiers to be associational for the partialized models

$\underline{M}' = \langle M', f_1, f_2 \rangle$, where $M' = \{o \in M;\ f_3(o) = 1\}$.

Let g be the cardinality of $\{o \in M;\ f_3(o) = 1\}$. Prove that, if $g! \leq 1/\alpha$ then $\mathrm{Asf}_{\sim_\alpha^c}(\underline{M}) = 0$ for each conditional associational quantifier $\sim_\alpha^c$ based on $\sim_\alpha$, $\sim_\alpha^2$ or $\sim_\alpha^3$.

(9) Define ternary associational quantifiers corresponding to tests of 2X2X2 tables studied in [Anděl 1973]. Apply to this case Theorem 3 from [Anděl 1973]. [Pokorný 1975] proved that $\mathrm{Asf}_{\sim_\alpha^3}(\underline{M}) = 1$ implies $\mathrm{Asf}_{\sim_\alpha^2}(\underline{M}) = 1$, and there is a non-empty set of models for which $\mathrm{Asf}_{\sim_\alpha^3}(\underline{M}) = 0$ and $\mathrm{Asf}_{\sim_\alpha^2}(\underline{M}) = 1$. Hence the χ^2 - test is strictly simultaneously more powerful than the interaction test.

The lesson from this result is the following:
It is meanigful, from a statistical point of view, to investigate observational properties; we can obtain valuable pure statistical results.

(10) Remember that $\sim_\alpha$ denotes the Fisher quantifier. Does $\mathrm{Asf}_{\sim_{r\alpha}^2}(\underline{M}) = 1$ imply $\mathrm{Asf}_{\sim_\alpha}(\underline{M}) = 1$ or does $\mathrm{Asf}_{\sim_{r\alpha}}(\underline{M}) = 1$ imply $\mathrm{Asf}_{\sim_\alpha^2}(\underline{M}) = 1$ for an $r \in [1,2]$? (Open.)

(11) Are there similar relations between distinctive quantifiers based on rank tests? (Open.)

Postcript: Some Remarks on the History of The GUHA-Method and its Logic of Discovery

Let us distinguish (a) the principle of the GUHA method, (b) particular realizations, (c) the theory on which the method is based and to which it gives rise.

(a) The principle of the method was discussed in detail in Chapter VI, Sect.1: it can be briefly formulated as the principle of "everything important" or the principle of automatic listing of important observational statements. Obviously, it has two contrasting aspects: the principle of exhaustiveness (everything) and the principle of relevance and optimization (importance). The idea of using the formulational possibilities of logic for the automatic investigation of all assertions of a certain syntactic form as to truthfulness on given concrete material (the principle of exhaustiveness) is due to Metoděj K. Chytil. He performed his first experiments based on propositional calculus in 1964. When P. Hájek (the first author of the present book) met Chytil at the end of 1964, and saw his experiments, he **suggested making use of** the deductive possibilities of logic to find true formulae as powerful as possible (important, relevant, interesting) Cf. Hájek, Havel, Chytil [1966b] and/or [1966a]([1966a] is the English version of [1966b] and Hájek, Havel, Chytil [1967]). As I. Havel is an expert in computer data processing he was invited to take part in discussions on the possibility of a computer implementation and if possible to write a computer program. It was only in 1974 that our attention was drawn to Leinfellner's book and we realized that Leinfellner had in fact arrived at the same idea. He wants everything ("wahllos einfache Hypothesen bilden") but only everything important ("auf keinen Fall ohne nachherige Selektion") . However, Leinfellner considered an "Induktionsmaschine" to be merely fictitions at that time("heute noch fiktiv"). In contrast, [Hájek and al. 1966b] already contains a particular (although primitive) method which existed at that time in the form of a computer program.

(b) Particular methods. A particular GUHA-method is divided into the method of determining inputs, the method of machine processing (the core method), and the method of interpreting results. This division was formulated by

Chytil; and the fact remains that as yet it was almost exclusively the core methods which were theoretically discussed in publications. The following is a summary of implemented methods:

(1) Method with true disjunctions(Hájek-Havel-Chytil [1966b] , Havel is the author of the MINSK 22 program).

(2) Almost true disjunctions (Hájek-Havel-Chytil [1967], a MINSK 22 program and an IBM 7040 FORTRAN program by Havel).

(3) Fisher association (Hájek [1968], Part II, MINSK 22 and IBM 7040 programs by Havel).

(4) Statistical modification of (1) and (2)(Havránek [1971] , a program for CELLATRON by Havránek).

(5) Three-valued modification of (2) and (3)(Hájek, Bendová, and Renc [1971], a program by Rauch in FORTRAN - it has never been in practical operation).

(6) Associational and implicational quantifiers (Hájek [1973a, 1974], Part III, programs by Havel and Rauch for IBM 370, in FORTRAN).

The last method contains and generalizes all the previous ones. It is a particular case of the method in Chapter VII of the present book (it is not possible to restrict oneself to incompressible formulae). For method(6) a textbook has ben written (Hájek, Havel, Havránek, Chytil, Rauch, Renc) which is, in essence, a "directions for use" - instruction on the application of a particular method with the necessary theory.

A number of informal ideas on the methodological elaboration of the inputs (i.e., of procedures suitable for deciding whether, in a particular situation, the application of the method is adequate and in what manner it should be applied) and of the interpretation i.e., of the forms of the communication of the computer results and their utilization , are contained in numerous unpublished comments by Chytil; however, up to now they have not been systematically elaborated with the exception of [Chytil 1969] and some parts of the above mentioned textbook. The chapter on interpretation in it, written by Havránek and Renc, is also partially based on Chytil's ideas.

(c) Theoretical background. The original version of the GUHA method was based throughout on elementary logical theory (in[Hájek, Havel, Chytil 1966a] we read: "From the mathematical point of view, the method does not contain any innovations" .)

However, later it became clear that the investigation into the possibilities of realizing the above mentioned principle requires a specific autonomous theory. Now it seems that this autonomous Theory of Automated Discovery possesses its own importance and its own sphere of problems even independently of the GUHA-methods. Below, we give a summary of the existing theoretical development.

(1) Logic. A first attempt aiming at a general theoretical framework is found in [Hájek 1968] (in the language of second order logic). The foundations of the logic of automated discovery were laid down in a series of papers [Hájek 1973a, 1974]: the formulation and the study of the concepts of the semantic system, problem and solution, formal definition of a GUHA-method, function calculi, particularly cross-nominal calculi, the application of the concept of a generalized quantifier (operator) in the Mostowski - Lindström sense, the introduction of the concepts of associational and implicational quantifiers. See also [Hájek 1973c, 1975c]; [Hájek 1974 and Hájek 1975b] are written from the point of view of "pure" logic.

(2) Statistics. Developed by Havránek, the second author of the present book. The points are (a) the introduction of particular statistically motivated quantifiers in [Havránek 1971] although not in the terminology of quantifiers , and (b) the development of the theory of error rates with respect to the problems of local and global interpretation [Havránek 1974]. It can be said that (a) is an application of statistics to logic while (b) is an application of logic to statistics.

(3) Methodology (philoshphy of science). Besides Chytil's comments, mentioned above, there is a modest contribution in [Hájek 1973a]. Methodological aspects are studied in more detail in Chapter III (Hájek-Havránek-Chytil) of the prepared textbook. Note that the thesis of Buchanan [1966] contains numerous considerations that are of basic importance in the development of the philosophical logic of automated discovery.

(4) Computer science. We believe **that** all points mentioned in 1 - 3 are relevant to Artifical Intelligence, especially to Hypothesis Formation; we have here a metatheory of Hypothesis Formation. Moreover, it can be seen that investigation of the relations between the logic of observational calculi and the problems of computational complexity is worthwhile (Hájek [1975b], [Pudlák] ; see Chapter III Section 5, of the present book).

EXAMPLE OF AN INDUSTRIAL APPLICATION

We illustrate the theory presented here an example of a particular application of the GUHA method with associational quantifiers to a problem from industry. The task was to analyse possible causes of simultaneous overflashing of the generator and motor of dieselelectric locomotives of a certain type. (The **application** of the GUHA method was done by E. Pavlíková - Technical University Žilina, in collaboration with M. Rabiška and I. Šulcek - ČKD Prague, with I. Havel - Mathematical Institute, ČSAV Prague, Z. Renc - Dpt. of Mathematics, Charles University Prague and the present authors. The example has been slightly simplified.) *)

Objects: locomotives in the moment of **overflashing on the generator.**

Attributes as follows:

		Remark:
1/ Velocity (km/hour)	$V_1 = \{0,1,2\}$	0 : $\leq$ 60 1 : between 60 nad 80 2 : $\geq$ 80
2/ Kilometer performance	$V_2 = \{0,1,2\}$	0 : (0,100 000] 1 : (100 000,200 000] 3 : (200 000,300 000]
3/ Throttle position of the **master controller**	$V_3 = \{0,1,2\}$	
4/ Load	$V_4 = \{0,1\}$	
5/ Change of the position	$V_5 = \{0,1\}$	
6/ Exciting of traction motors	$V_6 = \{0,1,2,3\}$	
7/ Switching of the relay	$V_7 = \{0,1\}$	
8/ Weather	$V_8 = \{0,1,2\}$	0 : dark 1 : fog 2 : rain

---

*) E. Pavlíková, Aplikácia metod automatizovaného výskumu v doprave a spojoch, VŠD Žilina(Czechoslovakia), research report P04-533-081-00-03 (1975).

9/ Air temperature	$V_9 = \{0,1\}$	0 : under 5°C
10/ Track character ; gradient	$V_{10} = \{0,1\}$	
11/ Track character; curve	$V_{11} = \{0,1\}$	
12/ Type of train	$V_{12} = \{0,1\}$	0 : local train 1 : fast train
13/ - 18/ Descriptions of various break-downs of the generator	$V_{13}\text{-}V_{18} = \{0,1\}$	
19/ Overflashing of the motor	$V_{19} = \{0,1\}$	

The model was selected by random sampling using tests of representativeness. One selected 33 objects - locomotives in the moment of overflushing on the generator. It should be clear that such a model can serve only for systematic inspiration (cf. Chapter VIII).

Now we can specify our r-problem in terms of Chapter VII.(Details of implementation are disregarded here.)
Remember that the parameter of the GUHA method used decomposes into three parts, $p = \langle CALC, QUEST, HELP \rangle$. Here $CALC = \langle CHAR, KQUANT, PQUANT \rangle$.
In our example we have the following:

(a) CHAR = (aa) number of function symbols - 19,

(ab) for each function symbol, its set of regular values - as given above.

(ac) our model has complete information.

(b) KQUANT = SYMNEG ; our quantifiers will satisfy the rules SYM and NEG (cf. 3.2.)

(c) PQUANT = the Fisher quantifier with $\alpha = 0.05$

Next we specify $QUEST = \langle KRPF, FORQ, SYNTR \rangle$:

(a) KRPF = CPF (conjuctive pairs of formulae),

(b) FORQ = SIMPLE (relevant questions have the form $\varphi \sim \psi$ where $\langle \varphi, \psi \rangle$ varies over relevant pairs),

(c) SYNTR = (ca) The succedent is fixed as (1) F_{19}.

(cb) F_{19} must not occur in any antecedent.

(cc) Only one-element coefficients are allowed.

(cd) Maximal number of function symbols occurring in our antecedents is 3.

HELP: helpful quantifiers are not used.

This completes the specification of parameters.

The model was processed on the MINSK 22 computer on September 9, 1975. (Duration under one hour.)

Output:

8 sentences of the form $\varphi \sim (1)\,F_{19}$ where φ is a literal,

64 sentences of the form $\varphi \sim (1)\,F_{19}$ where φ is a conjunction of two literals,

125 sentences of the form $\varphi \sim (1)\,F_{19}$ where φ is a conjunction of three literals.

For each output sentence $\varphi \sim (1)\,F_{19}$ we have the table

	$1\ F_{19}$	$0\ F_{19}$	
φ	a	b	r
$\neg\varphi$	c	d	s
	k	l	

.

Note that k is constant since the succedent is fixed; in our example k = 20 and k + 1 = r + s = 33 . For each output sentence the following numerical characteristics are printed:

α_{crit}, a, r, (a/r) 100 .

Let us present some examples of results:

	$\sim (1)F_{19}$	α_{crit}	a	r	(a/r) 100
1/	$(2)\,F_1$	.04681	8	9	88
2/	$(3)\,F_6$	.04712	10	12	83
3/	$(1)\,F_{12}$	.00059	17	20	85
4/	$(0)\,F_4 \& (2)\,F_1$	. 03499	6	6	100
5/	$(1)\,F_{12} \& (0)\,F_4$	.00029	12	13	90

6/	(1) F_{12}&(0) F_9	.02558	9	10	90
7/	(1) F_{12}&(1) F_{18}	.00907	8	8	100

The computer produced a long list of such **results, 197 output sentences.** They serve as a source of hypotheses for further investigations (cf. Chapter VIII). Observe the dependence of (1) F_{19} (overflashing on the motor) on the property that the train hauled by the locomotive is a fast train. This property occurs in output sentences both alone (in a sentence with one-element **antecedent) and in** conjunction with other factors (low air temperature, low load, some particular break-downs etc.). One must be careful in interpreting the results since e.g. low load may imply a fast train. Similarly for (2) F_1 (high velocity).

Further, the most important factors occurring in output sentences are of technical character and give little information to a layman. The core of the results consists in the combination of some functional states of the machine and the complex fast train - low load - high speed. The importance of the fact that some factors do not occur in the results (e.g. the kilometer performance) should not be overlooked.

Index

Bibliography

ANDĚL J. [1973]: On interactions in contingency tables, Aplikace matematiky 18, 99-109

[1974]: The most significant interaction in a contingency table, Aplikace matematiky 19, 246-252

BALAAM L.N., FEDERER W.T.[1965]: Error rate basis, Technometrics 7, 260-262

BENDOVÁ K. [1975]: Tříhodnotové logiky a observační kalkuly Three-valued logics and observational calculi , Thesis in preparation

BENDOVÁ K., HAVRÁNEK T. [1973]: Výběr regresních modelů Choosing regression models, Sdělení MSBÚ ČSAV, 2(1), Praha

BISHOP YVONNE M.M., FIENBERG S.E., HOLLAND P.W. [1975]: Discrete multivariate analysis, MIT Press

BUCHANAN B.G. [1966]: Logics of scientific discovery, Thesis, Michigan state university

BÜCHI J.R.[1960]: Weak second order arithmetic and finite automata, Zeitschrift für mathematische Logik und Grundlagen der Mathematik 6, 66-92

BURRIL C.W. [1972]: Measure, integration and probability, McGraw-Hill, New York

CARNAP R. [1935]: Testability and meaning, Philosphy and Science, vol. III, Williams and Wickins, Baltimore

[1950]: Logical foundations of probability, Chicago

CHANG C.C., KEISLER H.J.[1966]: Continuous model theory, Princeton

CHURCH A. [1956]: Introduction to mathematical logic, Volume I., Princeton

CHYTIL M. [1969]: Zadání semantického modelu pro zpracování metodou GUHA (On constituting of semantic models for GUHA - methods), Československá fyziologie 18, 143-147

[1974]: Decomposition calculi and its semantics, Studia logica 33, 277-282

CLEAVE J.P. : The notion of logical consequence in the logic of inexact predicates, Zeitschrift für mathematische Logik und Grundlagen der Mathematik (to appear)

COX D.R. [1965]: A remark on multiple comparisons methods, Technometrics 7, 223-224

COOK S.A. [1971]: The complexity of theorem proving procedures, Proceedings of Third Annual ACM Symposium on Theory of Computing, 151-158

CRAIGH W. [1953]: On axiomatizability within a system, Journal of Symbolic Logic 18, 30-32

ČUDA K. [1973]: Contributions to the theory of semisets III, Zeitschrift für mathematische Logik und Grundlagen der Mathematik 19, 399-406

DAVIS M. [1958]: Computability and unsolvability, New York

EDWARDS A.W.F. [1963]: The measure of association in a 2x2 table, Journal of the Royal Statistical Society, ser. A 129, 109-114

FABIAN V. [1968]: Statistische Methoden, Deutscher Verlag der Wissenschaften, Berlin

FAGIN R. [1973]: Contributions, the Model Theory of Finite Structures, Thesis, University of California, Berkeley

[1974]: Generalized first-order spectra and polynomial time recognizable sets, Complexity of Computations ed. R.Karp , SIAM - ACM proceedings, vol. 7, 43-73

[1975a]: Monadic generalized spectra, Zeitschrift für mathematische Logik und Grundlagen der Mathematik 21, 89 - 96

[1975b]: A spectrum hierarchy, ibid., (to appear)

FERGUSSON T.S. [1967]: Mathematical statistics, a decision theoretic approach, Academic Press, New York

FINE T.L. [1973]: Theories of probability, Academic Press, New York

FISHER M.S., RABIN M.O. [1974]: Super-exponential complexity of Presburger arithmetic, MAC Technical Memmorandum 43, MIT

FISHER R.A. [1933]: The design of experiments, Oliver and Boyd, London

FLUM J. [1975]: First order logic and its extensions, ISILC-Logic Conference, Lect. Notes in Mathematics 499, Springer Verlag

FRAISSÉ R. [1965]: A hypothesis concerning the extension of finite relations and its verification in certain special cases, The theory of models, ed. J.W.Addison, L.Henkin, A.Tarski , 364-375

FREIBERGER W., GRENADER U. [1971]: A short course in computational probability and statistics, Applied mathematical sciences 6, Springer, New York

FUHRKEN G. [1972]: A remark on the Härtig quantifier, Zeitschrift für mathematische Logik und Grundlagen der Mathematik 18, 227-228

GABRIEL K.R. [1969]: Simultaneous test procedures - some theory of multiple comparisons, Annals of Mathematical Statistics 40, 224-250

GOOD I.J. [1963]: Maximum entropy for hypothesis formulation, especially for multidimensional contingency tables, Annals of Math. Statist. 34, 911 - 934.

HÁJEK J., ŠIDÁK Z. [1967]: Theory of rank tests, Academia, Prague and Academic Press, New York

HÁJEK J., VORLÍČKOVÁ D. [1967]: Neparametrické metody (Nonparametrical methods), SPN, Praha

HÁJEK P. [1968]: Problém obecného pojetí metody GUHA (The question of a general concept of the GUHA method), Kybernetika 4, 505-515

[1972]: Contributions to the theory of semisets I., Zeitschrift für mathematische Logik und Grundlagen der Mathematik 18, 241,248

[1973a], [1974]: Automatic listing of important observational statements I - III, Kybernetika 9, 187-205, 251-271 and 10, 95-124

[1973b]: Why semisets? , Commentationes Mathematicae Universitatis Carolinae 14, 397-420

[1973c]: Some logical problems of automated research, Proceedings of the 1973 Symposium of Mathematical Foundations of Computer Science,High Tatras, Czechoslovakia

[1974]: Generalized quantifiers and finite sets, Proceedings of the Autumn School in Set Theory and Hierarchy Theory, Wroclaw (to appear)

[1975a]: On logics of discovery, Mathematical Foundations of Computer Science 75, ed.J. Bečvář, Lecture Notes in Computer Science 32, Springer-Verlag,30-45

[1975b]: Projective classes of models in observational functor calculi, Preprint of the V.International Congress on Logic, Methodology and Philosophy of Science, London Ontario, Canada

[1975c]: Observationsfunktorkalküle und die Logik der automatisierten Forschung, Elektronische Informationverarbeitung und Kybernetik 12 (1976), 181-186

HÁJEK P., BENDOVÁ K., RENC Z. [1971]: The GUHA - Method and the three valued logic, Kybernetika 7, 421-435

HÁJEK P., HAVEL I., HAVRÁNEK T., CHYTIL M., RAUCH J., RENC Z.: Metoda GUHA, Dům techniky ČVTS, České Budějovice, 1975

HÁJEK P., HAVEL I., CHYTIL M. [1966a]: The GUHA Method of automatic hypotheses determination, Computing 1, 293-308

[1966b]: GUHA - metoda systematického vyhledávání hypotéz (GUHA - a method of systematic hypotheses searching), Kybernetika 2, 31-47

[1967]: GUHA - metoda systematického vyhledávání hypotéz II GUHA - a method of systematic hypotheses searching II. , Kybernetika 3, 430-437

HAVRÁNEK T. [1971]: The statistical modification and interpretation of the GUHA Method, Kybernetika 7, 13-21

[1974]: Some aspects of automatic systems of statistical inference, Proceedings of the European Meeting of Statisticians, Prague (to appear)

[1975a]: The approximation problem in computational statistics, Mathematical Foundations of Computer Science 75, ed.J.Bečvář , Lecture Notes in Computer Science 32, Springer-Verlag, 258-265

[1975b]: Statistical quantifiers in observational functor calculi: an application in GUHA - methods, Theory and Decision 6, 213-230

[1975c]: Statistics and Computability, Research report nr. 1 - AISC, Center of Biomathematics, Czechoslovak Academy of Sciences, Prague

[1975d]: A note on simultaneous inference in contingency tables, (submitted)

HEMPEL C.G. [1965]: Aspects of scientific explanation, Free Press, New York

IVÁNEK J.: Master thesis, Department of Mathematics, Charles University, Prague

JENSEN R.B. [1965]: Ein neuer Beweis für die Entscheidbarkeit des einstelligen Prädikaten kalküls mit Identität, Archiv für mathematische Logic und Grundlagenforschung 7, 128-138

KARP R.M. [1972]: Reducibility among combinatorial problems, Complexity of computer computations, Plenum Publishers 85-104

KENDALL M.G. [1951]: Regression, structure and relationship, Biometrika 38, 11-25

1955 : Rank correlation methods, Griffin, London

KEISLER H.J. [1970]: Logic with the quantifier "there exist uncountable many", Annals of Mathematical Logic 1, 1-93

KLEENE S.C. [1952]: Introduction to metamathematics, Van Nostrand, Princeton

KORNER S. [1966]: Experience and theory, Routledge, London

KOWALSKI R. [1974]: Logic for problem solving, Memo 75, Department of Computational Logic, School of Artificial Intelligence, Edinburgh

KRIPKE S.A. [1959]: Completeness theorem in modal logic, Journal of Symbolic Logic 24, 1-14

LEHMANN E.L. [1959]: Testing of statistical hypotheses, J. Wiley, New York

LEINFELLNER W. [1965]: Struktur and Aufbau wissenschaftlicher Theorien, Physica Verlag, Wien

LINDSTROM P. [1966]: First order logic with generalized quantifiers, Theoria 32, 186-191

[1969]: On extensions of elementary logic, Theoria 35, 1-11

MAKOWSKI A. [1973]: Notes for lectures on model theory at the logical semester, Warsaw (mimeagraphed)

MATIASEVIČ Ju.V. [1970]: Diofantovost perečislimych množestv, Doklady Akademii Nauk SSSR vol. 191, 279-282.(English translation: Enumerable sets are diophantine, Soviet Mathematics 11 no 2 [2970]), 354-357.

MCCARTHY J., HAYES P. [1964]: Some philosophical problems from the point of view of Artificial Intelligence, Machine Intelligence 4

MELTZER B. [1970 a]: Generation of hypotheses and theories, Nature, 225, 972

MELTZER B. [1970 b]: Power amplification for theorem provers, Machine Intelligence 5, 165-179

MENDELSOHN E. [1964]: Introduction to mathematical logic, Van Nostrand, New York

MEYER A.R. [1973]: Weak monadic second order theory of successors is not elementary recursive, MAC Technical Memorandum, MIT

MILLER R.G. [1967]: Simultaneous statistical inference, McGraw-Hill, New York

MINSKY M. [1974]: A framework for representing knowledge, AI memo nr.305, MIT

MORGAN C.G. [1971]: Hypothesis generation by machine, Artificial Intelligence 2, 179-187

MOSTOWSKI A. [1957]: On a generalization of quantifiers, Fundamenta mathematice 44, 12-36

NARIAKI S., MASANORI O. [1973]: Approximate distribution of the maximum of c - 1 χ^2 - statistic 2 x 2 derived from a 2 x C contingency table, Communications in Statistics 1, 9 - 16

NILSSON N.J. [1971]: Problem solving methods in Artificial Intelligence, McGraw-Hill, New York

O'NEIL R.O., WEITHERIL G.B.[1971]: The present state of multiple comparisons methods, Journal of the Royal Statistical Society, ser.B, 33, 218-250

PEARSON E.S., HARTLEY H.O.[1972]: Biometrika tables for statisticians, vol. II, Cambridge University Press

PLOTKIN G.D. [1971]: A further note on inductive generalization, Machine Intelligence 6, 101-124

POKORNÝ D. [1975]: Asociační kvantifikátory v nominálních kalkulech (Associational quantifiers in qualitative calculi), Master thesis, Dept. of Math., Charles University, Prague

POPPER K.R. [1966]: Logik der Forschung, J.c. B. Mohr, Tübingen

POUR-El M.B., CALDWELL J. [1975]: On a simple definition of computable functions of a real variable with applications to functions of a complex variable, Zeitschr. f. math. Logik und Grundlagen d. Math. 21, 1-19

PUDLÁK P. [1975a]: Observační predikátový počet a teorie složitosti. Observational predicate calculi and complexity theory, Thesis, Charles University, Prague

[1975b]: The observational predicate calculus and complexity of computations, Commentationes Mathematicae Universitatis Caroliae 16, 395-398

[1975c]: Polynomial complete problems in the logic of automated discovery, Mathematical Foundations of Computer Science 75, J. Bečvář, Lecture Notes in Computer Science 32, Springer-Verlag, Heidelberg

RABIN M.O.[1974]: Some impediments to Artificial Intelligence, Information Processing 74 Proc. of IFIP Conf. , Stockholm, 615-619

RACKOFF C.W.[1975]: The computational complexity of some logical theories, MAC Technical report 144, MIT

RAO C.R. [1965]: Linear statistical inference and its applications, J.Wiley, New York (sec.ed.1973)

RAUCH J. [1975]: Ein Beitrag zu der GUHA Methode in der dreiwertigen Logik, Kybernetika 11, 101-113

REEKEN A.J. VAN [1971]: Report of the Dutch working party on statistical computing, Applied Statistics (JRSS-C), 20, 73-79

RESCHER N. [1962]: Plurality quantification, Journal of Symbolic Logic 27, 373-374

RÖDDING D. [1967]: Primitiv-rekursive Funktionen über einem Bereich endlicher Mengen, Archiv für Mathematische Logik und Grundlegenforschung, 10, 13-29

ROGERS H. [1967]: Theory of recursive functions and effective computability , McGraw-Hill, New York

ROSSER J.B., TURQUETTE A.R. [1952]: Many-valued logic, North-Holland, Amsterdam

SCOTT D. [1964]: Measurement structures and linear inequalities, Journal of mathematical psychology 1, 233-247

SCOTT D. S., KRAUSS P. [1966]: Assigning Probabilities to logical formulas, in: Aspects of Inductive Logic (J. Hintikka and P. Suppes, eds.) pp. 219-264. North Holland, Amsterdam.

SHOENFIELD J.R. [1967]: Mathematical logic, Addison-Vesley, [1971]: Degrees of unsolvability, North-Holland, Amsterdam

ŠIDÁK Z. [1962]: Rectangular confidence regions for the means of multivariate normal distributions, Journal of the American Statistical Association 62, 626-633

SIEFKES D. [1970]: Büchi's monadic second order succesor arithmetic , Lecture Notes in Mathematics 120, Springer - Verlag, Heidelberg

SLOMSON A. [1968]: The monadic fragment of predicate calculus with the Chang quantifier, Lecture Notes in Mathematics 70, Springer-Verlag, Heidelberg 279-302

SUPPES P. [1962]: Models of data, Logic, Methodology and Philosophy of Science; Proceedings of the 1960 Interantional Congress ed. E. Nagel, P. Suppes, A. Tarski , Stanford University Press, 252-261

SUPPES P. [1965]: Logics appropriate to empirical theories, The theory of models, ed. J.W. Eddinson, L. Henkin, A. Tarski, 364-375

THARP L.H. [1973]: The characterisation of monadic logic, Journal of Symbolic Logic 38, 481-488

TONDL L. [1972]: Scientific procedures, Reidel, Dortrecht

TRACHTENBROT B.A. [1950]: Nevozmožnost algoritma dlja problemy razrešimosti na konečnych klassach (Impossibility of an algorithm for the decision problem on finite classes), Doklady Akademii Nauk SSSR vol. 70, 569-572

VOPĚNKA P., HÁJEK P. [1972]: The theory of semisets, North-Holland, Amsterdam and Academia, Prague

YANAGIMOTO T., OKAMOTO M., [1969]: Partial orderings of permutations and monotonicity of a rank correlation statistics, Annals of the Institute of Statistical Mathematics (Tokyo) 21, 488-507

YASUHARA A. [1971]: Recursive functions, theory and logic, Academic Press, New York

ZINDEL P. [1966]: Gamma rays on man in the moon marigolds, Gerhard Pegler-Verlag, München

ZELEN M. [1970]: Exact significance tests for contingency tables embedded in 2^n classification, Proceedings of the 6th Berkeley Symposium in Probability and Mathematical Statistics, University of California Press, 737 - 757.

Perspectives in Mathematical Logic

In recent years interconnections between different lines of research in mathematical logic and links with other branches of mathematics have proliferated. The subject is now both rich and varied. It is the aim of this series to provide, as it were, maps or guides to this complex terrain as seen from various angles. The group is not committed to any particular philosophical program. Nevertheless, the critical discussion which each planned book undergoes ensures that it will represent a coherent line of thought; and that, by developing certain themes, it will be of greater interest than a mere assemblage of results and techniques.

The books in the series differ in level: some are introductory, some highly specialized. They also differ in scope, some offering a wide view of an area while others present more specialized topics. Each book is, at its own level, reasonably selfcontained. Although no book depends on another as prerequisite, authors are encouraged to fit their book in with other planned volumes-sometimes deliberately seeking coverage of the same material from different points of view.

J. Barwise

Admissible Sets and Structures

An Approach to Definability Theory

1975. 22 figures, 5 tables. XIV, 394 pages
ISBN 3-540-07451-1

P. G. Hinman

Recursion-Theoretic Hierarchies

1978. Approx. 500 pages
ISBN 3-540-07904-1

A. Levy

Basic Set Theory

1978. 2 tables. Approx. 350 pages
ISBN 3-540-08417-7

Springer-Verlag
Berlin
Heidelberg
New York

K. Schütte **Proof Theory**

Translation from the German by J.N. Crossley

1977. XII, 299 pages
(Grundlehren der mathematischen Wissenschaften, Band 225)
ISBN 3-540-07911-4

Contents: Pure Logic: Fundamentals. Classical Predicate Calculus. Intuitionistic Predicate Calculus. Classical Simple Type Theory. – Systems of Arithmetic: Ordinal Numbers and Ordinal Terms. Functionals of Finite Type. Pure Number Theory. – Subsystems of Analysis: Predicative Analysis. Higher Ordinals and Systems of Π^1_1-Analysis.

This book was planned as a new edition of Schütte's *Beweistheorie* (Grundlehren der mathematischen Wissenschaften, Band 103). However, in view of the development of the subject, the book was virtually completely rewritten, and translated into English. Intuitionistic predicate logic and simple type theory are studied as well as classical predicate logic, and proofs of cut elimination are provided. The Gödel interpretation of number theory is presented in full detail. Various sub-systems of analysis (including Π^1_1-analysis) and predicative systems of of Δ^1_1-analysis and ramified analysis are considered. The delimitations of deducible transfinite induction for all systems are studied. The required theory of ordinal numbers is provided first classically and then constructively. With the basic notions of positive and negative part of a formula, a clear and simple formalization of many parts of proof theory is provided.

Textbooks in Logic

D.W. Barnes, J.M. Mack
An Algebraic Introduction to Mathematical Logic
1975. 5 figures. IX, 121 pages
(Graduate Texts in Mathematics, Volume 22)
ISBN 3-540-90109-4

H. Hermes
Introduction to Mathematical Logic
Translator: D. Schmidt
Universitext
1973. XI, 242 pages
ISBN 3-540-05819-2

Y.I. Manin
A Course in Mathematical Logic
Translated from the Russian by N. Koblitz
1977. 1 figures. XIII, 286 pages
(Graduate Texts in Mathematics, Volume 53)
ISBN 3-540-90243-0

J.D. Monk
Mathematical Logic
1976. X, 531 pages
(Graduate Texts in Mathematics, Volume 37)
ISBN 3-540-90170-1